# WHAT ARE W

We begin life with a [...] fertilization we weig[...] gram. In this miraculous spark are contained our sex, our appearance—all the potentials inherited from a far-reaching line of ancestors.

*What happens to these potentials is to a considerable extent a question of environment. And environment begins in the womb, the moment after conception.*

*In this closely detailed study, the author shows the basic flaw in arguments that separate the interrelated effects of heredity and environment on the individual. He discusses how congenital disorders such as cretinism and many congenital malformations, which have been considered hereditary by fearful families, are in many cases environmental. Through a thorough description of the genetic mechanism, he outlines traits that are inherited . . . relates specific facts to larger issues of "race," sterilization, increased population, and the alarming dangers of fallout.*

The text is supported by tables and illustrations, and includes a unique, comprehensive list of inherited disorders in man, a list of heredity clinics where parents can be advised, and a glossary. Easy to read, yet full of information, here is a valuable guide for people who want to learn more about their origin and more about the prospects for their children.

*"Comprehensive and well-documented."*
—N.Y. Times

# SIGNET and MENTOR
## Books of Related Interest

**Man: His First Million Years** *by Ashley Montagu*
A vivid, lively account of the origin of man and the development of his cultures, customs, and beliefs.
(#P2130—60¢)

**Man In Process** *by Ashley Montagu*
The distinguished anthropolgist tells how man can learn to live in a more rational, social and co-operative manner. (#MT474—75¢)

**Heredity, Race and Society (revised)** *by L. C. Dunn and Th. Dobzhansky*
Group differences, how they arise, the influences of heredity and environment. (#MP532—60¢)

**The Genetic Code** *by Isaac Asimov*
A study of the discovery and significance of DNA, the chemical compound which determines the nature of all living organisms. (#P2250—60¢)

TO OUR READERS: If your dealer does not have the SIGNET and MENTOR books you want, you may order them by mail, enclosing the list price plus 5¢ a copy to cover mailing. If you would like our free catalog, please request it by postcard. The New American Library of World Literature, Inc., P.O. Box 2310, Grand Central Station, New York, New York, 10017.

# HUMAN HEREDITY

by **Ashley Montagu**

*Second Revised Edition*

*A SIGNET SCIENCE LIBRARY BOOK*
*Published by The New American Library*

To May Sarton

COPYRIGHT © 1959, 1960, 1963, BY ASHLEY MONTAGU

All rights reserved. No part of this book may be reproduced in any form without written permission from the publisher, except for brief passages included in a review appearing in a newspaper or magazine. For information address The World Publishing Company, 119 West 57th Street, New York 19, New York.

Published as a SIGNET SCIENCE LIBRARY BOOK by arrangement with The World Publishing Company, who have authorized this softcover edition.

FIRST PRINTING, OCTOBER, 1960
SECOND PRINTING (REVISED), OCTOBER, 1963

This SIGNET SCIENCE LIBRARY edition has been brought up to date by the addition of new material. The author is grateful to the publisher for the opportunity to make these changes.

The "Fifth Philosopher's Song," quoted on page 23, is from *Leda*, by Aldous Huxley, published by Harper & Brothers.

Most of the illustrations in this book were made by I. N. Steinberg.

SIGNET TRADEMARK REG. U.S. PAT. OFF. AND FOREIGN COUNTRIES
REGISTERED TRADEMARK—MARCA REGISTRADA
HECHO EN CHICAGO, U.S.A.

*SIGNET SCIENCE LIBRARY BOOKS are published by The New American Library of World Literature, Inc. 501 Madison Avenue, New York 22, New York*

PRINTED IN THE UNITED STATES OF AMERICA

# CONTENTS

| | |
|---|---:|
| PREFACE | xi |
| PREFACE TO THE SECOND EDITION | xiv |

## Part I. THE MEANING OF HEREDITY

| | | |
|---|---|---:|
| 1. | Introduction | 17 |
| 2. | In the Beginning | 23 |
| 3. | The Laws of Inheritance | 57 |
| 4. | "Blood Is Thicker than Water | 67 |
| 5. | Heredity and Environment | 76 |
| 6. | The Effects of Environment upon the Developing Human Being in the Womb | 86 |
| 7. | Environment After Birth | 113 |
| 8. | Twins, Genes, and Environment | 141 |
| 9. | Crime—Genes or Environment or Both? | 156 |
| 10. | Genes and Constitution | 164 |
| 11. | Sex | 178 |
| 12. | Heredity and "Race" | 217 |

## Part II. THE FAMILY ALBUM

| | | |
|---|---|---:|
| 13. | The Pigmentary System of the Skin, Eyes, and Hair | 235 |
| 14. | The Features | 256 |
| 15. | The Body | 266 |
| 16. | Heredity and Blood | 276 |
| 17. | What Do We Do About Heredity? | 297 |
| 18. | The Bomb, Radiation, and Human Heredity | 325 |

## APPENDIX A.

Inherited Disorders of Man — 346

## APPENDIX B.

The Chance of the Offspring of Marriage Between Cousins Exhibiting a Defective Condition — 375

## APPENDIX C.

Gene and Genotype Equilibria, Populations and Genes, the Hardy-Weinberg Law — 376

## APPENDIX D.

Location, Name of Institution, and Principal Counselor of Some Heredity Clinics (by State) — 380

## APPENDIX E.

Glossary — 382

## APPENDIX F.

Genetics in the U.S.S.R.: A Political "Science" — 393

Books for Further Reading — 396

Bibliography — 401

Index — 410

# FIGURES

| | | |
|---|---|---|
| I | Diagram of a Cell | 36 |
| II | Diagram of Protein Synthesis | 38 |
| III | The Sixty-Four Trinucleotides, or Triplets | 39 |
| IV | The Synthesis of an Amino Acid | 40 |
| V | The Elements of the Genetic Code | 43 |
| VI | Meiosis | 51 |
| VII | Crossing-Over | 52 |
| VIII | Meiosis: Sperm Formation | 54 |
| IX | Meiosis: Egg Formation | 55 |
| X | Independent Assortment in Chromosomes | 64 |
| XI | Hartsoeker's Homunculus | 79 |
| XII | Ideogram of Chromosomes | 92 |
| XIII | Cigarette Smoking and Premature Births | 109 |
| XIV | Cephalic Index of Immigrants and Their Descendants | 125 |
| XV | Transmission of the Hemophilia Gene | 170 |
| XVI | Hemophilia in Queen Victoria's Descendants | 172 |
| XVII | Inheritance of Color Blindness | 190 |
| XVIII | Family Pedigree of Balding | 193 |
| XIX | The Chromatin Body in Relation to the X Chromosomes | 202 |
| XX | Non-Disjunction | 208 |
| XXI | World Population | 214 |
| XXII | Major Ethnic Groups of Man | 223 |
| XXIII | Inheritance of Albinism | 237 |
| XXIV | An Albino Mating | 238 |
| XXV | Inheritance of Skin Color | 242 |
| XXVI | Inheritance of Hair Form | 252 |
| XXVII | Inheritance of Form of Earlobe | 264 |

| XXVIII | Inheritance of Tendency to Nasal Sinus Infection | 274 |
| --- | --- | --- |
| XXIX | Blood Groups in Europe | 279 |
| XXX | Inheritance of Blood Groups | 281 |
| XXXI | Linkage of Blood Groups with Nail-Patella Syndrome | 294 |
| XXXII | A Standard Pedigree Chart | 312 |
| XXXIII | Pedigree of Egyptian Rulers | 316 |
| XXXIV | Short Pedigree of the Darwin Family | 320 |

# PLATES

*Plates appear as a complete section between pp. 216 and 217.*

1. The DNA Molecule
2. Hybrid Basuto-English Mating
3. Color in Andalusian Fowl
4. Human Chromosomes in Diploid Number
5. Cretins
6. Sextuplets
7. Four Sets of Quadruplets
8. The Diligenti Quintuplets
9. Stature Difference in Identical Twins
10. Teeth of Identical Twins
11. Harelip and Cleft Palate in Identical Twins
12. Premaxillary Diastema in the Gorilla
13. Tongue Rolling
14. Bushman-White Hybrids
15. Patterns of Balding
16. Inheritance of Baldness
17. Dwarfs
18. Dwarfs and Giants
19. The Chromatin Body in Female Somatic Cells
20. X-Ray-Induced Chromosome Breakage

# TABLES

| | | |
|---|---|---|
| I | Chromosome Numbers in Some Animals and Plants | 29 |
| II | Pregnant Women and Their Diets | 102 |
| III | Nutrition of Mothers and Weight of Children | 103 |
| IV | Weight and Height of Children in Relation to Socioeconomic Status | 121 |
| V | Children Under Average Height | 122 |
| VI | Measurements of Immigrants' Children | 123 |
| VII | Life Expectancy in Selected Countries | 127, 128 |
| VIII | Test Scores Related to Social Status | 138 |
| IX | Schizophrenia Expectancy in Co-Twins | 154 |
| X | Criminal Behavior of Twins | 157 |
| XI | Mortality Rates | 182, 183 |
| XII | Sexual Differences in Disease Susceptibility | 186 |
| XIII | Human Traits and Sex-Linked Genes | 188 |
| XIV | Conditions Due to Lethal Genes | 201 |
| XV | Ability to Taste Phenylthiocarbamide | 262, 263 |
| XVI | Agglutinogens and Agglutinins of Blood Groups | 276 |
| XVII | Determination of Blood Groups, with Anti-A and Anti-B | 277 |
| XVIII | Determination of Blood Groups, Corpuscles A and B | 278 |
| XIX | Compatibility of Blood Types | 280 |
| XX | Gene Combinations in Blood Groups | 280 |
| XXI | Genetic Constitution with Regard to Blood Groups | 282 |

| XXII | Heredity of Blood Types M, N, and MN | 284 |
| XXIII | Inheritance of M-N Blood-Group System | 284 |
| XXIV | M-N-S-s Blood-Group System | 285 |
| XXV | Nomenclature of M-N-S-s Types | 285 |
| XXVI | Rh Genotypes and Phenotypes | 287 |
| XXVII | The Rh Series of Allelic Genes | 289 |
| XXVIII | Distribution of Rh Blood Types | 290, 291 |
| XXIX | States and Territories with Sterilization Laws | 303 |
| XXX | First-Cousin Marriages and Hereditary Disorders | 317 |
| XXXI | Diagnostic Radiology Compared with Radium Wristwatches | 330 |
| XXXII | Diagnostic Radiology Compared with Wearing Trousers | 335 |
| XXXIII | Spontaneous Mutation Rates of Some Human Genes | 337 |

# PREFACE

THERE ARE several good popular books on human heredity, and the outstanding among these is Amram Scheinfeld's *The New You and Heredity*. That book is highly recommended to readers of the present volume, as is Dunn and Dobzhansky's paperback, *Heredity, Race, and Society*, Goldstein's *Genetics Is Easy*, and Crow's *Genetics Notes*. The present volume, intended for the general reader, while setting out the basic facts of heredity, represents something of a departure from the conventional pattern of such books. While the usual book very properly concentrates on giving an account of the biological principles and the mechanisms involved in inheritance, the reader is all too often left with the impression that the genes and chromosomes, in their various permutations and combinations, are largely if not entirely responsible for heredity. The reader knows that there is also an environment, but somehow this merely seems to play the role of a setting within which the chromosomes and the genes pursue the even tenor of their ways. Biologists are, of course, agreed that the expression of genetic potential is greatly influenced by environment. But somehow in textbooks and more popular works on genetics the environment as a factor in influencing heredity manages to get slighted. This seems to be unavoidable in works written by biologists. Their emphasis is biological, organic, and much as they may try to avoid a biologistic bias, they cannot help, a great deal of the time, leaving the reader with the impression that genes and chromosomes are all. Since the facts are quite otherwise, this is an impression that requires correcting, and this I have attempted to do in this book.

There also exists a widespread belief that the mere demonstration that a trait is hereditary somehow means that it is not alterable through environmental agencies. The truth is that not only does the environment regulate the expression of genetic potentials but also it is in every instance capable of affecting the basic structure of those potentials.

Much of the confusion that exists concerning the nature

of heredity is easily correctable. I have attempted to reduce this confusion without, I hope, adding to it or, I trust, falling into the error of extreme environmentalism as opposed to the extreme hereditarianism I desire to correct. In the development of the individual, genes are extremely important, but environment is important too. In concentrating so much attention upon environment, I should not like to leave the reader with the impression that genes are of secondary importance—they are not. Their importance is at least as great as is that of environment. The power of the genes should not be underestimated—nor should that of the environment.

I have endeavored to help the reader understand the facts of heredity clearly so that he might be in a position to work out simple problems in heredity for himself. I have also provided a census of inherited disorders, which should enable the reader to look up any condition in which he may be interested and determine the possible or probable manner of its transmission, as well as other details.

As a rule, in the pages that follow, when an unusual term is used for the first time, it will be defined. Where that is not the case it will generally be defined a little later. But whenever the reader desires to check on the meaning of a term not immediately defined, he should turn to the glossary at the back of the book.

One of the things I regret, for which I must apologize to the reader, is that I have frequently not produced chapter and verse for many of the statements made in this book. The reason for this is that had I done so, the volume would have been of considerably greater dimensions than it now is. I desire, however, to assure the reader that for every statement claimed as a fact in this book the evidence exists, and where I have not cited the evidence in detail, the reader may find it in the works listed in the bibliography.

To Professor Conway Zirkle of the Department of Botany, University of Pennsylvania, I owe many thanks for his kindness in reading the manuscript of this book and for his many suggestions that have served to improve it. To Professor Sheldon C. Reed of the Dight Institute of Human Genetics, University of Minnesota, I am much indebted for his helpful reading of the galleys and for providing me with a list of counseling centers in human genetics in the United States. To Professor Th. Dobzhansky of the Department of Zoology, Columbia University, I am grateful for his help in reading the galleys. Professor R. Alexander Brink of the University of Wisconsin kindly read my account of his discovery of paramutation, and Professor H. J. Muller read my brief account of Russian genetics, for which I herewith express my thanks. For a similar service in connection with his own and his collaborators' work I owe thanks to Dr. W. L. Russell of the Oak Ridge National Laboratory, Oak Ridge, Tennessee. Dr. Richard Levins of the Department of Zoology, Columbia University, and Professor Max Levitan of the Department of Anatomy, Woman's Medical College of Pennsylvania, read page proof, much to the book's advantage. I thank them.

I am also most grateful to Doctor Michael A. Bender and Doctor E. H. Y. Chu of the Oak Ridge National Laboratory for their kindness in providing me with photographs of human chromosomes.

To Miss Elizabeth Corddry I owe many thanks for her searching criticism of the book while it was still in manuscript, and to Mr. Jerome Fried, my editor, I am deeply obliged for his constructive assistance at every stage of the making of a book that has not been easy to produce.

This book owes its being to William Targ, of The World Publishing Company, my publishers, who challenged me to write it, knowing full well that I am not a geneticist but an anthropologist. All complaints may therefore appropriately be laid at his door. I am greatly indebted to him for the suggestion, for I have enjoyed writing the book.

ASHLEY MONTAGU

*Princeton, N. J.*

# PREFACE TO THE SECOND EDITION

SINCE THE appearance of this book, in October, 1960, the breakthrough in our understanding of the basic mechanisms of heredity, as well as the increase in our knowledge of the relation between chromosomal aberrations and developmental disorders, has been so considerable that, without any exaggeration, they may be described as epoch-making. There have also been other advances. Without unduly increasing the size of this book or complicating the text, I have endeavored to bring the reader abreast of recent developments.

Every part of the book has been revised, new illustrations have been added, and additional material has been incorporated into the appendixes.

I am indebted to Dr. Kurt Hirschhorn, Department of Internal Medicine, New York University, who kindly read the book in page proof and saved me from some last-minute errors.

I am also most grateful to Miss Joan Meinhardt of The New American Library for the excellence of her editorial work on the book.

ASHLEY MONTAGU

*Princeton, N. J.*
*22 August 1963*

# PART ONE

# THE MEANING OF HEREDITY

*"The teachings of genetics may be summarized in Aristotle's saying—'the nature of man is not what he is born as, but what he is born for'—paraphrased perhaps into the form 'the inheritance of man is not alone what he is born with, but what he can develop.'"* HERBERT S. JENNINGS, *Prometheus*, 1925

# 1. INTRODUCTION

APART FROM IDENTICAL TWINS, have you ever seen any two persons who looked exactly alike? You haven't, and you never will. You never will because of the fact of variability —no two things are ever exactly alike—a fact that applies to all living things. That variety is the spice of life is not only true of human experience but constitutes a principle inherent in the very nature of nature. It is, indeed, the material of natural selection and the basis of social choice. For man the fact of natural variability is the best assurance against the dulling effects of uniformity and the nightmarish threats of totalitarianism.

What is the meaning of all this variation? Look about you in your own group. Observe how different people are in their appearance and in their behavior. In your own family consider how each member differs from the others, and yet how remarkably they resemble each other in some of their features and even in their mannerisms. The resemblance may be in the shape and form of the nose, the color of the eyes, body build, the walk, little mannerisms, and the like. One is very fond of music, another cannot carry a tune. One is allergic to pollens, the other is not, and so on. How do all these likenesses and differences come about in the same family and between different families? And what about the differences and likenesses among so-called races? Why are some people black and others white? Why do most Chinese and Japanese have "almond-shaped" eyes? Why are Pygmies pygmies? Why do people vary in intelligence so much? Which is stronger, heredity or environment? Should cousins marry? Should persons with such and such traits marry? What are "bluebloods"? What is—in short, what is heredity?

The questions just asked and others of a similar sort are all the concern of the science of heredity, genetics—and of the special branch of this science that deals with human heredity, human genetics, which is what this book is about. Genetics is the branch of biology concerned with the manner in which inherited differences and resemblances come into being between similar organisms.

In a world in which many human beings, for one reason or another, have been caused to feel insecure, it is a common response to want to cling to one's kind and to reject anything that is different. Hence, many people tend to commit the error of concentrating on likenesses without realizing the great importance and value of differences. Differences are what make the world go round, and if it were not for differences, there would be no likenesses—nothing endures but mutability. Indeed, it is becoming increasingly evident that each of us carries within him a great deal of the potential variability of the whole human species. As we learn to understand the causes of difference, variation, we learn to appreciate the meaning of differences as we see them among human beings of every kind and variety; far from utilizing those differences as pegs upon which to hang our prejudices or private peeves, we begin to perceive them as points of interest that, through the greater understanding we have acquired, serve to humanize and enlarge our interest in the world in which we are living. This is a world in which there exist not only physical differences and biological differences between individuals and between groups but also cultural differences, that is, differences in the ways of life of peoples. There are social differences, caste differences, class differences, political differences, national differences, and international differences. Somehow most of these differences have gotten themselves all mixed up in the minds of many people.

It is thought and believed by many that what a person is born with is somehow related to what he later does in life. To a certain extent this is, of course, true, but when that belief is extended to mean that what a person is born with *determines* what he will later do in life, and that therefore heredity is fate, a kind of predestination, such a belief becomes stultifying and damaging. Fortunately it can be shown to be quite unsound. Yet there are millions of people who subscribe to the belief that if you are born a member of one "race" or ethnic group, you are unlikely to achieve as much or achieve it as well as the member of another "race." There are millions of people who believe that race and society are inseparably connected, that some civilizations are highly developed and others less highly developed because the heredity of the groups involved differs in a manner significant enough to account for the differences in civilization. Millions believe that the differences in intelligence among individuals are due to heredity, and that therefore there is very little one can do about improving an individual's intelligence, which is set and determined by heredity. And then there are the plays like Clemence Dane's *A Bill of Divorcement,* which has been seen by millions of people in its stage and movie versions, its tragic theme being that because a father was insane his daughter could not marry, for fear she would pass on his insanity to her children or manifest it herself. More recently we have had William March's novel *The Bad Seed,* made into both a play and a movie, which demonstrates how the "bad" heredity of the mother, whose mother, in turn, was a murderer, shows up in her daughter, who has inherited her grandmother's disposition to engage in the not-so-fine art of mass murder. The novels, short stories, collective myths, and loose talk about these matters together constitute a formidable miseducative force concerning the facts of heredity. Add to this the confusion that reigns in the minds of many persons who are regarded by the public as the kind of people who *should* know, such as doctors and even biologists, and we have a quite sizable body of epidemic confusion about what heredity is and how it works. Finally, there is the word itself. It has more than once been suggested that we would be better off if the word "heredity" could be dropped altogether. And there is sound sense in the argument that the long-standing abuse of the meaning of a word constitutes the best reason for its total exclusion from common usage. Unsound words make for unsound ideas, and unsound ideas tend to result in unsound action. The word "race" is a horrid example. But most of the trouble is not with a word but with its users. And the difficulty with those who misuse "heredity" is that they simply do not give the word its full meaning, but use it rather in a too limited sense. That sense is usually far too fateful, too gloomy, and unhopeful.

On the other hand, when heredity is understood in the light of the scientifically discovered facts, it is seen to be a science that yields findings that enable us to contribute to the welfare and general happiness of mankind. No science is in itself either hopeful or unhopeful. A science is simply a system of knowledge derived from observation, study, and experiment calculated to determine the nature or principles of what is being studied. The truths revealed by such systematized knowledge make it possible for us as human beings to apply them to the solution of human problems. It is when we acquaint ourselves with the truths revealed by the science of heredity that we are then able to see what needs to be done and what may be done with human beings as such, what may be done with human beings as biological *and* social organisms.

Knowledge is not the end of wisdom but only the beginning of it. A knowledge of the facts of heredity should form part of the intellectual equipment of every citizen for the simple reason that the very process of being human requires an adequate knowledge of the principles of heredity if one is to be successfully human. Knowledge of the facts of heredity is indispensably necessary if one is to have the beginnings of an understanding of the nature of human nature. Had such understanding been widespread among the German people, for example, they could not possibly have fallen for the racist theories of Hitler and his minions. If they had been able to identify those theories for the poppycock they were, they might have been able to go on from that point to recognize Hitler's political programs and prognostications for what *they* were. False biology and fantastic theories of heredity, often with the full consciousness of their falsity, were foisted upon a whole nation because the greater part of that nation had not been educated in the facts, an education they could easily have received during their school days, for as we shall see, there is nothing intrinsically difficult or complicated about the essential facts of heredity.

In the name of fantastic theories of heredity millions of human beings were deliberately exterminated by the Nazis. The human tragedies, the suffering, the losses to the human species as a whole thus produced, can never be totaled—they add up to something beyond computation. Politics and heredity are indeed closely linked. It was Aristotle who defined politics as the complete science of human nature. Our politics will be greatly influenced by the view we take of human nature, and often it works the other way round, our view of human nature being influenced by our politics. It has been shown, for example, that scientists working in the field of heredity tend to divide up into two camps: the environmentalists, who believe that the environment is a more powerful influence than heredity, and the hereditarians, who believe that one's biological heredity is a more powerful influence than the environment. When inquiry was made into the political beliefs of a sample of such scientists, it was discovered that the lineup was pretty much as one might have expected: the environmentalists were almost invariably Democrats, the hereditarians almost invariably Republicans. And the amusing thing was this: two of these scientists, who in their younger days were socialists and environmentalists, in their older days tended toward Republicanism and became hereditarians! Dr. Nicholas Pastore has most interestingly presented these facts in his book *The Nature-Nurture Controversy*. Dr. Pastore's conclusion was that in general the political allegiances of the scientists were significantly related to their position on nature-nurture questions, and that in most cases it was probable that they were unaware of the specific impact of their political loyalties upon their scientific thinking. The opposite, however, is also true, namely, that some scientists modify their political views as a consequence of the development of their scientific ones—as was the case with several of the individuals mentioned in Dr. Pastore's book. Thus, when scientists tell us in a discussion that they are themselves uninvolved, we should take particular care to check on this and then allow for any involvement we discover in evaluating their assertions. This is a very necessary precaution, especially when discussing so important a matter as human heredity.

Scientists with political axes to grind—some of them of the highest scientific distinction—have been responsible for some most unfortunate social changes. Such scientists are likely to be anxious to introduce "reforms" in society wherever they can. As a consequence of their activities, our unfortunate immigration laws were first put into the form they have had ever since, except that they have gone from bad to worse, since politicians uneducated in the facts of heredity began to modify them. Many of the sterilization laws enacted by many states of the Union were also largely the work of politically minded scientists and pseudoscientists (see pages 299–303). The immigration laws were designed to keep out the "unfit," and the sterilization laws were calculated to prevent those judged to be "unfit" from reproducing their kind. In fact, both laws achieve nothing of the kind, but, in the case of the immigration laws, serve to work great injustice upon thousands of human beings who seek to become Americans and from whose admission to these shores America would greatly benefit, and, in the case of the sterilization laws, put a most dangerous and unwarranted power into the hands of the State. Politics and heredity are far from unrelated to each other.

Unquestionably there are certain individuals who should be discouraged from reproducing. But it is extremely necessary to have a thorough knowledge of the hereditary processes involved in particular cases before one may make such a judgment. Always the genetic processes involved in the

## 22 HUMAN HEREDITY

transmission of traits in cases of this kind must be diagnosable. And, of course, such diagnoses will be possible only on the basis of earlier studies made of similar cases. It is one of the purposes of this book to acquaint the reader with the grounds upon which such diagnoses are made.

What we know about heredity in human beings has been largely learned through observation, though a good deal has also been learned by experiment with human beings. What we have learned in the laboratory by experiment upon other animals we have found, upon further study and comparison, is applicable to virtually all living things, plant and animal. The same basic biological laws of heredity apply to plants, and flies, and cats and dogs, and monkeys, apes, and men. In the laboratory, using the experimenter's favorite animal, the fruit fly *(Drosophila melanogaster)*, it is possible to study about twenty to twenty-five generations in a year, owing to the rapid development of this fly. Counting thirty years to a human generation, this means that it would take 600 to 750 years to study the heredity of the number of human generations that the fruit fly produces in only a year. We have been helped in the understanding of human heredity by many other lowly animals—the mouse, the guinea pig, and man's oldest friend, the dog. But from the standpoint of heredity, man continues to be the principal object of his own interest and study, and in this book we will report not on fruit flies but on human beings—without forgetting our debt to the fruit flies and other animals from whom we have learned so much about ourselves.

## 2. IN THE BEGINNING

A million million spermatozoa,
  All of them alive:
Out of their cataclysm but one poor Noah
  Dare hope to survive.

And among that billion minus one
  Might have chanced to be
Shakespeare, another Newton, a new Donne—
  But the One was Me.

Shame to have ousted your betters thus,
  Taking ark while the others remained outside!
Better for all of us, froward Homunculus,
  If you'd quietly died!

SO WROTE Aldous Huxley's "Fifth Philosopher." The view is somewhat extreme. There has been only one Shakespeare, and Newtons and Donnes are excessively rare. The conditions that go to make a Shakespeare, a Newton, and a Donne occur by chance, and the chances of their occurring in any generation of human beings are vanishingly small. So it is far better that our "froward Homunculus" (forward little man) came through, for if we were to wait for the Shakespeares, the Newtons, and the Donnes to make their appearance, the human species would have ceased to exist long ere this—there just aren't that many geniuses in the makeup shared by all humanity. What is that makeup?

Every human being starts off as a fertilized egg. Every female at birth already has in each ovary about 250,000 undifferentiated eggs (oogonia). Many of these will develop into fully differentiated ova. The half million primordial egg cells in the female ovaries at birth were all produced by the time she was a fetus of five to six months of age. After that age no more of these egg cells are propagated. As the female, the potential mother, grows (increases in size) and develops (increases in complexity), the primordial follicles, the undeveloped ova, grow and develop into fully differentiated and mature eggs, or ova (Graafian follicles). Shortly after the female reaches puberty, the ova begin to pass singly out of one ovary at a time, approximately each twenty-eight days, and into one of the tubes, or oviducts *(fallopian tubes)*, the fimbriated openings, or commencements, of which are situated just above the ovary on each side, and from thence into the womb *(uterus)*. When the female pairs with a male, he introduces into her reproductive tract (at each mating) something over 250 million of his germ cells *(spermatozoa)*. These spermatozoa develop in his testes and are ejaculated at copulation.

After a human ovum has passed into the fallopian tube of the female, it has a life of a little over twenty-four hours. After this time, if it has not been fertilized by a spermatozoon, a process that occurs in the first third of the fallopian tube (that is, nearer the ovary than the uterus), the unfertilized ovum breaks down and dies. Only one spermatozoon fertilizes the egg out of the more than 250 million that have been introduced into the female reproductive tract.

The human ovum is a spherical cell approximately one-seventh of a millimeter, or 1/175 inch, in diameter, and weighing about one 20-millionth of an ounce; in other words, it would take about 20 million ova to weigh an ounce. If you want to visualize the actual size of the human ovum— and you can just barely see that single cell with the naked eye—its size is about one-half of the diameter of the period at the end of this sentence. If you think the ovum is small, it is a giant compared with the spermatozoon. You cannot see a spermatozoon without a microscope, for its volume is 85,000 times smaller than that of the ovum. It is shaped something like a very young tadpole, with a head, neck, body, and tail, and its total length is about 1/20 millimeter, or 1/500 inch, the tail making up the greater part of this length and the head being only 1/200 millimeter, or 1/5000 inch, in length.*

---

* Expressed in the customary language of science, we say that the diameter of the human ovum is between 130 to 140 micra (about 0.140 millimeter) where a *micron*=1μ, or 0.001 millimeter (one-thousandth of a millimeter, therefore approximately 25,400 micra in an inch), and the spermatozoon is about 0.05 to 0.06 millimeters long, or about 50 to 60 micra, the head being 5 micra long and 3 micra wide, the neck 5 micra long, and the tail about 50 micra long. The spermatozoon is, in fact, by far the smallest cell in the body. Some idea of its size may be gained by considering that if we laid the sperm heads like a pavement on the top of the period just a little larger than the one at the end of this sentence it would take

It has been estimated that since man's first appearance on earth, there have been born something of the order of 100 billion human beings. At the present time there are about 3 billion human beings. Allowing one sperm and one egg to each of them as being responsible both for their existence and for their genetic heredity, we have a total of 5.78 billion germ cells involved, a number that could fit into a container of about two and a half quarts. The sperm cells would occupy less space than an aspirin tablet. In fact, the chromosomes, the actual bearers of the hereditary particles, the genes, within the cells of this huge number would occupy less space than half an aspirin tablet! Reflect upon that! All the hereditary materials—the heredity of the whole human race—of all those now living could be contained within the space of half an aspirin!

Why is the egg so much bigger than the sperm? Presumably because it carries so much of the nutriment necessary for the fertilized egg to feed on. The sperm consists almost entirely of a *nucleus* in which the *chromosomes*, which carry the hereditary particles, the *genes*, are situated. The egg consists of both a nucleus and a large quantity of more or less transparent, viscous fluid called the *cytoplasm* (Greek *cyton*, cell; *plasm*, fluid). The nucleus of the egg carries gene-bearing chromosomes. That what is contained in the nucleus, rather than in the cytoplasm, is chiefly responsible for one's genetic heredity is suggested by the fact that children resemble their mothers no more often than they resemble their fathers. Mulattoes, who are the offspring of whites and Negroes, do not resemble their mothers more than they do their fathers. The nuclei of egg and sperm are, however, strikingly similar. When these are brought together at fertilization, through the union of a male with a female, and the sperm fuses with the egg, the hereditary materials are combined and the process of development commences. From the two cells thus combined into one, two will develop, then four, then eight, then sixteen, then thirty-two, and so on until the total number of body cells at birth has been reached, a total altogether of about 200 billion, and in the adult about 10 trillion, containing about $7 \times 10^{26}$ atoms. It has been calculated that only about forty-four divisions are necessary to bring into being the number of cells present at birth, and

---

about 2,500 to cover it. It would take only twelve eggs to cover the same area. The big egg is immobile, but the small spermatozoon is very active and can move at the rate of about an inch in three minutes.

only four more divisions are required to bring about the adult number. At fertilization you weighed about fifteen 10-millionths of a gram; at birth (if you were a seven-pound baby) you weighed 3,175 grams. In the nine months from conception to birth you increased your weight 2 billion times. Adults weigh about 50 billion times as much as they weighed at fertilization. Perhaps a more striking way of stating what this means is to say that it is as if a hen's egg, which weighs about two ounces, were to grow in the same period to a mass of 33,000 tons. Where did all this material that makes up so much of the bulk of the individual originate? Where did enough matter come from to get the baby up to the point of birth? And where did enough come from in order to turn him into an adolescent and then an adult? In both cases the answer is the same: from the environment; from the food that the mother took in and was able to pass on in chemical form to her child in the womb, and after its birth, as nature intended, through the milk from her breasts, and subsequently through the foods that human beings obtain from their environment. So we perceive at the outset that environment constitutes an indispensably necessary condition of development, for without the nutriment necessary for the energizing of the developing organism there could be no development.

What is environment? Anything apart from genes that can act upon or influence the genes or the organism is environment. Cells, among other things, constitute one another's environment. Small accidental variations in the environment of the egg may have a decisive effect upon development. The complexities of the mother's body constitute the baby's environment while it is in the womb. But these are matters with which we shall deal very shortly in some detail. Let us proceed to the discussion of the nature of the bearers of the hereditary materials.

## The Chromosomes and the Genes

In the nucleus of every cell there lie a number of threadlike structures, which are called *chromosomes* (Greek *chroma,* color; *soma,* body). The chromosomes are so called because they readily take up the color stain the biologist uses in order to distinguish these bodies more clearly. In the unstained state they are clear and transparent and therefore not easy to see. In man there are forty-six chromosomes in every body cell,* and half that number, twenty-three, in the germ cell, the egg or the sperm. When egg and sperm unite, they each contribute their twenty-three chromosomes, so that the fertilized egg from which all the body cells are derived contains the total of forty-six. Animals and plants of different kinds have different numbers of chromosomes, some, like the intestinal worm *(Ascaris megalocephala)*, having as few as two, and others, like certain ferns, having more than five hundred. The potato has two more chromosomes than man. The single-celled rhizopod *Castinidium variable* is estimated to have between 1,500 and 1,600 chromosomes! Number of chromosomes does not appear to be associated with the degree of complexity of an organism. In Table I is presented a list of some representative animals and their chromosome numbers.

Since the diameter of the nucleus of the cell is not much more than a thousandth of an inch, the chromosomes contained in it are extremely small. The stretched-out chromosome (it is usually coiled in the nucleus) is between two and three thousandths of an inch long and about a ten thousandth of an inch wide. The longest human chromosome is about 7 and the shortest about 1.4 micra in length. (See Plate 4 and Figs. 12a and 12b.) Because chromosomes are so small, their activities can be followed only with the aid of an extremely powerful microscope.

It is principally through studying the behavior of the chromosomes of plants and lower animals and observing the association between differences in the behavior and structure of chromosomes and differences in the behavior of plants and animals that we have learned to understand something of the mechanism of heredity. Throughout the plant and animal kingdoms that mechanism is basically similar.

The chief importance of the chromosomes is that they afford the means by which potentialities may not only be transmitted to offspring but also the means by which the variation and change may be transmitted, for it is in the chromosomes that are lodged the physicochemical packages that are themselves the basic hereditary materials, namely, the *genes*. Recently, scientists working in the laboratory have succeeded in partially following the action of the gene both structurally and chemically. This is one of the most marvelous of scientific achievements. To know how a gene produces a

---

* Until 1956 it was believed that the number was forty-eight. It has now been established that there are only forty-six.

## 28 HUMAN HEREDITY

hereditary change seemed to many scientists not so long ago the most difficult of all problems to solve. Today that problem has been solved for at least one human trait, and the way is now open for the solution of the manner of action of many other genes.

What is the size of a gene? It is ultramicroscopically small, and therefore no one has ever seen one. Estimates of size range between 4 to 50 millimicrons in diameter—a millimicron is one-millionth of a millimeter—estimated gene size is therefore between 1/250,000 and 1/20,000 millimeter, or about one half-millionth to one forty-thousandth of the size of the period at the end of this sentence.

How many genes are there in each of the twenty-three chromosomes? Estimates range all the way from about 500 to about 5,000, and for the gene content of all twenty-three chromosomes the estimated number runs from about 10,000 to 100,000, with about 30,000 as the estimate most in favor, that is, with about 1,300 genes in each chromosome (with the exception of the male Y chromosome). These estimates are for the genes in the sex cells; the genes in the body cells, having a full complement of forty-six chromosomes, would have double the number of genes given in the above estimates. In the fruit fly, *Drosophila melanogaster,* in which there are four pairs of chromosomes in each cell, it has been estimated that there are between 5,000 and 10,000 genes in each cell. Man has twenty-three pairs of chromosomes, so if we tentatively award him the same number of genes *Drosophila melanogaster* is believed to have on at least one chromosome, that is 1,250, then man would have at least 28,750 (1,250 × 23) genes in the chromosomes of just his sex cells, and trillions throughout his body. In a single mating the possible combinations between the twenty-three chromosomes of the male and those of the female are 8,388,608, or 2 raised to the twenty-third power, and the chance of any one such combination being repeated more than once is 1 in approximately 70 trillion, or $2^{23} \times 2^{23}$. The different combinations that a 30,000-gene system can assume reach a stupendous figure, that is, $2^{30,000}$ This is on a purely quantitative basis. When the physicochemical factors and those of the environment are introduced as modifying agencies, the possible differences in human development within the limits of the species become practically infinite.

A gene is a small region in a chromosome composed of a giant nucleic acid molecule or part of such a molecule and

## TABLE I
### CHROMOSOME NUMBERS IN SOME ANIMALS AND PLANTS IN THE DIPLOID NUMBER

| Common Name | Scientific Name | Diploid Number (2N) of Chromosomes |
|---|---|---|
| Intestinal worm | *Ascaris megalocephala* | 2 |
| Small crustacean | *Cyclops viridis* | 4 |
| Fruit fly | *Drosophila willistoni* | 6 |
| Fruit fly | *Drosophila melanogaster* | 8 |
| Fruit fly | *Drosophila obscura* | 10 |
| Fruit fly | *Drosophila virilis* | 12 |
| Opossum | *Didelphis virginiana* | 22 |
| Bullfrog | *Rana catesbiana* | 26 |
| Salamander | *Amblystoma tigrinum* | 28 |
| Beetle | *Tirirhabda canadense* | 30 |
| Bat | *Plecotus auritus* | 32 |
| Lizard | *Anolis carolinensis* | 34 |
| Minnow | *Fundulus heteroclitus* | 36 |
| Cat | *Felis domestica* | 38 |
| Mouse | *Mus musculus* | 40 |
| Baboon | *Papio papio* | 42 |
| Rat (albino) | *Rattus norvegicus* | 42 |
| Squirrel monkey | *Saimiri sciureus* | 44 |
| Gibbons | *Hylobates* (all) | 44 |
| Marmoset | *Callithrix chrysoleucos* | 46 |
| Red titi monkey | *Callicebus cupreus* | 46 |
| Man | *Homo sapiens* | 46 |
| Gorilla | *Gorilla gorilla gorilla* | 48 |
| Chimpanzee | *Pan satyrus* | 48 |
| Orangutan | *Pongo pygmaeus* | 48 |
| Reptiles | Class *Reptilia* | 48 |
| Siamang | *Symphalangus syndactylus* | 50 |
| Slow loris | *Nycticebus coucang* | 50 |
| Wood rat | *Neotoma floridana* | 52 |
| Capuchin monkey | *Cebus capucinus* | 54 |
| Tantalus monkey | *Cercopithecus tantalus* | 60 |
| Guinea pig | *Cavia cobaya* | 62 |
| Donkey | *Equus asinus* | 62 |
| Mule | ♀ horse × ♂ donkey | 63 |
| Hinny | ♂ horse × ♀ donkey | 63 |
| Horse | *Equus caballus* | 64 |
| Armadillo (Texan) | *Dasypus novemcinctus* | 64 |
| Herring gull | *Larus argentatus* | 68 |
| Arabian camel | *Camelus dromedarius* | 70 |
| Dog | *Canis familiaris* | 78 |
| Crayfish | *Cambarus virilis* | 200 |
| Paramecium | *Paramecium* | Estimated at several hundreds |
| Rhizopod | *Castinidium variable* | 1,500–1,600 estimated |
| Plants | | |
| Some fungi | | 4 |
| Pea | *Pisum sativum* | 14 |
| Onion | *Allium cepa* | 16 |
| Primrose | *Primula sinensis* | 18 |
| Corn | *Zea mays* | 20 |
| Tomato | *Lycopersicon* | 24 |
| Wheat, cultivated | *Triticum vulgare* | 42 |
| Potato | *Solanum tuberosum* | 48 |
| Cherry | *Prunus laurocerocerasus* | 170–180 |
| Mulberry | *Morus nigra* | 224–308 |
| Fern | *Ophioglossum vulgatum* | 512 |

## 30 HUMAN HEREDITY

believed to consist mainly of a chemical substance called deoxyribonucleic acid, or DNA for short. The stuff out of which life is made consists of protein (Greek *proteios*, prime), and nucleic acids constitute the blueprints that direct the manufacture of proteins. DNA is so structured that it carries within itself a kind of Morse code. Each chromosome contains hundreds of genes, possibly thousands—it is not known exactly how many. In the last paragraph we settled for about 1,250 genes on each chromosome. The genes are strung together in a row. A gene relating to a particular response or enzyme has a specific position, or locus, in the string. Nucleic acids are the blueprints that the proteins must follow if the body is to be properly built. If the blueprint isn't followed for some reason, something is bound to go wrong at the point of departure from it, and the result may be a slight or a serious disorder. The evidence strongly indicates that the nucleic acids control the making of the body's characteristic living substances—its proteins—and that DNA carries the master plans, or code. The information contained in the code is there in a particular order, and this order determines the order in which amino acids (chemical compounds that are the building blocks, or structural units, of proteins) fall into place in the protein molecule for which it is responsible. There are about twenty different amino acids. (See Fig. 5, p. 43.) A protein molecule may be made up of hundreds or thousands of amino-acid units. Each protein owes its uniqueness to the specific sequence of its amino acids.

It was a known defective protein in man that threw the first light on the probable manner of action of genes. There is a disorder of the blood that affects mostly individuals of Negroid origin. The condition is genetically inherited, and it is called sickle-cell anemia because if one takes a small sample of blood from such a person and places it on a slide under a cover glass to cut off the oxygen, and then examines it under the low power of a microscope, the red blood cells are seen to assume bizarre shapes frequently resembling a sickle. This is due to a difference in the chemical structure of the oxygen-carrying part of the red blood cell, the *hemoglobin* (*hema,* blood; *globus,* globe). By a most ingenious systematic chemical and electrical analysis it was found that the difference between normal hemoglobin and that which causes sickle-cell anemia is due to a difference in a single amino acid—in 1 out of the 574 amino acids constituting the hemoglobin molecule. The person afflicted with sickle-cell anemia is most unfortunate, and yet this severe disease is due to the simple fact that a glutamic acid unit has been replaced by a valine unit!

The theory of the nature of chromosome and gene is, then, that the chromosome consists of a chain of long DNA molecules and that the genes are segments of the chain. The DNA molecule, of which there are thousands in each chromosome, is believed to be made up of two sugar-phosphate chains that are twined around each other in a spiral structure thousands of turns long, linked together by certain basic nitrogenous substances (adenine and thymine or guanine and cytosine). The nucleic acids of DNA are made up of three components, which together form a small molecular unit called a *nucleotide*. Each nucleotide is a storehouse and supplier of energy, one of the most important forms of which is adenosine triphosphate (ATP). Each nucleotide consists of one molecule of a *nitrogenous base*, one molecule of a *sugar*, and one molecule of *phosphoric acid*. The nitrogenous bases are *adenine* and *guanine*, which are *purines*, and *cytosine* and *thymine*, which are *pyrimidines*. Thus there are four different nucleotides. All nucleotides are linked together in long chains, in which the purines and pyrimidines occupy well-defined and ordered positions. Were these positions always to be uniform, things would be very uncomplicated, repetitious, and uninformative. All the molecules of DNA would then be very much the same in all living things, and if DNA is the code-carrying material of heredity, all living things would then be of similar heredity, which is obviously not the case. The nucleotides of nucleic acid must therefore be arranged in a very irregular sequence indeed. This, in fact, has been demonstrated to be the case by Dr. E. Chargaff and his coworkers. These investigators found that the four nucleotides are present in nucleic acids in very unequal amounts. Furthermore the proportions of individual purines and pyrimidines were found to be different in the DNA of different species—which is precisely what we should expect if different DNA's are the carriers of different hereditary codes. If distinctive differences are to be maintained and transmitted, then the DNA molecule must be both complex and variably irregular in its structure. Such variability can be achieved simply by variation in the position of the purine and pyrimidine bases in relation to each other in complex, varying patterns. Such sequences of variant patterns constitute the genetic code—the code of heredity. Simply by changing the sequences of the bases in

## 32 HUMAN HEREDITY

one DNA molecule, the number of possible patterns resulting from this alone would amount to $10^{10,000}$! This is more than sufficient to account for all the variability that has ever existed among human beings from the very beginning of the human species. The number of human beings who have ever existed cannot have exceeded $10^{14}$. The number is probably less than $10^{11}$.

At a molecular weight of about 10 million, the DNA molecule contains about 300,000 nucleotides. With the double helical structure of DNA, this would yield a molecular unit of information about 150,000 nucleotides long, which represents a molecular coding capacity of quite considerable dimensions. Different DNA molecules may represent different genetic units, and the approximately 400,000 DNA molecules in the nucleus of the 46-chromosome cell clearly contains more than enough information for the development of the most complex of organisms. Dr. J. R. Goldacre, of the University of London, has calculated that the number of bits of information a fertilized human egg must contain to produce all the thoughts and deeds that will finally be thought and done by the person that egg will become is about $10^{12}$, that is, about 1 trillion.

In 1953 Dr. M. F. H. Wilkins and his colleagues reported the results of their studies on the scattering of a beam of X rays by specially prepared threads of DNA. They showed that DNA was a highly crystalline substance, and that calculations from the angle of scattering suggested that the DNA molecule was arranged as a coiled two-stranded helix.

Following this lead, and with the knowledge that the DNA molecule contained nucleotides of adenine, guanine, cytosine, and thymine, and that the ratio of all purines and pyrimidines in any DNA was found always to be equal to one, as if one purine always formed a pair with one pyrimidine, doctors J. D. Watson and F. H. C. Crick proposed the now famous Watson-Crick model of the DNA molecule.

Watson and Crick, in 1953, suggested that the DNA molecule is composed of two complementary sugar-phosphate chains, or strands, each consisting of a series of nucleotides, as shown in Plate 1, in which each spiral (chain, strand, or helix) is unwound and appears to be somewhat flattish rather than helical. Each spiral consists of five-carbon sugar molecules (deoxyribose) alternating with phosphate groups. These are situated toward the outside of the spirals, whereas the bases are situated toward the inside, facing the complementary bases of the other spiral, a purine always facing a pyrimidine in the other. Adenine always faces thymine and guanine always faces cytosine, or vice versa, to form a complementary pair of bases. A molecule of DNA may have as many as 10,000 such purine-pyrimidine pairs. Each strand is, therefore, also the complement of the other strand. A specific sequence of purines and pyrimidines of one strand can pair only with a very specific sequence of such bases on the other strand, that is, one with the complementary, but not the same, sequence of bases.

Each strand, then, determines the sequence of the bases in its opposite member, and each carries the complete information necessary for the direction of the numerous processes that interact in function and development to produce the organism. Reproduction is achieved by the unwinding and separation of each strand, and by each strand replicating its missing complement. Each strand can accomplish this because it carries the necessary information, the genetic instructions for itself as well as for its complement. The strands are similar, in principle, to the magnetic tapes used in electric computers. In the organism the "tapes" provide the sets of instructions that operate during development.

The genetic instructions are spelled out by the sequence of the nucleotides, of which there are about 10 billion in the forty-six chromosomes of the fertilized egg, not counting those in the sex chromosomes. A gene is a nucleic acid molecule made up of a thousand or so nucleotide pairs of DNA. Each gene directs the functioning of a particular process or group of processes in the functioning and organization of the body. For example, a particular gene is responsible for the selection and assembly of the proper sequence of amino acids, out of the twenty kinds available, that will provide an essential protein constituent of hemoglobin. Should, for some reason, an error occur in the selection of a single acid, the right one being replaced by a wrong one, the wrong message will be delivered to the hemoglobin and a deficiency, such as sickle-cell anemia, will result. The DNA molecule has been likened to an enormously long winding staircase, with the stair treads corresponding to the cross-linking chemical bases. The various segments of these DNA molecules each contain their specific codes or blueprints for the plan to be followed in the manufacture and organization of the basic materials out of which the organism is to be developed.

It has long been clear that such a master controlling system must be present in the chromosomes (code centers) or genes since the materials that enter into the making of a living organism are so numerous and so complex that if it were left to chance, only disorganization would result instead of organization. That, in fact, is precisely what the word "organism" means, an organized living entity, moreover, an organized living entity that functions to maintain organization and is opposed to all states of disorganization. Disorganization represents a breakdown in the system. Orderly and organized development is brought about as a result of the communication of the coded information contained in the living "tapes," the chains of DNA molecules. The genes are, as it were, the paragraphs in which the words or sentences are the nucleic acids (DNA). Each of the four bases contains a different bit of information. If a molecule, for example, contained only one base, it would either be A or T or G or C—in other words, there would be four possible kinds of information available. If a molecule contained only two bases, the possible kinds of information would then be $4^2$ (4 × 4), or 16, as AA, AG, AC, AT, GA, GG, GC, GT, CA, CG, CC, CT, TA, TG, TC, AND TT. If the molecule contained only three bases, there would be $4^3$ (4 × 4 × 4), or 64, possible kinds of information, as AAA, AAG, AAC, AAT, etc. A molecule containing four bases in sequence would yield $4^4$, or 256, possible kinds of information, and one with 10 bases will have $4^{10}$, or 1,048,576, possible kinds of information, while one with 20 bases would have $4^{20}$, or 10,995,111,627,776 possible kinds of information. When these bases are paired, as they are in the double helix of the DNA molecule, the number of possible kinds of information conveyable staggers the imagination.

Or put in another way, just as it is possible to form thousands of different words by using only four letters, so it is possible to make thousands of different genes and DNA out of the four bases. The genetic message of DNA is coded by the manner in which the four "letters" follow one another in making up a chain (strand, spiral, "tape") in relation to the "letters" on the opposite chain (see Plate 1). All this information is contained within the nucleus of the cell, information that, written out in words, would fill well over five hundred volumes of this size.

The *genetic code* represents the way in which a sequence of twenty amino acids is determined by a sequence of four bases.

Each gene is now thought to be constituted by hundreds

of bases arranged in a unique sequence within the DNA molecule. Sometimes, during the process of reduplication of the bases, one or more of them is erroneously replaced by one of the other three possibilities, and in this manner a *mutant* gene is produced, resulting in a transmissible hereditary modification in the expression of a trait. The effects of most such mutations are not usually detectable in any one generation. Such "mistakes" in the basic self-copying process of the gene provide the raw materials, the building blocks, of the evolutionary process. Most mutations occur at the rate of about one in a million copies. Without mutations evolution would be impossible. A mutation, then, is a transmissible alteration of the sequence of nucleotides in the genic DNA. This may take the form of the replacement of a single nucleic-acid molecule by a different purine or pyrimidine.

The particular order of the base pairs at a particular place on the chromosome is thought to determine a specific pattern of information. This is the gene. Genes not only instruct but also control the execution of their instructions. The genetic information is transmitted from the chromosomes in the nucleus by *ribonucleic acid* (RNA), which has been synthesized in the nucleus, where it has received the patterns determined by the DNA base-pair sequences. The RNA then migrates outside the nucleus into the cytoplasm of the cell and there imprints corresponding patterns on the enzymes, globulins, and other large molecules in the process of being synthesized. Specific patterns are transmitted from DNA to RNA and from RNA to the protein by a process resembling the manner of making a cast, or template. At least, this is the *template hypothesis*. It is further discussed below.

DNA is the model for RNA (ribonucleic acid). RNA is in the service of DNA and amounts to less than 10 percent of the total quantity of nucleic acid in the chromosomes. RNA is formed at the DNA sites in the chromosomes and stored in the nucleolus (or "little nucleus") within the cell nucleus. The function of RNA is protein synthesis. Whereas DNA is limited to the cell nucleus, RNA is for the most part found in the cytoplasm, that is, in the cell protoplasm outside the nucleus. While the cell is in its "resting phase," that is, while it is engaged in the processes of growth and development (increase in size and in complexity) between divisions, the chromatin material in which the cell's entire complement of DNA is located lies diffusely distributed in the nucleus. In this way the DNA makes maximum surface contact with the other materials in the nucleus, from which it puts together the molecules of RNA and replicates itself. Preparatory to division, the chromatin coils up tightly to form the chromosomes, which will then be equally distributed to the daughter cells.

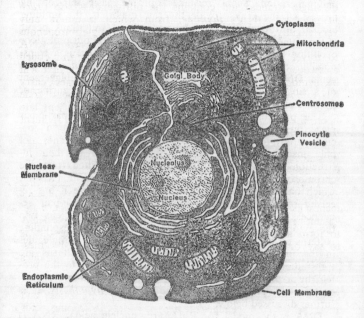

FIG. 1. Diagram of a Typical Cell. The ribosomes are the dots lining the endoplasmic reticulum, the sites of protein synthesis. The mitochondria are the power plants providing the cell with energy. Lysosomes are the bodies containing the digestive enzymes that break down larger molecules into smaller constituents that can be oxidized by the enzymes of the mitochondria. The centrosomes, shown both in longitudinal section and in cross section, move away to form the poles of the apparatus that separates two duplicate sets of chromosomes. Pinocytic vesicles represent areas of the cell membrane that invaginate to incorporate external materials into the cytoplasm. The function of the golgi body is unknown.

DNA-coded messenger-RNA carries instructions from the nucleus to the cytoplasmic ribosomes, the minute particles situated in great number on the margins of the endothelial reticulum in the cytoplasm (Fig. 1). Ribosomes are the assembly lines, the machinery, for the production of proteins. At the ribosome, messenger-RNA supervises the formation of protein. RNA differs from DNA in that it is composed of single strands, and contains the nitrogenous base *uracil* in place of DNA's *thymine,* from which it differs only very slightly. It may be this difference in a single nitrogenous base that keeps DNA within the nucleus and permits RNA to travel unimpeded into the cytoplasm. It is the coded messenger-RNA that determines the type of protein that will be synthesized, and not the ribosome. Ribosomes carry RNA, but ribosomal-RNA does not carry the genetic code —this is carried by messenger-RNA, which acts as a template upon the ribosomes. Ribosomes contain about 50 percent RNA and an equal amount of protein.

Messenger-RNA, or messenger-template-RNA, as it may be called, is synthesized with great rapidity by DNA replication, and may carry an ordered arrangement of as many as 1,500 nucleotides, the genetic code of the gene that produced it. Moving into the cytoplasm, messenger-RNA attaches itself as a template to a vacant ribosome. There the "blank" ribosomal-RNA is "keyed in" with messenger-RNA to form a specific protein. Another way of putting this is to say that the ribosomal "factories" are "hired" by messenger-RNA to make any protein for which RNA provides the instructions.

In the cell are free amino acids, for which specific enzymes are present. Activated by their specific enzymes and energized by ATP (adenosine triphosphate), the amino acids are picked up by the freely soluble fragments of RNA, which are distributed throughout the cell. These fragments are called transfer-RNA. Enzyme and phosphate are released and the amino-acid transfer-RNA units are transferred to the ribosome, where each sticks to a special place on the messenger-template-RNA according to the code. In other words, in the ribosome the transfer-RNA amino-acid units are aligned on the template of messenger-RNA, and messenger-RNA directs the linkage of amino acids into specific sequences.

One end of each amino acid attaches to one end of its particular transfer-RNA. The other end of that transfer-RNA strand sticks to the template like a foot. The free end of the amino acid is left dangling in proximity to its adjacent amino acid, and in that order these amino acids are then enzymatically linked together into a single specific polynucleotide chain that peels off the ribosome as a complete protein. This whole process, from beginning to end, takes no more than a minute and a half. It is diagrammatically illustrated in Fig. 2.

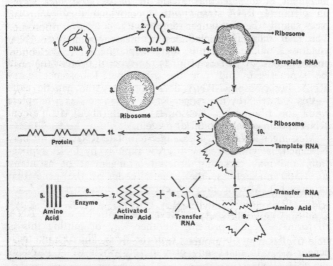

FIG. 2. Diagram Showing Steps in Protein Synthesis. Within the nucleus of the cell (1) is contained the chromosomal DNA with its master set of templates coded for the synthesis of proteins. Complementary templates are constructed out of nuclear RNA molecules (2), which then move into the cytoplasm and as messengers carry the genetic instructions to the ribosomes (3), where the RNA strands are laid down as a mold or template (4). Meanwhile, free amino acids (5), activated by specific enzymes (6) and energized by adenosine triphosphate (7), are picked up by and combined with specific transfer-RNA molecules (8); the transfer-RNA-amino-acid complexes thus formed (9) proceed to the ribosomes (4), where each fits into a special place on the template-RNA according to the code. The dangling amino acids, now arranged in a specific sequence, are enzymatically linked together (10), forming a polynucleotide chain that peels off the ribosome as a completed protein (11), containing about 200 of all or most of the 20 amino acids in various combinations. To function properly, a living cell requires many hundreds of proteins.

There are several varieties of transfer-RNA, each of which will attach itself to the adenylic (adenine) portion of a specific activated messenger-RNA amino acid at a specific point, and to no other. As we have already said, transfer-RNA molecules, one for each of the twenty amino acids, pick up and transfer the latter to the messenger-template-RNA at the ribosome. This is achieved by the attachment of a particular group of bases, in *triplet* form, on a transfer-RNA strand with a complementary group of bases along the messenger-template-RNA in the ribosome. The four different nucleotides in messenger-template-RNA yield $4 \times 4 \times 4$, that is, sixty-four different trinucleotide combinations, or triplets (Fig. 3). Most of the sixty-four pos-

| AAA | ACA | GAA | GCA | CAA | CCA | UUA | UCA |
|-----|-----|-----|-----|-----|-----|-----|-----|
| AAG | ACG | GAG | GCG | CAG | CCG | UAG | UCG |
| AAC | ACC | GAC | GCC | CAC | CCC | UAC | UCC |
| AAU | ACU | GAU | GCU | CAU | CCU | UAU | UCU |
| AGA | AUA | GGA | GUA | CGA | CUA | UGA | UUA |
| AGG | AUG | GGG | GUG | CGG | CUG | UGG | UUG |
| AGC | AUC | GGC | GUC | CGC | CUC | UGC | UUC |
| AGU | AUU | GGU | GUU | CGU | CUU | UGC | UUU |

FIG. 3. The Trinucleotides, or Triplets. **U** Uracil (Uridylic acid), **C** Cytosine (Cytidylic acid), **G** Guanine (Guanylic acid), **A** Adenine (Adenylic acid).

sible triplets may by grouped into twenty groups yielding the twenty amino acids, although it is probable that more than one triplet codes each amino acid, and more than one amino acid may be determined by a single triplet. In general it is the order, or sequence, of the bases in each triplet that codes a single amino acid. The set of bases that codes one amino acid is called a *codon*. The evidence suggests that all codons consist of a triplet of bases. What is the sequence of the bases in each triplet that determines the nature of the codon and the amino acid that it codes? In other words: What is the genetic code? The triplet dictionary?

The breakthrough in deciphering the genetic code came when doctors Marshall W. Nirenberg and J. Heinrich Matthaei, of the National Institutes of Health, in 1961, added to a cell-free extract of the colon bacillus, containing all the components necessary for protein synthesis, a synthetic RNA molecule consisting of polyuridylic acid, that is, of repeating units of uracil. The resulting protein molecule was found to be made up of repeating units of the amino

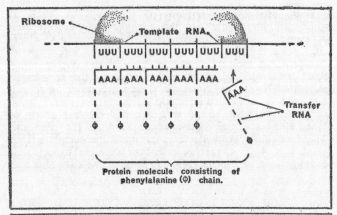
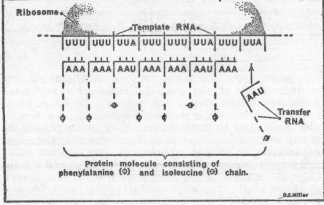

FIG. 4. Diagram of Protein Synthesis. Synthetic RNA containing only uracil (polyuridylic acid, or poly-U) added to mixtures of the 20 amino acids showed that (as in upper diagram) only those transfer-RNA's whose attachment sequence of poly-A, that is, polyadenylic acid containing only adenine, requiring that they carry phenylalanine, could match up, and a protein entirely of that amino acid result. The code letters for phenylalanine, therefore, consist only of U's. Poly-U template RNA can be "doped" with varying quantities of a single base, such as adenine (lower diagram), to direct the incorporation, for example, of isoleucine into protein, indicating that the code for that particular amino acid would be two uracils and one adenine, or UUA.

acid phenylalanine, UUUUUUUU . . . . . . . —which could only mean that the triplet codon for phenylalanine was UUU.

The triplets (but not the codons) for the twenty amino acids as we now know them to be are shown in Fig. 3. Though not yet proved, there is some evidence that the codon for tyrosine is AUU and for cysteine GUU. It will not be long before the exact codon has been worked out for each amino acid. At the present time the evidence strongly suggests that the code is universal, that it works in the same way in all living things, that is, as sequences of triplets or amino acids linked together in trinucleotide chains, and acting in the manner already discussed in the building of proteins.

From what has been said above it is evident that there are three main forms of RNA. What is their fate once they have performed their respective tasks? The facts are interesting. Messenger-RNA, being an unstable intermediate form of nucleic acids, as soon as it has delivered its message, breaks down into its individual nucleotides, which then become available for other uses within the cell. Transfer-RNA is stable, and once it has performed its task of aligning the amino acids on the template in the ribosome, it returns to its former state within the cell. Ribosomal-RNA is also stable, and when its work has been done it returns from its template, or "keyed," form to its blank, or neutral, state in readiness for any other coded message that may be brought to it for the formation of protein.

The facts of molecular genetics discussed above are set out in the box below in summary form.

## Brief Summary of Molecular Genetics

Sequence of AMINO ACIDS in PROTEIN MOLECULE is determined by sequence of BASES in region of a particular NUCLEIC ACID MOLECULE

The 4 BASES in the DNA (deoxyribonucleic acid) MOLECULE are: ADENINE, THYMINE, GUANINE, and CYTOSINE

In RNA (ribonucleic acid), URACIL is substituted for THYMINE

There are approximately 20 AMINO ACIDS, which constitute the BUILDING BLOCKS OF PROTEIN

The sequence of BASES that determine the structure of the 20 AMINO ACIDS occurs in TRIPLETS, that is, in the form of THREE LINKED BASES, or TRINUCLEOTIDES.

There are a total of 64 TRIPLETS

TRIPLETS in the RNA MOLECULE are situated on SHORT RNA STRANDS called MESSENGER- or TEMPLATE-RNA

A TRIPLET codes one AMINO ACID and is known as a CODON

It is possible that some CODONS may contain a small multiple of 3 BASES, such as 6 or 9, but this is not very likely

Most of the 64 possible TRIPLETS may be grouped into 20 groups

In general, more than one TRIPLET codes each AMINO ACID

The code is probably UNIVERSAL, being much the same in all living organisms

There are 3 main forms of RNA:

1. MESSENGER-TEMPLATE-RNA. Unstable-intermediary. Replicated in DNA in nucleus with substitution of *Uracil* for *Thymine*. Stored in nucleolus, but found mostly in cytoplasm. Carries instructions that are laid down as template in RIBOSOME, the PROTEIN-manufacturing assembly line, where RIBOSOMAL-RNA is organized into TEMPLATE form and where MESSENGER-RNA directs linkage and specific sequence of amino acids.

2. TRANSFER-RNA. Stable. Freely soluble, occurring as fragments in cytoplasm. Picks up free cellular amino acids, which, activated by specific enzymes and ATP (adenosine triphosphate), are carried by TRANSFER-RNA to RIBOSOMES

3. RIBOSOMAL-RNA. Stable. Assists in formation of RIBOSOMAL TEMPLATE for production of PROTEIN. This completed, resumes its former state and is ready for new instructions from MESSENGER-TEMPLATE-RNA

Fig. 5. Elements of the Genetic Code

| Amino Acid | RNA Bases or Code Word | | | |
|---|---|---|---|---|
| Alanine | CCG | UCG ■ | | |
| Arginine | CGC | AGA | UCG ■ | |
| Asparagine | ACA | AUA | | |
| Aspartic acid | GUA | | | |
| Cysteine | UUG △ | | | |
| Glutamic acid | GAA | AGU ■ | | |
| Glutamine | ACA | AGA | AGU ■ | |
| Glycine | UGG | AGG | | |
| Histidine | ACC | | | |
| Isoleucine | UAU | UAA | | |
| Leucine | UUG | UUC | UUA | UUU □ |
| Lysine | AAA | AAG ● | AAU ● | |
| Methionine | UGA ■ | | | |
| Phenylalanine | UUU | | | |
| Proline | CCC | CCU ▲ | CCA ▲ | CCG ▲ |
| Serine | UCU | UCC | UCG | |
| Threonine | CAC | CAA | | |
| Tryptophan | GGU | | | |
| Tyrosine | AUU | | | |
| Valine | UGU | | | |

**U** Uracil (Uridylic acid), **C** Cytosine (Cytidylic acid), **G** Guanine (Guanylic acid), **A** Adenine (Adenylic acid).

The sequence of the "letters" in the code "words," that is, of the bases in relation to one another in each triplet, or codon, has not been established; the order of the bases is therefore arbitrary. Several distinct triplets may code a single amino acid, and some triplets may code more than one amino acid, but each amino acid has its unique code word. Various combinations of AAC apparently code for asparagine, glutamine, and threonine. UUU appears to be the exception to the rule since it codes for phenylalanine and, somewhat less effectively, for leucine. Eighteen of the amino acids can be coded by only two different bases.

△ Uncertain whether UUG or GGU. ■ Need for U uncertain. □ Codes preferentially for phenylalanine. ● Need for G and U uncertain. ▲ Need for U, A, G, respectively, uncertain.

As we have seen, the reference system that revealed the manner in which the genetic code works is protein because the manufacture of protein is directed by nucleic acid. When the double-stranded DNA molecule unwinds, it separates into two strands. Each strand can then replicate itself and so produce more DNA, or it can direct the manufacture of ribonucleic acid (RNA). It is the single-strand RNA that governs the production of protein. A mutation, or change, in DNA is passed on as a change in RNA, which transmits the alteration to the protein it makes. The change resulting from such a mutation reflects an alteration in DNA, and the size and nature of the alteration probably define a single gene.

Since RNA governs its own replication as well as protein production, working with RNA alone, doctors A. Tsugita and H. Fraenkel-Conrat, of the University of California, produced a mutation in tobacco-mosaic virus by stripping the protein coat from the RNA and then treating the naked nucleic acid with nitrous acid. When the virus particles were reconstituted and rubbed on a tobacco plant, the viral protein now produced a kind of infection different from the usual. The mutant change in RNA had produced a corresponding change in the viral protein. The change in the protein was identified as having occurred near the end of the molecular chain, so that for the first time it has become possible to associate a specific alteration in the composition of a molecule, that is, its code, with a specific change in its effect.

In 1958 DNA was for the first time synthesized in the laboratory by doctors S. M. Beiser, A. Bendich, and H. Rosenkranz at Columbia University. They found that artificial DNA inhibits the natural substance's ability to cause heritable changes in pneumococci, the organisms that cause pneumonia. Some pneumococci are resistant to streptomycin. DNA taken from a resistant strain can make a sensitive strain resistant, *and this* change is subsequently inherited. Artificial DNA, however, stops the reaction. This discovery may lead to similar results in such conditions as cancer, in which the cells are thought to have an altered or abnormal DNA.

In bacteria it has been possible to transfer the DNA of one organism to the body of another bacterium, and in this way to transfer the heredity of one bacterium to another. From such experimental advances we may some day go on to achieve the correction of functioning in a defective gene by exposing the tissue cells in man to DNA derived from normal individuals or DNA synthesized in the laboratory, and in this manner restore the individual to normal health.

In May, 1961, Dr. Austin S. Weisberger, of Western Reserve University, reported the apparent transformation of cell function in human material by DNA. Weisberger took hemoglobin-producing cells from patients with pernicious anemia (megaloblastic) and with sickle-cell anemia. He put extracts of megaloblastic marrow and nutrients into two flasks. He added sickle-cell DNA to one, and to the other, the control flask, he added physiological saline solution. Ten hours later a completely new kind of hemoglobin began to appear in the first flask. This was neither sickle-cell nor megaloblastic hemoglobin. What it is is yet to be determined. Possibly it is a form of hemoglobin on its way to becoming sickle-cell hemoglobin as directed by the DNA from the bone marrow of the patient with that disorder.

In April, 1960, doctors Paul M. Doty and Julius Marmur, of Harvard University, succeeded in "unzipping" the two stranded DNA molecule into two distinct strands and "rezipped" it without destroying its biological activity. Hybrid molecules have been built by using half the DNA from two different strains of bacteria. "Crosses" were also produced by building hybrid molecules with DNA from two different species.

From the example of sickle-cell anemia we have gained a general idea as to how a gene probably acts. A gene may be regarded as a unit of reproduction, responsible for one or more specific steps in development. We have now to ask the question, How does a gene reproduce itself? And how can all the complexity that a human being represents (or an earthworm, for that matter) be reproduced as happens in the process so well called *reproduction?* The two questions are not quite the same, but they are very closely related.

Reproduction is just another word for *reduplication*. However, the word "reduplication" seems to describe rather more nearly the manner in which a gene reproduces itself. We have good reason to believe that a gene reproduces itself by making a duplicate, or replica, of itself. But how does a gene make a duplicate of itself?

The human species, *Homo sapiens* (Latin *homo*, man; *sapiens*, wise), as well as every other living species, owes its existence to the fact that living organisms are able to make copies of themselves. These copies reproduce themselves into other copies, and in this way the species maintains its specific character, generation after generation. We say then that there is an inbuilt specificity in the basic structure of every organism, that the manifest characteristics of an organism, the phenotype, consist of an exact sequence of amino acids in protein—the genetic constitution, the genotype, consisting of a corresponding sequence of nucleotides in DNA. The little girl who when asked by her teacher, "What is it that an elephant has that no other animal has?" replied, "Little elephants," was, of course, perfectly correct in terms of genetics, although the teacher wanted an answer in terms of anatomy. The reason that elephants have little elephants and not little animals of other sorts is because the information

## 46  HUMAN HEREDITY

contained in the elephants' genes is "coded" in such a manner as to produce copies of themselves. Precisely how this is done we do not know, but it is generally agreed that the process of copying probably resembles something like the process of making a mold, or template. When a sculptor wishes to make exact copies of his work, he uses his original piece of sculpture from which to make a mold, from which he can then make innumerable copies. The mold he makes is generally a negative cast, and it is from the negative casts that the positives are made. This is done by pouring the liquid material into the mold and allowing it to flow into every nook and crevice until it has hardened and a positive cast emerges.

A template is a pattern, usually in the form of a thin plate, used for testing accuracy of form in woodwork, machine parts, and the like, and in another form it is used in the mass production of all sorts of industrial and commercial products, such as the parts of automobile bodies and all sorts of familiar objects. It is this latter form of the template that many biologists believe represents a crude model of the living template by which the organism reproduces its molecular structure. The theory is that the DNA molecule by its very structure constitutes a perfect template. Consider the structure of the DNA molecule as shown in Plate 1. It has already been stated that the DNA molecule has been likened to a long winding or spiral staircase. Let us, for diagrammatic purposes and clearer understanding, straighten it out and unwind the staircase so that in the diagram it now resembles something like a spiral ladder with two uprights and a lot of rungs (treads). The uprights consist of sugars bound together by phosphates, and the rungs constitute the nitrogen bases, consisting of (1) adenine, (2) thymine, (3) guanine, and (4) cytosine. The rungs of the ladder always consist of one of the two pairs, 1–2, adenine-thymine, or 3–4, guanine-cytosine. The two uprights of the ladder are perfect complements of each other, and the bases 1–2 and 3–4 are always linked together in a pair on each rung, either in the order 1–2, 2–1, 3–4, or 4–3. There is, therefore, one base forming the left side of the rung and another forming the right side of the rung. Now suppose we cut the rungs clear down the middle of the ladder. We should then have two independent uprights with half a series of rungs, the left series oriented to the east and the right series oriented to the west, and each complementing the other. Thus separated, each half ladder immediately picks up from the protoplasm in the cell the chemical substances that form the complementary uprights (sugar and phosphate) and bases, base 1 picking up base 2, base 2 picking up base 1, base 3 picking up base 4, and base 4 picking up base 3. In this manner the two separated halves of the DNA molecule become two separate, identical DNA molecules.*

## Information and Cell Functions

The structure of all the protein molecules, which constitute the tissues and fluids of the body, is largely determined by the genetic information carried in the fertilized ovum and transferred by cellular division to all descendant cells. Coded in molecules of DNA, every living cell contains in its chromosomes all the genetic information necessary for its replication. Yet not even the simplest unicellular organism requires all the information at its disposal in the DNA at every stage of its life cycle. In multicellular organisms the cells become specialized and therefore require only a small amount of the total available information. How is this information turned on and off? A possible mechanism has been suggested by doctors Ru-chih C. Huang and James Bonner, of the California Institute of Technology.

These investigators found that chromatin (chromosomal substance) consists of about one-third DNA, one-third of the protein histone, one-sixth of another protein, and one-sixth RNA. The nonhistone protein includes the RNA enzyme polymerase, which facilitates the synthesis of messenger-RNA. Instructions for the synthesis of specific proteins are transmitted from DNA by messenger-RNA. When

---

* Dr. R. L. Sinsheimer has shown that the dwarf virus, Phi X, which infects and destroys sewer bacteria, consists of but a single strand of DNA wrapped in a skin of protein. The DNA is composed of one molecule made up of "beads" of nucleotides. Apparently a single strand of DNA is sufficient for the dwarf virus to reproduce itself. Twenty minutes after Phi X invades a cell, it forces the cell to manufacture about 300 Phi X viruses, each capable of infecting a new cell.

Whether or not the dwarf virus is unique with its single strand of DNA is not at present known, but the single strand of DNA is built on the same pattern as the double one. The virus being a somewhat primitive form of life, it is not surprising that it shows the simplest form of DNA.

An excellent and instructive DNA model kit for preparing three-dimensional models of the Watson-Crick DNA double helix, devised by R. V. Potter, can be obtained for one dollar from Burgess Publishing Company, 426 South 6th Street, Minneapolis 15, Minnesota.

two isolated, suitable amino-acid building blocks (riboside triphosphates) are added, messenger-RNA is synthesized, but at a rather slow pace. Removal of the histone fraction from the chromatin immediately quintuples the rate of RNA synthesis. The histone apparently functions to bind DNA and block the expression of particular genes, according to the requirements of particular cells at particular times.

The process of doubling and copying described in the preceding pages is at present still a hypothesis. But there are at present many scientists at work endeavoring to test this hypothesis, to discover whether or not this is how genes, the hereditary materials, really do reproduce themselves. In any event, whatever the manner in which the process of doubling may be achieved, it is believed to occur during the process when the cell is reproducing itself, a process called *cell division*. The doubling occurs long before the actual division of the cell, so that the daughter cells will have as many genes as the mother cell. During one of the phases of cell division the threadlike chromosomes are seen to duplicate themselves, so that where formerly there were a certain even number there are now double that number. It is probably at this phase of cell division that the genes are also duplicated in something like the manner described above.

The twenty-three pairs of chromosomes are of different sizes, as shown in Fig. 12 (page 92). The large chromosomes replicate at a different period of time from the smaller ones. The two homologues of most chromosome pairs also duplicate at different times. The two arms of the X chromosome replicate at different times, and the Y chromosome replicates at a time different from that of the smaller chromosomes. The interval between DNA synthesis and the beginning of chromosomal replication varies between four to five hours.

It should be quite clear that if chromosomes duplicate themselves, and the genes are duplicated during cell division, there must be some mechanism that holds the number constant in each body cell. Every cell in the human body contains forty-six chromosomes, or twenty-three pairs. The only exception to this statement being that the germ cells contain a total of only twenty-three single, or unpaired, chromosomes. And this fact gives us the answer, at least in part, as to how it comes about that the number of chromosomes and genes is not in fact doubled at every cell division. It is this process of cell division that we must now discuss.

## Cell Division

"Cell division" is another name for reproduction. When we speak of cell division we mean something very definite, namely, the series of continuous changes that occur during the process of duplication, in which all the elements of the cell are identically reproduced.

The process of body cell division is known as *mitosis* (Greek *mitos*, thread; *-osis*, condition, hence, the condition of forming threads). The word "mitosis" refers, then, to the essential characteristic of the process, which is the duplication of the threadlike chromosomes. The duplicated chromosomes then separate from each other and become members of separate nuclei in separate cells in this way: The thin, long, threadlike chromosomes remain quite separate from one another and gradually thicken into short rodlike bodies. The nuclear membrane of the nucleus within which they lie dissolves, so that the chromosomes now lie freely within the general protoplasm of the cell, the cytoplasm. A part of the protoplasm forms a spindle-shaped body, which encloses the chromosomes. The duplicated chromosomes take up a position at the widest part, at the center of the spindle. The two lines of duplicated chromosomes then separate and pass in opposite directions to the tips of the spindle. Here a new nuclear membrane develops around each group of chromosomes, and the cell then divides between the two new nuclei, and there are now two complete and completely separate cells that, at least so far as their chromosomal components are concerned, are completely identical. The chromosomes then become gradually invisible, until the cell is ready to go through this whole process again. Thus, from one cell there are produced, at varying rates in a geometric progression, two cells, then four, then eight, sixteen, thirty-two, sixty-four, and so on until in the human body there are many billions of cells—the estimates range from 26 billion to more than 100 trillion.

How much time does this whole process of mitosis take? It depends upon the demands of the body. The process is often completed in less than thirty minutes. In organs that are being constantly renewed, such as the skin, the intestines, and the blood-forming tissues, cell division is rapid in order to replace cells that are being worn out and sloughed off. In kidney and liver cells the process is slower. Nerve cells are not replaced at all. Hence, damage done to cells in the brain is not reversible.

## 50 HUMAN HEREDITY

By this time it may have occurred to the reader to ask himself the question, If all the genes in all these cells are identical, how then can one account for the development of the many different parts of the body? It is a good question, and one that happens to be among the most difficult in the whole realm of biology. No one is at the present time in a position to answer it adequately, but it is possible to discuss the question in the light of such evidence as we have and thus suggest something of the probable nature of the answer. But before we come to that we have a little way to go yet.

### Meiosis: The Reduction to Half the Number of Chromosomes in the Germ Cells

The cellular changes by means of which the full, or double (*diploid*), number of chromosomes is reduced to half the number (*haploid*) in the germ cells is variously known as maturation or reduction division or *meiosis,* from the Greek word meaning "to make smaller." Meiosis occurs only in the reproductive cells and is characterized by *two* nuclear divisions in rapid succession.

In meiosis the chromosomes uncoil into slender threads, but instead of doubling, as in mitosis, each chromosome arranges itself in longitudinal contact throughout its length with the one originally derived from the opposite parent. (Remember that twenty-three chromosomes were derived from the mother and twenty-three from the father.) These two chromosomes become closely entwined about each other, lengthwise (see Fig. 6). In this way each gene is brought into place opposite its corresponding gene. This process is called *conjugation* or *synapsis*.

Each chromosome of a synapsed pair now doubles, just as it does in preparation for mitosis. There are now four *chromatids*, which is the name for each of the four strands of new chromosomes (the four together are called a *tetrad*). Each pair of strands, or chromatids, that arose in the doubling process is called a *bivalent* or *dyad*. In each group of four synapsed strands the pairs (dyads) pull away somewhat from one another. The original two synapsed chromosomes then separate, and it is found that two of the four strands (one from each dyad), in crossing each other, have broken in one or more identical places, and the broken end of each strand has united with an end of the other strand. This is a very important fact to understand, for what it means is that there has been an interchange of corresponding genes from the

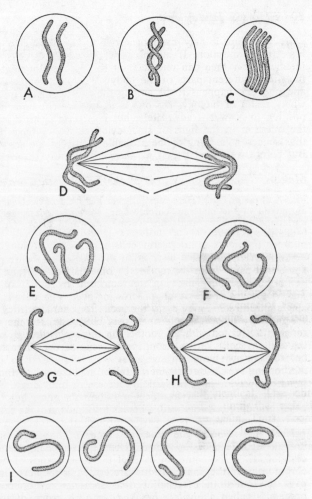

FIG. 6. Diagram of Meiosis. A. Nucleus of primary meiocyte—one pair of homologous chromosomes. B. Homologous chromosomes pairing. C. Each chromosome split and forming two chromatids. D. First Meiotic Division—chromatids separating in pairs. E and F. Nuclei of two secondary meiocytes—each with a pair of chromatids. G and H. Second Meiotic Division—chromatids of each chromatid pair separating. I. End of second division—four nuclei with one chromatid each. (After J. B. and H. D. Hill, *Genetics and Human Heredity*.)

52  HUMAN HEREDITY

father and the mother (see Fig. 7). By this process, called

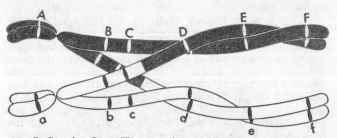

FIG. 7. Crossing-Over. Here crossing-over has occurred between a chromatid of the maternal chromosome, shown in white, and a chromatid of the paternal chromosome, shown in black. (After McLeish and Bryan.)

*crossing-over,* genes that were originally located in a chromosome that came from the mother become part of a strand of genes that originally made up a chromosome from the father, and vice versa. As a result of crossing-over, one part of a single chromosome may carry the genes for characteristics of the father, whereas the other part may bear genes for characteristics associated with the mother. In this way, through our sex cells, we pass on to our offspring some of the characteristics of our own parents.

Crossing-over is yet another mechanism responsible for a great deal of the kind of variation we see in individuals. We say that an individual has one grandfather's nose, but the other grandfather's forehead, whereas his hands are like those of one grandmother. It is partly through the crossing-over process that some of the genes and traits of both the male and the female ancestors come to be represented in the individual.

But let us return to the meiotic process. Following crossing-over, the four strands (tetrad) become shorter and more compact, and each group of four breaks up into two pairs (dyads), each pair then separating and passing to the middle of the division spindle. This is the first meiotic division. In the second meiotic division, or *reduction division,* each pair of strands separates into two single strands, each of which passes to opposite poles. There is no duplication of the chromosomes in this reduction division. In the manner just described the double, or diploid, number of chromosomes is halved, or becomes haploid, in the germ cells, and the numerical constancy of the chromosomes is thus maintained in the species from generation to generation. All this is clearly set out in Fig. 6.

The process of meiosis results in the formation of four nuclei, each carrying one chromatid from each of the four strands into which the two original chromosomes had split. But we have seen that as a result of crossing-over, the chromosomes may no longer be like those derived from the maternal and paternal chromosomes. Now, since, in addition, the paired strands (dyads) arrange themselves in relation to each other in a random manner, and the crossing-over can have occurred at hundreds of different places, each of the germ cells will be different from the other germ cells derived from the single original cell (mother cell) and will differ from all other germ cells derived from other mother cells. This means that every egg and every sperm carries an independent set of genes, which differs from those carried in every other egg and sperm. And that is one reason why no two individuals are ever exactly alike, unless there has been no change in the structure of the pairing chromosomes in the germ cells of the parents. For the same reason—that is, because of the differences existing in the germ cells—the exact type that the parents represented can never be restored or reproduced.

And so we see how it is that by means of the process of sexual reproduction (1) variation is achieved and maintained from generation to generation, (2) characteristics are rearranged and redistributed in such a manner as to maximize the improbability of any two things ever being exactly alike, and (3) the uniqueness of the individual is ensured by such rearrangement.

By means of the two consecutive cell divisions in meiosis, the number of chromosomes is reduced, so that the resulting sex cells have exactly half the number of chromosomes, twenty-three, as compared with the double number of chromosomes formed by mitosis. The four cells that result from this division of one male cell are called *spermatids*. The spermatids change into *spermatozoa* by developing a head. The spermatozoon is composed almost entirely of a nucleus and a neck, and is made motile by a long tail.

Heller and Clermont have recently shown that the whole process of spermatogenesis in man, from just before primary spermatocyte formation to the completion of the spermatids' metamorphosis into spermatozoa and their extrusion into the lumen of the seminiferous tubules, takes sixty-four days.

## 54 HUMAN HEREDITY

The primordial eggs, or ova—the oocytes—produced by the female, develop in the same way. However, in females, three of the four cells resulting from meiosis, the *polar bodies,* as they are called, are very small and soon disintegrate. Most of their materials have gone to form the cytoplasm of the giant mature ovum. The mature egg or sperm is called a *gamete* (marrying cell), the fertilized egg is called a *zygote* (yoked, or married, cell). At fertilization the double, or diploid, number of chromosomes is restored, and the processes of mitosis and meiosis are repeated in the formation of the body (somatic) cells on the one hand and in the formation of the germ, or reproductive, cells on the other.

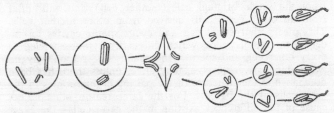

FIG. 8. The Formation of the Sperm. Commencing with two pairs of chromosomes, each member duplicates to form a pair of tetrads, or four-strand configurations. Two successive nuclear divisions result in the formation of four sperm, each with one member of each pair of chromosomes.

Out of the millions of body and germ cells, only a few germ cells will be directly engaged in transmitting the spark of life to the next generation. But this does not mean that the millions of other cells of the body are not involved—they are. They provide the environment for the germ cells. As Samuel Butler, a nineteenth-century thinker and writer, put it, "A hen is only an egg's way of making another egg."

### How Sex Is Determined

At fertilization, that is, the union of the male sperm with the female ovum, the twenty-three chromosomes contributed by the sperm and the twenty-three contributed by the ovum restore the number to forty-six in the zygote. How, in what manner, is the sex determined of the organism that will eventually develop from this coming together of chromosomes? The answer is provided by the fact that out of the twenty-

# IN THE BEGINNING 55

three chromosomes contributed by the sperm, one chromosome differs in character and structure from the twenty-two others, and the same is true of the chromosomes contributed by the ovum. These chromosomes—which differ from the other chromosomes, the *autosomes*—are the sex chromosomes, or *gonosomes*. There are two kinds of sex chromosomes: a large, well-nourished-looking sex chromosome known as the *X chromosome* and one that is a third to a fifth of the size of the X, known as the *Y chromosome*. Half the sperm cells carry a single Y chromosome in each of their heads; the other half carry a single X chromosome in each of their heads. Ova, however, all carry a single X chromosome and never carry a Y chromosome.

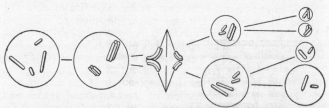

FIG. 9. Meiosis in the female results in only one functional egg—the three smaller *polar bodies* eventually disintegrate. At the first meiotic division one of the cells receives very little cytoplasm (this is the polar body); the other cell, the *secondary oocyte*, receives most. At the second meiotic division the first polar body divides into two, and the secondary oocyte divides into the ovum with most of the cytoplasm and another polar body with very little.

It is a beautifully simple arrangement. Any fertilized egg, or zygote, resulting from a union of sperm and egg, will have received either an X or a Y chromosome from the sperm and always an X chromosome from an egg.

If the zygote receives a Y chromosome from the sperm, the child will be a male because it has received a Y chromosome from the sperm and an X chromosome from the ovum, and XY sex combinations always yield a male. If the zygote receives an X chromosome from the sperm, the child will always be a female because the XX sex-chromosome combination always yields a female.

From these facts it is easy to see that the sex of a child is determined "accidentally." That is to say, it would appear that the ovum has a fifty-fifty chance of being fertilized by an X-bearing or a Y-bearing sperm and of thus producing females (XX) in 50 percent of cases and males (XY) in the other 50 percent of cases. Actually we know that more males are conceived and born than females. Between 120 and 150 males are conceived for every 100 females. We shall examine the possible reasons for this later.

Until April, 1959, it was believed that although the X and Y chromosomes are *associated* with sex they do not determine sex in themselves, that they do not in themselves carry genes for femaleness and maleness. The Y chromosomes came to be thought of as virtually empty of genes and, therefore, as having probably no influence on the determination of sex.

It was considered that sex is determined by the interaction of the genes on the twenty-two autosomes with the genes on the X chromosome. The best evidence indicated that the autosomal genes are oriented in the direction of maleness, that a single X chromosome is not sufficient to overcome this maleness orientation, but that a double dose of X chromosomes is. Hence it was thought that the sex of the zygote depends upon the relative *number* of X-chromosome genes that it receives, but that always the sex of the offspring is the result of the interaction of the autosomal genes with either one or two X chromosomes.

While these conclusions remain true for the fruit fly, *Drosophila,* from which creature they were extrapolated to other animals and to man, researches first published in April, 1959, and since now render it certain that the X and Y chromosomes play a much more significant role in the determination of sex than has been previously supposed by geneticists.

Because of the role that the X chromosome plays not only in influencing the development of sex but also in the distribution of the genes that it carries, it has greatly helped us to understand much concerning the nature of heredity that would otherwise have remained obscure. This, again, is a matter we shall discuss more in detail in the chapter on sex.

## 3. THE LAWS OF INHERITANCE

IN AMERICA we have perhaps better opportunities for observing how the biological laws of heredity work among human beings than in any other land—with the exception always of Hawaii. In America whites and Negroes have been mating for several centuries. Their offspring have been mating with others like themselves, backcrossing to Negroes or whites, and entering into all sorts of other conjugal combinations. Individuals belonging to the Mongoloid major group have mated with whites and with Negroes. American Indians have mated with whites, with Negroes, and with Mongoloids. In Hawaii Polynesians have formed a basic ingredient of many such intermarriages. Indeed, all the world over, such bringing together of diverse heredities has been going on before our eyes, and it is to be feared we have not always understood what was going on and only too often drawn the wrong conclusions.

Let us in a practical way, then, illustrate the basic laws of heredity as they are exhibiting themselves before our eyes.

We have already learned that genes constitute the basic particles or packages of heredity, and we know something of the manner in which they are transmitted to the body cells and to the germ cells. We have now to observe how those genes distribute themselves in the observable form, the *phenotype* (Greek *phainein,* to show, appear; *typos,* form; hence, the visible type), of human beings.

When a white and Negro mate, their offspring are known as mulattoes. The appearance of mulattoes, as everyone knows, is intermediate between their white and Negro parents. Their skin color seems to be a blend of the white and the black of the parents. The hair is not as kinky as it is in the Negro parent, the nose not as flat, and so on. Such facts have given support to the ancient idea that what is inherited is a blending of two separate heredities, a mixture of the two. If such matings stopped with or at mulattoes, it would be difficult to prove that their intermediate physical traits were not due to a blending of their "bloods." But since

## 58 HUMAN HEREDITY

mulattoes do go on to mate in all sorts of possible combinations, we can see at once, if we pay careful attention to the facts, that something is wrong with the blending theory, that, in fact, it won't hold up in the light of the observed facts.

When mulattoes mate, their offspring are not mulattoes. What, then, are they? Remembering that many different genes are involved and that these will be independently assorted, let us take a specific case. Two mulattoes marry—both the offspring of different white fathers and different Negro mothers. They have nine children. Two of these children have a black skin, kinky hair, broad noses, small ears with no lobes, and are in every respect indistinguishable from unmixed Negroes; occasionally one or two of the children may appear in every respect to be indistinguishable from American whites; the other five or six children will show intermediate shades of skin color and other characteristics. Many other families of a similar sort that have been studied show the same kind of segregation of traits. What such studies prove is that since skin color, to deal with only one trait at a time, is undoubtedly influenced by genes, the genes do not blend with one another but retain their independence generation after generation. But the same is true of all other traits. Take, for example, the case of the mother and child in Plate 2. Here is the case of a woman who is the offspring of a Basuto (South African) Negro mother and a white English father. She has brown-yellow skin, brown eyes, and kinky hair. She married a white man who had brown eyes and straight brown hair. Their two sons both had white skin, straight brown hair, and eyes like the father.

Such cases as these can only be explained on the assumption that the genes do not blend but maintain their integrity, generation after generation, and express themselves as such entities whatever the form of mating. Hence, the distribution of skin colors we encounter in the offspring of mulattoes is explicable in terms of the fact that since mulattoes carry genes for Negroid and white traits in their chromosomes, and since these will be distributed in a random manner in the sex cells, some of the sex cells will carry mostly Negro genes, some will carry mostly white genes, and some will carry varying proportions of these genes. Now, if two germ cells that carry mostly white genes unite, the chances are great that the offspring will be mostly white in its traits. If the genes carried in the germ cells are mostly Negroid, the chances are great that the offspring will be Negroid. And if the genes carried in the conjugating germ cells are of varying varieties, this will express itself in an apparent blending of traits.

In the instance of the Basuto-Negro hybrid woman mating with a white man, the explanation here is that the woman, who was the offspring of a Negro and a white, and therefore a mulatto, carried both white and Negro genes in her ova. Her husband, who was white, carried only the white genes in his sperm cells, and it so happened that his sperm united with ova that carried mostly white genes; hence, the two boys turned out to be whites, like their father and grandfathers. But it might have happened that the father's sperm united with ova that carried mostly Negroid genes, in which case the boys (or girls, as they might have been) would have been less white in their traits.

Facts such as these, which were first experimentally derived from work on plants and later on animals, gave us the first of the laws of heredity.

## The Law of Segregation

The first law of heredity states that in the zygote the paired genes derived from the two parents, which influence the development of traits, do not blend but retain their individuality and segregate unaffected by each other to pass into different gametes, and are thus able to enter into new combinations when they unite to form a new zygote. This first law (of Mendel) is known as the *law of segregation*. In man it is not possible to set up experiments to test the law of segregation, but this is possible in other animals. Let us, therefore, illustrate the manner in which the law of segregation works in a different animal, so that you may clearly understand it and also so that you may clearly understand how it works in man.

Speaking now entirely of physical traits or characters, it is known that any trait or character is under the control of pairs of genes, one gene being derived from one parent and the other gene from the other parent. Such corresponding pairs of genes are situated opposite each other on a single chromosome, and each member of such a pair is known as an *allele* (Greek *allelon*, of one another). Actually, they are not limited to acting on single characters but have many different effects and influence the development of many different characters. Genes, then, have multiple effects. When each of an allelic pair of genes produces the same effect, the genes

## 60 HUMAN HEREDITY

are said to be in a *homozygous* state (Greek *homo,* same; *zygotos,* yoked). When each of the genes of an allelic pair produce different effects, they are said to be in the *heterozygous* state (Greek *hetero,* different). For example, a person with two blood groups of A is homozygous for A, a person with one gene A and one gene B is heterozygous for A and B.

To trace the action of a gene we generally take a particular trait and observe how it behaves in inheritance. By this means and by experiment we can discover what gene or genes may be involved.

As a practical example let us cite an experiment on birds, focusing our interest on color, and examine what happens to the color of the feathers in such a breed as the Andalusian, a chicken similar to the leghorn. When a black bird is crossed with a splashed white, all the offspring in the first filial generation ($F_1$) are blue (actually they are gray, but the bird fanciers prefer to call them blue). When bred together the blues produce in the second filial generation ($F_2$) one-fourth black, which breed true, one-half blue, which breed like the $F_1$ blues, and one-fourth white, which breed true (see Plate 3). So we have in the second filial generation one black, two blues, and one white, the blacks when bred with each other yielding only blacks, the whites when bred with each other yielding only whites, and the blues when bred with each other yielding one black, two blues, and one white. The second filial generation is more variable than the first filial generation. The blacks and whites are in equal numbers, the blues twice as numerous as either blacks or whites, the ratio being 1:2:1.

As in the case of human skin color, such facts tell us that throughout the process of gene exchange and transmission the genes have not altered one bit but have retained their identity. In the first filial generation (corresponding to our mulattoes) there appears to have been blending since all the offspring are blue. But in the second filial generation there is a segregating out of the original colors as well as the appearance of the blended forms. Half are the original colors and half are blended. But the blends are not anything other than color blends; the genes remain unblended in these animals, which is proved by the fact that when they are mated, their genes combine to produce pure blacks, pure whites, and blues, and the interbreeding blacks always yield blacks and the interbreeding whites always yield whites.

The black Andalusians were produced, therefore, by one black gene from one parent and another black gene from the other parent. The white Andalusians were produced by a white gene from one parent and a white gene from the other parent, and the blue Andalusians were produced by a white gene from one parent and a black gene from the other. The parents contribute equally to the heredity of their offspring, and it doesn't matter whether it is the father or the mother who is one color or the other. This being so, it should be clear that when blues are mated, and they have offspring that are black, blue, and white, the genes carried by the blues must have remained distinct and must have separated from one another so that each gene entered different germ cells. Since in mating the chances are equal that each germ cell will unite with its own or with the other type—that is, black with black, white with white, black with white, and white with black—we may expect the ratio of color varieties we actually get, namely, one black, two blues, and one white. The black birds are produced by a pair of like genes, and so are the white birds—they are each homozygous for their particular color genes—but the blue birds are produced by a pair of unlike alleles, and are therefore heterozygous for the genes influencing the color of their plumage. It should be clear why blues can never breed true.

We conclude from these facts, then, that the genes influencing the development of characters are present in pairs (alleles), that each allele (member of the pair of genes) separates from its fellow allele during meiosis and enters a separate germ cell, which is the process of segregation. At fertilization the separated genes are brought together in a manner that is random, so that genes that find themselves paired do so by pure chance. Throughout this process the genes retain their identity.

Literally millions of observations that have been made on plants and animals support in every detail the facts described by the law of segregation.

This law and the other fundamental laws of heredity were discovered by an Austrian monk, Father Gregor Mendel (1822–1884), experimenting in his spare time in the Augustinian monastery garden at Brünn, of which he eventually became the abbot. After eight years of experiment and analysis Mendel presented his findings before the local Brünn scientific society in 1865 and published his results a year later. But the value of his work went completely unrecognized until sixteen years after his death. In the year 1900, three different scientists, Hugo de Vries of Holland, Karl Correns of Germany, and Erich von Tschermak-Seysenegg of

Austria, almost simultaneously brought the importance of his work to the attention of the scientific world. Mendel simply entitled his paper, "Experiments in Plant Hybridization." He did not announce it for what it was, namely, the discovery of the age-old mystery of the laws governing the inheritance of traits. Had he done so, it is just possible that he would have gained the attention for his work that was denied it for thirty-five years. "My time will come" were Mendel's last words. It has. Mendel would be astonished and delighted to know how important the science he created—*genetics,* the science of heredity—has indeed become.

What Mendel did was not what his predecessors had done. Instead of studying, as they had done, whole complexes of traits and lumping together parents, offspring, and their descendants, Mendel clearly saw that it was the inheritance of separate traits that must be studied, and that each of the members in each generation must be examined and recorded separately and as a member of its distinctive generation. Wrote Mendel in his original paper, commenting on the failure of previous investigators to do this, "It requires indeed some courage to undertake a labor of such far-reaching extent; this appears, however, to be the only right way by which we can finally reach the solution of a question the importance of which cannot be overestimated in connection with the history of the evolution of organic forms."

## The Law of Independent Assortment

In reporting the results of his eight years of experiments on the garden pea *(Pisum),* Mendel "left to the friendly decision of the reader . . . whether the plan upon which the separate experiments were conducted and carried out was the best suited to attain the desired end." The friendly readers, and we know that there were more than one, apparently entirely failed to see the point.

Mendel studied seven pairs of characters in peas. These were seed form, color of seeds with coat, color of seeds without coat, form of ripe pods, color of unripe pods, position of flowers, length of stem.

Studying the offspring of a cross between a pea plant possessing round yellow seeds with one characterized by wrinkled green seeds, he found that the plants of the $F_1$ generation all had round yellow seeds. When the $F_1$ hybrids were crossed, the $F_2$ generation showed not only the two original combinations of characters, round-yellow and wrinkled-green, but also two new combinations, namely, round-green and wrinkled-yellow, in the proportions of 9/16 round-yellow, 3/16 round-green, 3/16 wrinkled-yellow, and 1/16 wrinkled-green.

By this experiment (and others) Mendel discovered the two additional laws of heredity. The second law is that the pairs of alleles conditioning the different pairs of characters in the offspring are distributed independently of each other in the way they recombine in the formation of characters. This is the *law of independent assortment,* or *the law of free recombination.* New combinations as well as old ones of genes and traits will occur, and it will be seen that every trait is inherited independently of every other trait.

Two important modifications of this law that have been discovered since Mendel's day are the effects of the *linkage* of genes, that is, of genes situated on the same chromosome, and of *pleiotropy,* the effects that a single gene is capable of exerting upon more than one character.

## The Law of Dominance and Recessiveness

In Mendel's experiment described above, the cross between round-yellow and wrinkled-green seeds yielded a first filial generation that in every case was round-yellow. Yet when these hybrids were crossed, the original combination of characters plus two new combinations showed up in the $F_2$ generation in the combinations 9/16 round-yellow, 3/16 round-green, 3/16 wrinkled-yellow, and 1/16 wrinkled-green. Evidently the two color factors differed markedly in their capacity to express themselves. One seemed to dominate the other. The same was true of the texture factors. Factors or characters that dominate over others are called *dominants,* and the characters over which they prevail are called *recessives.* Thus in Mendel's experiments with peas he found round-yellow to be dominant and wrinkled-green (which remained unexpressed in the hybrid $F_1$ generation) to be recessive.

Mendel's third law states that every character is represented by two genes, one derived from each parent. When these genes are different, one may dominate over or cover up the other. This is the *law of dominance and recessiveness.* It should be clearly understood that dominance and recessiveness refer only to the relative power of alleles to express themselves in the individual. Dominance or recessiveness in no way determines which of the alleles or traits become the

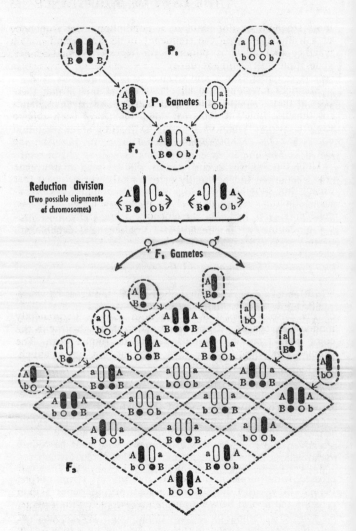

FIG. 10. Independent Assortment of Two Pairs of Chromosomes. (After Sinnott, Dunn, and Dobzhansky, *Principles of Genetics*.)

most frequent or dominant in a population. The frequency of a trait depends on the frequency of its allele, and not on whether that allele is dominant or recessive. Furthermore, genes that are dominant under some conditions may act as recessives under others, and vice versa.

Mendel's laws apply to all plants and animals. With these laws at their disposal, students of heredity had in their hands for the first time the tools by which they have been able to explore the workings of heredity. The result is a vast and complex body of knowledge that has grown up during the last sixty years.

Until the beginning of 1959 the individuality of the gene was considered to be largely unalterable. It was believed that except for spontaneously originating factors and such artificial factors as radiation, formaldehyde, mustard gas, colchicine, and the like, which alone were capable of producing mutations, genes remained constant and unchanged, transmitting their information unaltered from generation to generation. In January, 1959, The National Science Foundation reported the findings of Professor Alexander Brink, of the University of Wisconsin, which challenge the genetic axiom, hitherto prevailing, of the unalterability of the gene.

Brink found that the gene that produces color in the kernel of the purple variety of Indian corn can be permanently modified by bringing it into combination in a hybrid with the color gene from the stippled variety of Indian corn. The stippled gene produces the same pigment, but not as much, as the purple-corn gene.

In crossbreeding the stippled with the purple corn variety, the stippled color gene, or an element closely associated with it, produces a heritable change in the purple color gene. When the latter is removed by outcrossing from association with the stippled gene, the purple gene is found to have permanently lost its capacity to produce purple seed pigment in normal amounts.

Purple pigmentation is dominant, so that when stippled is crossed with purple corn, all the hybrids ($F_1$) are purple, though they carry both stippled and purple genes. Upon crossing the hybrid between purple and stippled with a white, or colorless, variety of corn, it was found that instead of half the offspring being purple and half being stippled, the half that should have been purple were now almost wholly white, retaining only the faintest quantity of pigment, whereas the stippled remained unchanged.

When a similar test is made of a cross between colorless

with normal purple corn which has not been previously associated with the stippled color gene in a hybrid, the conventional result is observed, namely, half the offspring are white and half purple.

The $R^r$ gene, conditioning color, is stable in homozygous condition, but is invariably changed to a weakly pigmenting form in the progeny of heterozygotes carrying the stippled allele $R^{st}$. Something in $R^{st}$ brings about a permanent change in $R^r$, $R^{st}$ is not affected by $R^r$ in the heterozygotes. $R$ is now changed to a new allele termed $R^{r:st}$. Test crosses on plants ($r^r r^r$ or $r^g r^g$) with colorless seed granules (aleurone) showed that $R^{r:st}$ will regularly revert partially, but not completely, toward the normal pigment-producing level of standard $R^r$. Thus $R^{r:st}$ is not entirely stable, and since it does not occur at random, it cannot be regarded as a mutation. Professor Brink has suggested the term *paramutation* to describe the fact that the phenomenon is distinct from but not unlike mutation.

Paramutation implies an extraordinary reactivity at some stage of cellular development of some genetic component of a chromosome when the latter is present in the nucleus with the modifying chromosome.

Professor Brink's important discovery will almost certainly be followed by corroborative findings on other forms. Meanwhile our conception of the means by which genetic variability is produced has been deepened and enlarged.

The question may be asked, What relation does Professor Brink's discovery have to the claims of the Soviet geneticist T. D. Lysenko that environmental influences such as soil, nutrition, and climate can change the hereditary properties of the gene? The answer to that question is that there is no relation between Professor Brink's discovery and that of Lysenko's claims, for what Brink's discovery proves is that a specific gene can alter the hereditary properties of another specific gene. The change is a purely genetic one, and is not an effect of external, or environmental, factors. (For a discussion of Russian genetics, see Appendix E.)

All this is by way of preliminary preparation for what is to follow, after we have considered some of the erroneous ways in which many people still think of the nature of heredity today.

## 4. "BLOOD IS THICKER THAN WATER"

"BLOOD IS THICKER THAN WATER." So it is. Blood is also stickier than water, and nowhere nearly as clear—which is exactly what may be said about a good deal of the thinking that has been left over in our society from pre-genetics days. The phrase "blood is thicker than water" is a relic of the days when the blood was thought to be the carrier of heredity. Many people still think this is so. They believe that at conception the blood of the mother mixes with that of the father, and it is in this way that the offspring come to exhibit a mixture of the traits of both parents.

How the idea came about that blood was the carrier of the hereditary traits that were transmitted to one's descendants is readily understandable. How ancient the idea is no one knows, but it must be very old. The fact that blood is closely related to survival is a conclusion that must have been drawn quite early in the history of man. The weakening effect or demise of the individual following an appreciable loss of blood has led all peoples of whom we have knowledge to identify blood with the vital principle itself. Among most peoples blood is regarded as a richly powerful element endowed with strength-conferring qualities as well as many others of every conceivable kind. To exchange blood or to drink blood from a common source has, among many peoples, meant the establishment of an unbreakable bond between those involved, for in this way they have become "blood brothers."

If the blood contains the life force, it was probably reasoned, then it must also contain the materials out of which the human being is built up and through which life is continued. Through such a path of reasoning, blood became the element through which not only the life-giving but also the hereditary qualities were transmitted. In this way all persons of the same family stock came to be regarded as of the same "blood."

The idea of "blood" as the carrier of the heritable qualities of the "race," or nation, has led to its application in such extended meanings as are implied in terms such as "blue blood," "blood royal," "pureblood," "full blood," "half blood," "good blood," "blood tie," "blood relationship," and "consanguinity." "Blood" comes to be identified with "race," with nation, culture, and even with religion, as in the terms "Negro blood," "German blood," "English blood," "Jewish blood," and "Islamic blood."

When we examine the usage of these terms, we begin to understand their meaning a little more clearly.

The term "blue blood," for example, refers to a presumed special kind of blood that was supposed to flow in the veins of members of ancient and aristocratic families. "Blue blood" is a translation from the Spanish *sangre azul*, the blood that was attributed to some of the oldest and proudest families of Castile, who claimed never to have been contaminated by "foreign blood." Many of these families were of fair complexion among a population that was dominantly brunet. In members of these exclusive families the veins would appear strikingly blue in comparison with those of the rest of the population. Hence, the difference between an aristocrat and a commoner could easily be recognized as a difference in "blood"; one was a "blue blood" and the other was not.

There is, in fact, no such thing as blue blood. The blood in the veins is dark red in color, and the veins themselves are creamy-white. The fact that the veins when seen through the skin appear to be blue is due to the refractive properties of the tissues through which they are seen, much as in the case of "blue" eyes.

The expression "blood royal," or "royal blood," refers to the widespread notion that only persons of royal ancestry have the "blood of kings" flowing in their veins. No person, it is held, however noble his ancestry, can be of the "blood royal" unless he has the blood of kings flowing in him. Kings have usually been held to belong to a special class of mankind, principally by virtue of the supposed unique qualities of their blood. In order to keep the "blood" of the royal house pure, marriages were arranged exclusively between those who were of the "royal blood." Even in ancient Egyptian times the Egyptian kings had so hypertrophied a sense of their own quality that in order not to adulterate their "blood" in any way they married their own sisters. Cleopatra, for example, was the wife of her brother Ptolemy XII, and after his early death she married her still younger brother Ptolemy XIII. As a queen, and as the pharaoh's daughter and therefore of divine origin, Cleopatra could not possibly have married anyone of a "blood" "lower" than her own family's.

In the manner in which they are usually used, terms like "full blood," or "pureblood," and "half blood" clearly illustrate the supposed hereditary character of the blood and the manner in which, by simple arithmetical division, it may be diluted. In this way "full blood" and "pureblood" are expressions alleged to define the supposed fact that a person is of unadulterated blood; that is, he is a person whose ancestors have undergone no admixture of "blood" with persons of "blood" considered undesirable. Within the last century these terms have come to be applied almost exclusively to persons who are not of the so-called white race, to persons, in short, whose place is alleged to be on the lower rungs of the "racial" ladder. It is possible that this restricted usage has been determined by the fact that these expressions have generally done most service in the description of native peoples or of slaves, as in "full-blooded Negro," "pure-blooded Indian," or merely "full blood" or "pureblood." Such an imputed lowly association would be sufficient to procure the nonapplication of the term to any member of the self-styled "superior races."

A "half blood," in contradistinction to a "full blood," or "pureblood," is supposed to be half of one "race" and half of another—for example, the offspring of an Indian and a white. What is actually implied is that whereas a full blood may claim relationship through both parents, a half blood may claim relationship through one parent only. For example, a mulatto, the offspring of a white and a Negro, is for all practical purposes classed with the group to which the Negro parent belongs, and his white ancestry is ignored. In practice, it often works out that the half blood is not fully accepted by either of the parental groups because of his "adulterated blood," and he becomes in the true sense of the expression a "half-caste," belonging to neither caste, for in the Western world the so-called different races are treated as if they belonged to different castes.

A person is said to be of "good" or "gentle" blood if he is of "noble birth" or of "good family." Here the assumed biological determinance of social status by blood is clearly exhibited; that is to say, a person's rank in society is assumed to be determined by his "blood," when, in fact, it is the other way round, "blood" actually being determined by rank. The ancestors of all noblemen were once commoners, plebeians. It was not a sudden metamorphosis in the composition of their blood that caused them to become noble; it was rather an elevation in social status which endowed them with supposedly superior qualities. Such qualities are not biologically superior in any sense whatever but rather belong purely to the ascriptive variety of things; that is to say, they have no real, but possess a purely imagined existence.

The statement that a person is of "bad blood," in the sense that he is of common or inferior character or status, is rarely encountered, for the reason, presumably, that those who use such terms have not considered the "blood" of such persons worth mentioning at all. For example, while there is an entry in the great *Oxford English Dictionary* for "blood worth mention," there is none for "blood *not* worth mention." In the sense in which blood is considered as the seat of the emotions, "bad blood" is taken to be the physiological or serological equivalent of ill feeling. In this sense, of course, "bad blood" may be created between persons of "good blood."

The terms "blood relationship" and its Anglicized Latin equivalent, "consanguinity," meaning the condition of being of the same "blood" or relationship, by descent from a common ancestor, enshrine the belief that all biological relationships are reflected in and are to a large extent determined by the character of the blood. This venerable error, along with others, requires correction.

This brief examination of the ways in which "blood" is used and understood in the English language, and in Western civilization in general, renders it sufficiently clear that most people believe blood to be the equivalent of heredity, and that blood, therefore, is that part of the organism that determines the quality of the person. By extension it is also generally believed that the social, as well as the biological, status of the person is determined by the kind of blood he has inherited.

These beliefs concerning blood are probably among the oldest surviving from the earliest days of mankind. Certainly they are found to be almost universally distributed among the peoples of the earth in much the same forms, and their antiquity is sufficiently attested by the fact that in the graves of prehistoric men red pigments are frequently found in association with the remains. These pigments, authorities believe, were probably used much as they are used among nonliterate peoples of the recent past and of today, namely, to represent the blood as the symbol of life and of humanity,

## "BLOOD IS THICKER THAN WATER"

a belief embalmed in the expression "he is flesh and blood" to signify humanity as opposed to deity or disembodied spirit. There in the grave was the flesh, and the pigment was introduced to represent the blood.

As an example of a myth grown hoary with the ages, for which there is not the slightest justification in scientific fact, the popular conception of "blood" is outstanding. It is a myth that enshrines so many errors that on that ground alone it would be desirable to outlaw this usage of the term, for it is false and extremely misleading. But what is far worse, these beliefs in "blood" are harmful in their effects and are capable of being severely damaging to human beings; hence, it is today more than ever necessary to set out the facts as science has come to know them.

In the first place it is necessary to say that blood is in no way connected with the transmission of hereditary characters or the bases for these. The transmitters of hereditary materials, out of which characters are developed, are the genes, which lie in the chromosomes of the germ cells—the ova of the mother and the spermatozoa of the father—and nothing else. The genes, carried in the chromosomes and cell substance of a single ovum and a single spermatozoon, are the only parts of the organism that transmit the genetic materials, which permit the development of the organism's characters. Blood has nothing whatever to do with the transmission of heredity.

The belief that the blood of the pregnant mother is transmitted to the child in the womb, and thus becomes a part of the child, is ancient and wholly erroneous. Blood does not normally pass from the mother to the fetus. The blood cells of the mother are far too large (0.007 millimeter, or 0.0003 inch, in diameter) to be able to pass through the placenta, and so are the blood cells of the fetus. The fetus manufactures its own blood, and the character of its various blood cells, both structurally and functionally, is demonstrably different from that of either of its parents. These facts should forever dispose of the ancient notion that is so characteristically found among nonliterate peoples, that the blood of the mother is continuous with that of the child. Even so astute a thinker as Aristotle subscribed to this belief, stating in one of his works, *The Generation of Animals* (i. 20), that the monthly periods, which fail to appear during pregnancy, contribute to the formation of the child's body. Modern scientific investigations have demonstrated that this and similar notions are quite false, and have completely disposed of a blood tie between any two persons, whether ther and child or even identical twins. Hence, any claims to kinship based on the tie of blood can have no scientific foundation of any kind. Nor can claims of group consciousness based on blood be anything but fictitious or at most a manner of speaking that is best avoided, since the character of the blood of all human beings is determined not by their membership in any group or nation but by the fact that they are human beings.

The blood of all human beings is in every respect the same, except for variations in the frequencies with which the blood-group factors, the Rh factors, the sickle-cell factors, and similar factors occur. But these blood groups, the Rh factors, the sickle-cell factors, and the like, are again matters of genes that happen to be differently distributed in different populations. This distribution is not a matter of quality but of quantity. There are some ethnic groups and some local populations in which certain chemical components of the blood are present that are absent in other groups, but these differences are themselves expressions of the genes that are differentially carried by such groups and populations. These differences in the chemical composition of the blood are in no way the carriers of hereditary traits themselves.

The facts about blood did not in the least serve to deter political propagandists such as the German Nazis from using the blood myths in order to set human beings against one another. The chief falsifier of the facts employed by Hitler to deceive millions of Germans and poison their minds was one Alfred Rosenberg, who was subsequently hanged by the Allies as a war criminal. He was the author of, among other works, a book entitled *Blood and Honor* (1934), and as an illustration of the nonsense that millions of Germans took seriously, we quote him:

> A nation is constituted by the predominance of a definite character formed by its blood, language, geographical environment, and the sense of a united political destiny. These last constituents are not, however, definitive; the decisive element in a nation is its blood. In the first awakening of a people, great poets and heroes disclose themselves to us as the incorporation of the eternal values of a particular blood soul. I believe that this recognition of the profound significance of blood is

now mysteriously encircling our planet, irresistibly gripping one nation after another.

Nonsense such as this led to the murder of millions of human beings. That is perhaps one reason why it is important to understand the facts about the falsities involved in the myths about blood and heredity, for, as Voltaire remarked, those who believe in absurdities will be in danger of committing atrocities. Few of us are guiltless in this respect. The extravagant and preposterous claims the Nazis made on the basis of the blood myth are paralleled by the superstitions that prevail among many other peoples. America is no exception. During World War II some of these superstitions were brought noisily out into the open when the Red Cross segregated the blood of Negroes for the purposes of transfusion. At that time the myth of "blood" seemed almost as strongly entrenched in the United States as it was among the Nazis.

The remarkable thing about the objection to Negro blood is not so much that it is based upon a misconception but that the same person who refuses to accept Negro blood may be willing to have his children suckled by a Negro wet nurse just as he had himself been as an infant. The same person will be ready to submit to an injection of serum derived from a monkey, a horse, a cow, or some other animal, and although he himself may have been suckled by a Negro wet nurse and may even entertain the greatest affection for Negroes, he will violently object to any "pollution" of his blood by the injection of Negro blood into his own bloodstream.

This is an irrational belief, a superstition for which there is no ground in fact. In reality the blood of the Negro is similar to that of all other human beings. For purposes of transfusion, or any other purposes, it is as good as any other blood.

The objection to Negro blood is, of course, based on the antique misconception that the blood is the carrier of the hereditary characters, and since the Negro is regarded—quite erroneously—as possessing "racially" inferior characters, it is feared that these may be transmitted to the receiver of the transfusion. Both ideas are false.

## Telegony

Left over from the days when heredity was regarded as something that was transmitted in the "blood" is the widely diffused belief that once a female has mated with a male, no matter how many other males she subsequently mates with, the heredity of her offspring by these later males will all be affected by the heredity of the first male with whom she originally mated. This belief in the supposed transmission of the hereditary characteristics of one sire to the offspring subsequently born to other sires by the same female is known as *telegony* (Greek *tele*, far, distant; *gonia*, progeny). Thus, it was believed, a pedigreed horse having once mated with a draft horse and given birth to young could never thereafter give birth to purebred offspring of her own kind. Similarly, a pedigreed bitch having once mated with a mongrel and given birth to young would thereafter be unable to give birth to purebred offspring when mated with a dog of her own pure breed. It is understandable how such an erroneous idea came into existence, for it *is* nothing more than an *idea* that is erroneous, as simple attention to the facts themselves might have demonstrated. The idea of telegony, however, is a striking example of the truth that ideas based on erroneous theories will make men insensible to the facts that are before their eyes and cause them to see facts not as they are but as their belief says they should be. To repeat, it is understandable how such an erroneous idea came into being when it was believed that heredity was carried in and transmitted through the blood, for if the first sire's hereditary qualities entered into the female's bloodstream, her subsequent offspring by subsequent sires would acquire some of the "heredity" of the first sire through its presence in their mother's blood. Today we know, on both theoretical and experimental grounds, that this idea is entirely erroneous.

Breeding experiments with animals have shown that telegony is a nonexistent phenomenon. Nevertheless, in pre-genetics days it was one way of explaining the fact that some offspring sometimes showed characteristics that were not present in their parents. This was especially true of many domestic animals in which traits would appear as a result of the fact that each of the parental animals carried the genes for them in recessive state while not exhibiting the traits themselves. But the simplest explanation was to attribute such traits expressed in the offspring to the effects upon heredity of some earlier sire.

The idea of telegony with respect to man has reappeared in modern times, principally in Germany. There, in 1919, immediately after the First World War, an anti-Semitic novel was published by one Herr Artur Dinter, entitled *Die Sünde wider das Blut* (*The Sin Against the Blood*), in which the author argued by implication that the blood of the German people was being "poisoned" by the aftereffects of admixture with Jews. This novel sold by hundreds of thousands, and there can be little doubt that it made Hitler's later promulgation of the same doctrine all the more easily acceptable to the German people.

Since many people still believe that the seed comes from all parts of the body and is carried in or is merely a specialized portion of the blood,* it should be stated once and for all that neither the male nor the female seed is ever carried in the blood or ever forms any part of the blood. The seed is carried in their germ cells, the egg of the female and the sperm of the male, in the one case from the ovary to the fallopian tube and in the other case from the testes through its ejaculatory ducts into the male organ. Blood is not in any way involved.

The erroneous belief in the inheritance of acquired characters, namely, the idea that changes in the body structures or functions of one or both parents could be transmitted to their offspring, was based on this false notion concerning the relation of the seed to the blood. If the blood gathered the seed from every part of the body, then it was reasoned that any modification of the body would be reflected in the seed and hence would be transmitted to the offspring. Just as the evidence is entirely contrary to this view of the relation of seed to blood, so the evidence is entirely contrary to the existence of such a phenomenon as the inheritance of acquired characters.

But enough, I hope, has been said about the myth of blood as the carrier of hereditary characters to exhibit it to the reader for what it is: an erroneous and harmful idea, perfectly understandable in its development, but belonging in the same class with the ideas that flies are generated from putrefying meat or that the touch of kings is capable of curing scrofula and other diseases. As a synonym for heredity the word "blood" should be abandoned.

---

* We find this view clearly stated by the Father of Medicine, Hippocrates (c. 460–377 B.C.), in his book *Airs, Waters, and Places*, xiii. 14.

# 5. HEREDITY AND ENVIRONMENT

HEREDITY OR ENVIRONMENT: Which Is Stronger? This was a very popular subject for debate during the school days of almost everyone. Such debates were usually quite exciting, and since the discussants were not generally too well informed, confidence and passion would tend to conceal the insecure grasp of the facts, and the words would fly. Nothing would be resolved except, possibly, the resolve to acquaint oneself with the facts. Not too many seem to have made such a resolution, for the debate still goes on in spite of the fact that the question Heredity or Environment: Which Is Stronger? is a spurious one—spurious in the sense that it opposes and treats as separate two things that, in the context in which they are generally discussed, are not naturally either opposable or separable.

In practice we often do behave *as if* things that are not separable are capable of being separated, and in this way we are often able to discover how much of one is involved in the other. This is a common daily practice. But it is quite another thing to assume that two things we have arbitrarily decided to treat as separate are in fact separate in their action. Usually what we are dealing with, especially when concerned with living things, is not so much *action* as *interaction*, and certainly it is of the greatest importance to grasp the fact that living things are not results of actions but of *interactions*. Things acting by themselves do not in living organisms produce developmental changes; they do so only in interaction with other things. For example, there is no such thing as a gene that acts as such, by itself. Genes interact with other genes and with the environments in which they occur. The environment interacts with the genes and the genes interact with one another. Thus we can recognize that the environment and genes exist separately as such, but with respect to the development of the organism, they never exist separately but are always in the process of interaction. Neither genes nor environment alone can produce development. Both genes and environment in interaction are necessary for development to take place. It is the *interaction* that is the important part of the process, for the genes and environment are not simply parallel processes. In the interactive process within the organism, the environment affects the action of the genes, and in turn the genes in interaction both with one another and with the environment produce changes in their own environment. This is a short statement of the genetic developmental process, the detailed expansion of which could easily occupy the space of a hundred books as large as this one. The point to grasp here is that development is an interactive process involving the hereditary packages of chemicals, the genes, and the environment, that is to say, whatever is capable of interacting with the genes.

Development consists of changes in size, complexity, kind, number, position, shape, composition, and functions of the parts of the organism. Every one of these facets of development is under the combined influence of the genes and environment. The organism represents the end effect of the interaction of the two. (Remember that when we say "of the two" we are referring to a vastly complicated chemical series of processes in interaction with an equally complex series of environmental processes.) In development from the germ cell to the adult a tremendous number of quantitatively and qualitatively different interactions of elements must occur, and during the progress of these interactions almost anything can happen along the way to modify their expression. The fact that this is so, often enables us to disentangle something of the nature of the factors, both genetic and environmental, that have entered into the formation of a trait or character at the other end of its developmental history. Much light can be shed on all of this by studying the actual process of development in specific examples.

The idea that development simply represents a sort of unfolding of what is already preformed in the egg or sperm or the combination of both is one that was held in the seventeenth and eighteenth centuries and has persisted down to the present day in the minds of many people in a more or less vague sort of way. This idea is known as the *preformation hypothesis,* and in its day was held by quite a number of respectable scientists. In the days when microscopes were very primitive, imagination had to make up for what the lens was unable to do, and so the sperm was seen as a homunculus, or "little man," all ready and preformed in all its parts, simply waiting for the opportunity to unfold like the elements in the bulb of a flower. Several of these investigators actually showed a figure of a little man seated in the head of a sperm (see Fig. 11).

The opposite of this theory, known as *epigenesis,* was developed by a twenty-six-year-old German, Kaspar Wolff, in 1759. In his book *A Theory of Generation* he suggests that organization of the embryo occurs absolutely spontaneously by means of an "essential force" acting on the undifferentiated germ, the embryo resulting from the coagulating power produced by the warmth and development of the blood vessels.

The mistake in the doctrine of preformation lay in supposing that the germ was a miniature organization of the adult; the error of epigenesis was in maintaining that the germ lacked any organization at all.

The "little man, what now?" theory, as the doctrine of preformation may be called, and the "spontaneous-combustion" theory, as we may describe the epigenetic theory, now belong to the realm of the history of science. No scientist subscribes to them today, but in one form or another these ideas linger on in the minds of some people. We can dismiss such ideas from ours.

## What Is Heredity?

It is extremely important to grasp the relationship between heredity and environment, and there is nothing in the least difficult about doing so. When we are able to perceive (1) what heredity is, (2) how development comes about and may be modified, and (3) what we human beings can do about both heredity and development, we can understand what the relationship between heredity and environment really is.

To the question What is heredity? the answer usually returned by many authorities is that it is the innate endowment or equipment of the individual. It is not what the individual is born with, but what he receives at fertilization, his genetic endowment or pattern of genes, the *genotype.* However, as soon as we begin considering the role the genes play in the development of the individual as a functioning organism, structurally, physiologically, and psychologically, it is evident that the genes can do none of this, and there can be no development, without the interacting environment.

The end product the organism represents at birth is the result of the interaction of the genes with one another and with the varying environments within the developing organism they are in the process of helping to form. In addition,

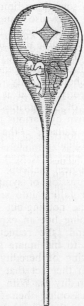

FIG. 11. Hartsoeker's drawing of a human spermatozoon, showing a homunculus, or miniature man, 1694.

there is the environment directly provided by the mother—we call this the *maternal* or *uterine environment*. This is of quite a complex nature and is only now coming to be recognized for the importance of the role it plays in the development of the organism. We shall return to this. Then there is the environment outside the mother, which is capable of affecting the developing organism within the womb. Radiation is an example. Loud sounds constitute another. These things are capable of reaching and affecting the development of the fetus directly without any intermediation on the part of the mother. And as we shall see in the next chapter, there is good evidence that many externally originating stimuli of various sorts are capable of seriously affecting the development of the organism.

## The Dangers of Overvaluing Genes

It is a vast and confusing oversimplification to think of an individual's heredity as something limited to the genotype or the innate endowment. It is also dangerous and harmful to think in this way, for if it is believed that heredity is what is given in the individual's genes, then the conclusion is likely to be drawn that what the individual is going to be, or what he is, is wholly determined by the genes, and therefore there is nothing one can do about heredity. This is a common way of thinking about heredity, and it is quite false. Unfortunately, quite a number of "authorities" have fallen into this error. This error is known by various names: "the biologistic fallacy," "the pathetic fallacy," "the nothing-but fallacy," and "the reductionist fallacy." "The biologistic fallacy" describes the erroneous belief that man is nothing other than the product of his biological makeup. "The pathetic fallacy" is used here as an expression of sympathy with those who should know better for falling into so pathetic an error as the biologistic trap; "the nothing-but fallacy" refers to the belief that man is nothing but an expression of his genetic or animal structure, and "the reductionist fallacy" is the error of reducing man to his innate structure.

The danger of thinking of heredity as equivalent to genetic endowment lies in the fact that one is likely to think that this is all that heredity is. With such a conception of heredity the line of reasoning then follows that since the genes determine heredity, there is nothing much one can do about the heredity of the individual once he is conceived or after he is born. This is the kind of belief that makes it possible for an uninformed doctor or anyone else to say, "Since this is a hereditary condition, there is nothing that can be done about it." Another word that is often used in a similar way, instead of "hereditary condition," is "constitutional." If a condition is "constitutional," the frequent assumption is that there is little or nothing that can be done to modify it. And if one wants to be very learned, one can call the condition a "constitutional idiopathy"—and that *really* closes the door on doing anything about it.

It has been truly said that while high walls do not a prison make, scientific terms frequently do. Many an individual and his family have been needlessly caused to abandon all hope by the pronouncement of the verdict that a disorder was due to heredity. It is, of course, perfectly true that there are some disorders that are due to the action of genes about which we can do little or nothing at the present time, but this does not mean that this is true of all conditions that are so caused. Nor does it mean that those about which we can do nothing at the present time must forever remain beyond remedy. On the contrary, there is every reason to believe that when we have more fully learned to understand the manner in which these conditions come about, we may be able to control their manifestations. In many instances, diabetes is a genetic disorder. When I was a youth, it was incurable, but since 1922, as a result of the discovery and isolation of insulin by two Canadian investigators, Banting and Best, the sufferer from diabetes has obtained relief by being injected with insulin, in which his body is genetically deficient. Today it is in many cases possible to substitute pills that stimulate the pancreas to produce insulin. Insulin does not cure or alter the defective heredity, but normal carbohydrate metabolism can be maintained by taking insulin. Hemophilia, or "bleeder's disease," a hereditary condition affecting some males, which takes the form of intractable bleeding from a slight wound, sometimes fatal, was until recently unrelievable. With the isolation of blood-clotting substances in recent years, it has for the first time become possible to help some of the sufferers from this unfortunate condition. We can already foresee that it will not be long before hemophilia becomes as relievable as diabetes. The treatment of pernicious anemia with liver extract is another case in point. Indeed, theoretically there is no chemical that genes are capable of producing that cannot be introduced into the body in some other way. Hence, inadequacies due to defective genes should be replaceable by chemicals originating outside the body. The point, of course, is to know what is the nature of the chemicals that are responsible for the defect.

There are many kinds of other hereditary conditions that may be similarly alleviated by environmental means. Tuberculosis, allergies, defects of vision and hearing—these are some of the common genetically conditioned disorders that have been greatly helped by control of the environment. In the case of hereditary defects of vision and hearing, the benefits conferred by eyeglasses and hearing aids can be at once appreciated.

At the same time attention must be drawn to the fact that there exist many genetic deficiencies that cannot be remedied in any environment. Furthermore, every time we do discover a means of compensating for the genetic deficiency, we thereby relieve the only control we have for *mutation pressure,* and thus increase the frequency of the gene and therefore the number of individuals who are affected by it.

## 82 HUMAN HEREDITY

The danger, then, of making an exclusive identification of the genes with heredity is that it leads one to believe that innate endowment is equivalent to fate, and that the pattern of genes an individual inherits determines his destiny. Such a belief is a modern form of the older doctrine of predestination. There is little more truth in the one than there is in the other. When, then, we think of the action of genes in development, it is desirable to remember that that action is an *interaction* with the environment, for development is a continual process of the adjustment of cells to their environment.

### The Action of Genes

Development is not something static; it is a dynamic condition that varies with the conditions. What any cell shall become, what part of the body it will form, is not determined solely by the genes it contains but depends on the conditions surrounding it, on its relation to other cells. This constitutes the answer to the question that arose on an earlier page, namely, How is it possible that cells that contain identical genes, the same materials, form the numerous different parts of the body?

Genes furnish the essential chemical substances necessary for development, and in interaction with other genes produce changes that further serve to constitute part of their environment. The cells, in their molecular structure, shape, stress, pressure, surface tension, the enzymes to which they give rise—that is, chemical substances that speed up the reactions in other chemical substances—are in a continuous state of adjustment and adaptation. Though they interact, the genes in the nucleus remain the same, but the composition of the protoplasm outside the nucleus (the cytoplasm) is differentially elaborated. Hence we see how very important the role of environment is in all heredity. The structure of the genes sets the specific limits of development; that is to say, their inherent composition is such that only an elephant can develop from elephant genes, and only a man can develop from the genes of human beings. The genes also set the maximal limits to which the various differentiated cellular masses can normally develop. Abnormal conditions can disturb each of these limits: In the case of specific characters, the organism develops in such a manner as scarcely to resemble the species that gave it birth; such organisms, generally known as "monsters," rarely survive to birth, and when they do they are likely to die shortly thereafter. In the case of the growth of cells into tissues, all sorts of abnormalities of growth and development may occur affecting either parts of the organism or the organism as a whole. Oversized or undersized parts of the body, or decreases or increases in size of the body, are familiar examples.

In many cases of dwarfs and giants it can be clearly shown that it is not so much the genes concerned with the size of the individual that are at fault as it is the environment in which these genes develop, an environment that is usually provided by certain secretions (*hormones*, or chemical messengers) from several of the ductless glands, notably the pituitary gland, situated at the base of the brain. We shall deal with these cases more elaborately later. Here we are engaged in discussing the general principles involved. From the analysis of cases of monsters, dwarfs, and giants we perceive how important the environment is in regulating the manner in which the genes respond in the process of development. As we shall see, no characteristic of man (or of any other living thing) is exclusively the product of genes or exclusively the product of environment, but the product of the interaction of each with the other.

The individual, then, does not inherit characteristics. When we speak of "inherited" eye color or "inherited" nose shape, this merely represents a figurative way of speaking. Such characteristics are no more present in the germ cells than are the parts of an airplane in the various materials out of which it is constructed. Those materials have to go through a vast number of complicated environmental processes before they can be put together to make an airplane. Exactly the same materials, with or without the addition of some others, can by other environmental processes be turned into an automobile or a cabin cruiser. The same materials in different environments may result in different end products. And so it is with heredity. The same sets of genes in different environments may result in the different expression of characters. Heredity does not transmit characters. What heredity transmits are the responses that the genes have made in the developmental process to the environments in which they have interacted. In a sense, genes may be compared with tools and the environment to the materials upon which they act. The genes can do only as much as (1) their particular limits enable them to do, and (2) the opportunities provided by the material. We can turn this around and regard the environment as the tools that work upon the genes, the material. Poor tools won't do justice to good material or to poor. Good tools will do as well by the material as the material permits, whether it be good or poor.

We come to understand, then, that genes are most usefully looked upon as chemical packages of potentialities. To what extent those potentialities are realized will depend upon the environmental conditions under which they develop. For this reason, whenever one thinks of heredity as a process it might be helpful to substitute the idea of potentialities for it, for this immediately suggests the idea of possible states that can be brought into being only through the stimulus of the environment.

The upshot of all this is of the greatest importance for us as human beings, for where we control the environment, we to some extent control heredity. Heredity, it has been said, determines what we can do, and environment what we do do. The limits of what we can do are determined by the genes, but it is the environment that determines the extent to which the potentialities within those limits are realized. We do not, therefore, stand helpless and impotent before the implacable fate that heredity is misconceived to be. On the other hand, through the intelligent management of the environment there is a great deal that we can do about it, as we shall see in the pages that follow.

To the question, therefore, that we set out to answer in this chapter, What is heredity? we may best make the answer that definitions are not properly meaningful at the beginning of an inquiry, but only at the end of one, and it is therefore only at the end of this book that we shall be able to give a reasonably complete answer, but at this juncture we may sum up our discussion thus far by saying what we have found heredity to consist of. First and foremost, heredity consists of the innate endowment of genes acquired by every individual at conception. It is these genes alone that are transmitted, and nothing else is transmitted in the genetic process. This, strictly speaking, is what is meant by "heredity." But as we have already seen, genes do not work their effects in a vacuum. Indeed, they are unable to work at all in the absence of environmental stimulation and they are to varying extents dependent for their very action upon environmental conditions. These environments are: (1) the intercellular environments in which the genes interact with one another, (2) the uterine environment, or environment within the womb, and (3) the extrauterine environment, that is, the environment of the external world outside the womb.

# HEREDITY AND ENVIRONMENT

Each of these environments is extremely complex, and we shall be discussing their nature throughout the remainder of the book.

One thing more: When, in this book, we sometimes speak of the "inheritance" of blue eyes or of this or that character or trait, or when we say that a character or trait is "inherited" as a dominant or a recessive, it should be remembered that these are only shorthand ways of speaking. Characters or traits are, of course, not inherited as such, but only the genes connected with their development. The relationship between heredity and characters or traits is not static; genes do not determine characters in a one-to-one, invariable manner. On the contrary, the relationship between heredity and traits or characters is a dynamic one, with genes influencing development through chains of physiological responses of the organism *to its environment*. The same genotype in different environments may give rise to quite different responses.

# 6. THE EFFECTS OF ENVIRONMENT UPON THE DEVELOPING HUMAN BEING IN THE WOMB

IT HAS BEEN KNOWN for many years that it is quite possible to change the form of various animals by altering the environment in which they develop. One of the early classical experiments illustrating this, involved the larval form of the small boy's delight, the common Atlantic Coast minnow *(Fundulus heteroclitus)*. It was shown that if a salt such as magnesium chloride is added to the sea water in which the eggs of this fish develop, many of the embryos and young fish develop a single eye instead of two normal eyes. In some individuals the eye is in the middle of the head, and in others it develops on only one side.

In ordinary sea water, minnows develop two eyes, but in sea water that is slightly higher in magnesium salts, minnows with exactly the same genetic constitution develop only one eye. This does not mean that such minnows will thereafter transmit only one eye to their offspring; on the contrary, they will always transmit the genes for two eyes and their offspring will develop the normal two eyes when they undergo development in normal sea water. The minnows have inherited a genotype for a certain set of characteristics that will go to form the visual apparatus, but it is clear that the form in which that genotype will be expressed is to a large extent dependent upon the environment with which it interacts. Genes, as we have seen, will respond to the environment only in the manner that that environment enables them to do so. What a given set of genes will produce, therefore, depends not upon themselves but upon their interaction with the environment. Under different environmental conditions the same genes will produce different results. It is therefore not true that because an individual has inherited a given set of genes he is thereby destined to develop certain characters and no others. The kind of characters he will develop will depend upon the kind of environments in which the building blocks of those characters undergo development.

# EFFECTS OF ENVIRONMENT IN THE WOMB

In the present chapter we shall deal with the evidence concerning the relation of development of the human fetus to changes in its environment.

For many years it was believed and taught that the human fetus is so well insulated within the womb from the rest of the world, including its mother, that what went on outside the womb was, on the whole, unlikely to affect the fetus. Anything capable of affecting the fetus, it was believed, had to be considerable. In recent years we have learned that this is quite untrue, that, on the contrary, the human embryo—the term we give to the human organism from conception to the end of the eighth week—is extremely sensitive to all sorts of changes both in the mother and in the external world, and that as it develops as a fetus—the term for the conceptus from the end of its embryonic period at the eighth week to the time at which the organism is born—it continues to be sensitive to innumerable environmental changes, though less so than the embryo. This follows a general law of embryological development, namely, the younger the developing organism, the more likely it is to be seriously affected in development by disturbing conditions. There are periods in the development of the organism that are known as critical developmental periods, so called because during these periods fundamental developmental changes are occurring that depend on perfect timing and the correlation of many different processes. Any disturbing factor introduced into the developmental process at such a time is likely to produce structural and functional disorders in varying degrees and of various kinds. It should be understood that the "developmental process" is the name we give to the orderly procedure by means of which the normal structures and functions are being built up. These processes involve not only the constructive ones of building but also certain limiting processes, which perform the tasks of guiding, checking, and channeling the building-up processes. All these processes can be affected by changes in the environment in which they are going on. And this is as true of man as it is of all other living things.

Since a great many physical and mental defects in children are believed to be due to disturbances occurring during the prenatal (before birth) development of the organism, it is important to know how these come about, for with the knowledge thus gained we will be in a better position to control the situation.

The effect of environmental factors upon the development of the prenatal organism may be considered under the following ten headings: (1) maternal age, (2) maternal parity, (3) maternal dysfunction, (4) maternal sensitization, (5) nutritional effects, (6) infections, (7) drugs, (8) physical agents, (9) emotional factors, and (10) other environmental factors.

## 1. Maternal Age

Conception takes place in a female. That female has had a history. That history has extended over a number of years. A female's past history as a growing and developing organism is closely related to her reproductive capacity. The age at which a female becomes capable of conception is, on the average, about three years after the first menstruation (menarche), which puts the average age at which a female becomes capable of conception at about sixteen and a half years. This does not mean that many females cannot become pregnant before that age. Many can and do, but most in fact cannot and do not. It is known that females who conceive at an early age experience a larger number of miscarriages and stillbirths than those who conceive at more efficient ages. Not only do more of their babies die at birth but also the youngest mothers have the highest mortality rates, and a large percentage give birth to defective children. The younger mothers (in the age group generally under seventeen) are normally physiologically incompletely prepared for the process of reproduction, hence the high frequency of unsatisfactory terminations of their pregnancies.

Studies of the reproductive development of the female show that from every point of view the best period during which the female may undertake the process of reproduction extends *on the average* from the age of twenty-one to about twenty-eight years of age. During these years, on the average, the female is at her optimum period of development. All the hormones necessary for the successful maintenance of the pregnancy are present at their optimum levels, and the organ systems are in their healthiest states. In mothers under seventeen years of age the hormones have not yet reached the optimum levels, nor have the reproductive organs—the organs serving the functions of bringing the pregnancy to a successful conclusion—reached their optimum growth and development. It is for these reasons that the younger mothers have a higher maternal- and infant-mortality rate than the mature mothers in the age group of twenty-one to twenty-eight years. The mothers between seventeen and twenty and their offspring do better than the mothers in the younger age group but not as well as the mothers and their offspring in the twenty-one to twenty-eight age group.*

Very gradually from the age of twenty-nine onward there is a rise in maternal- and infant-mortality rates; the stillbirth and miscarriage rates rise, and so do those for the number of defective children. From the age of thirty-five years onward there is a sudden jump in the number of defective children that are born, especially of the type until recently known as *mongoloids,* so called because of an alleged superficial but genetically quite unrelated resemblance to people belonging to the Mongoloid major group of mankind. At one time it was suggested that such defective children were "throwbacks" to a Mongoloid stage of development. This suggestion has no scientific merit or validity whatever. The anomalous condition from which these children suffer is now increasingly coming to be referred to as *Down's syndrome,* after Langdon Down, who first described it in 1866.

Down's children may or may not have the fold of skin over the inner angle of the eye (epicanthic fold) or the flat root of the nose that goes with this, but they do have smallish heads, fissured tongues, a transverse palmar crease, with extreme intellectual retardation. Their IQ ranges between 15 and 29 points, from idiocy to an upper limit of about seven years. Down's children are generally cheerful and have friendly personalities, with often remarkable capacities for imitation and memories for music and complex situations far superior to their other abilities. The expectation of life at birth is about nine years.

Half the known cases of Down's syndrome were born to mothers of thirty-five years or more, and 1 out of every 400 births or so occurring in the United States is a Down's. It has been estimated that once a Down's child has been born into a family, the risk of another such birth in that family is approximately fourfold (3.7). The causes of Down's syndrome were until recently quite obscure. Most authorities agreed that the physiological aging of the maternal organism, resulting in a progressive decline in the functions of her reproductive system, were almost certainly involved, and that this, among other things, takes the form of inadequacies in the secretions of the hormones of the ductless glands. Since these hormones are necessary for the proper develop-

* For a detailed discussion of the facts, see Ashley Montagu, *The Reproductive Development of the Female* (New York: Julian Press, Inc., 1957).

ment of her eggs, their failure to appear in sufficient quantity has a severe retardative effect upon the development of the fetus. An inadequate hormonal environment may occur in pregnant women at any age, in which event they are likely, in a high proportion of cases, to give birth to a Down's child. The mere accumulation of years is not in itself a cause of Down's syndrome or of any other disorder. There is promise that the knowledge yielded by the study of the endocrine status of women may enable us to reduce considerably the number of Down's children born. Such studies, by enabling us to recognize that some mothers are providing their own eggs with an inadequate hormonal environment, or by discovering the women who are likely to provide such an inadequate environment for their eggs even before they become pregnant, may lead to the remedy by providing such women with the necessary hormones. It is now believed that Down's syndrome is but one of many disorders principally due to a decline in the efficiency of cell division.

That genetic factors manifesting themselves in part through the endocrine system, especially in younger mothers, may be responsible for a certain number of children with Down's syndrome is strongly supported by the work of Ek, and of Rundle, Coppen, and Cowie, who found that mothers who had borne Down's children at a younger age exhibited a very significantly higher output of the steroid dehydroepiandrosterone (DHA) than mothers of Down's children above the age of twenty-seven. Since the older group of mothers had a normal output of DHA, clearly Down's syndrome cannot be due to a high output of this hormone. What is probably involved is an imbalance of hormones. In the younger mothers this would be a constitutional imbalance; in the older mothers the imbalance would be due to the effects of aging.

In Down's syndrome we have the tragic example of what may be an adequately sound genetic system being provided with an inadequate environment with resulting disordered development in the gametes. That defective genes are sometimes involved is a possibility. It is, however, generally agreed that in most cases defective genes are not primarily responsible, even though we have several records of Down's identical twins, and fraternal twins in which one child was a Down's and the other normal. In such cases it still remains a possibility that the defective uterine environment was at fault, and that in the case of the fraternal twins one child managed to obtain a sufficient quantity of hormones while the other failed to do so. (Apropos of twins, it is also a fact that two-egg twinning increases with maternal age; that is, the twins are derived from separate eggs and are therefore not identical but fraternal twins.) Down's syndrome in some cases may be due to a combination of a defective gene or genes with low penetrance and a defective environment. In an adequate uterine environment the gene would not express itself; in an inadequate one it may.

In January, 1959, Lejeune, Gautier, and Turpin reported that each of nine cases of Down's syndrome investigated by them possessed an extra autosome, making a total of forty-seven chromosomes instead of the normal forty-six. In April, 1959, a group of English investigators independently confirmed this discovery in six cases of Down's syndrome, in whom the chromosome count was also forty-seven instead of forty-six. The significance of the extra chromosome in the causation of Down's syndrome is not at present clear, but the fact that an extra chromosome is present is in itself highly significant. The extra chromosome (No. 21) reflects a failure in the normal process of mitosis, and this may be due to a number of conditions, aging of the ova being but one of them.

In April, 1960, with the publication of two separate communications, one by Polani and his coworkers and the other by Fraccaro and his collaborators, it became clear that Down's syndrome and, no doubt, many other disorders could also be associated with translocation of a portion of one chromosome to another. In such cases the number of chromosomes may remain forty-six, but their reciprocal

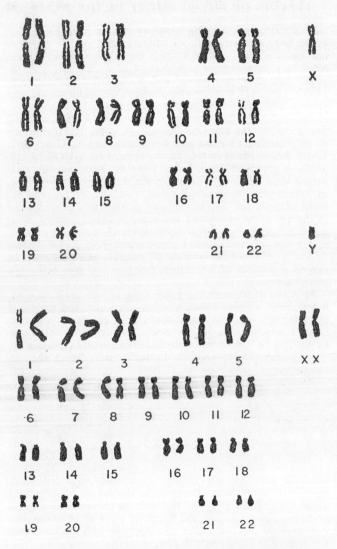

FIG. 12a. The Chromosomes of the Human Female. 12b. The Chromosomes of the Human Male. Arranged according to the Denver Classification. (Courtesy of Dr. Kurt Hirschhorn.)

## THE DENVER CLASSIFICATION OF HUMAN MITOTIC CHROMOSOMES

| | |
|---|---|
| Group 1-3 | Large chromosomes with approximately median centromeres. The three chromosomes are readily distinguished from each other by size and centromere position. |
| Group 4-5 | Large chromosomes with submedian centromeres. The two chromosomes are difficult to distinguish, but chromosome 4 is slightly longer. |
| Group 6-12 | Median-sized chromosomes with submedian centromeres. The X chromosome resembles the longer chromosomes in this group, especially chromosome 6, from which it is difficult to distinguish. This large group is the one which presents major difficulty in identification of individual chromosomes. |
| Group 13-15 | Medium sized chromosomes with nearly terminal centromeres ("acrocentric" chromosomes). Chromosome 13 has a prominent satellite on the short arm. Chromosome 14 has a small satellite on the short arm. No satellite has been detected on chromosome 15. |
| Group 16-18 | Rather short chromosomes with approximately median (in chromosome 16) or submedian centromeres. |
| Group 19-20 | Short chromosomes with approximately median centromeres. |
| Group 21-22 | Very short, acrocentric chromosomes. Chromosome 21 has a satellite on its short arm. The Y chromosome is similar to these chromosomes. |

relations will have been altered, with usually resulting alterations in development. Fraccaro and his coworkers have described a case of a Down's boy with forty-six chromosomes with a healthy father who had forty-seven chromosomes! The mother's chromosomes were perfectly normal. She was thirty-seven and the father was fifty-eight at the birth of their child. In this case the Down's child had only four short acrocentric centromeres (toward the top of the two chromosomal arms) instead of five, but an extra chromosome similar to that of the nineteenth and twentieth pairs of chromosomes was present. In the father the forty-seventh chromosome was similar to the nineteenth pair.

Hydrocephalus, or "water on the brain," a condition in which the head becomes greatly enlarged owing to the abnormal increase of the fluid in the lateral ventricles of the brain, is another condition that is associated with late maternal age at pregnancy. No adequate studies have been made on the mothers of such children that indicate causative factors in human beings, but experiments made on healthy female rats show that when a mother was fed a diet low in folic acid, more than 7 percent of her young were born with hydrocephalus, suggesting that similar nutritional factors may sometimes be responsible in man.

It is known that disturbances due to sensitization such as result from blood transfusion and other conditions become more marked with increasing age of the mother, and such sensitization may seriously affect the development of the fetus. The age of the father has not been found to be significant as a possible factor in the production of fetal malformation. In connection with this point Jalavisto has brought forward some evidence indicating that the offspring of older mothers do not, on the average, enjoy as many years of life as do the offspring of younger mothers.

## 2. *Maternal Parity*

It is now well established that parity, or the number of previous pregnancies a woman has experienced, is significantly related to the survival of the infant as well as to the frequency of fetal malformations. There is evidence that firstborn children, as well as those born at the end of a long series of pregnancies, are less viable (able to live) than those born in between, irrespective of maternal age. Fetal malformations are slightly more common in children of primiparae (mothers having their first pregnancy) than in children who are the result of a second or third pregnancy.

Average birth weight and linear dimensions increase with the order of birth, the first-born being least in both these respects. At this point it is necessary to underscore the fact that the findings we shall be mentioning in this section are all of a statistical nature, and that there are plenty of firstborn and last-born children, as well as children who were born well after their mother's thirty-fifth birthday, who are in every way perfectly healthy. Number of children or age of mother should never alone deter a woman from maternity. It is, however, always a good idea to see a competent doctor for a complete checkup before embarking upon such an important undertaking.

## 3. Maternal Dysfunction

The term "maternal dysfunction" describes noninfectious functional disorder or disease in the pregnant mother. Such dysfunctions in the mother may seriously affect the development of the fetus. Pregnant women suffering from hypertensive disorders (high blood pressure) show a very high rate both of fetal loss and of maternal mortality, as well as of other serious conditions. In one series, reported by Chesley and Annitto, of 301 pregnancies in 218 women suffering from essential hypertension (disease due to no known cause), the gross fetal loss in the first hypertensive pregnancy reached the staggeringly high figure of 38 percent, the higher the maternal blood pressure, the more fetuses dying. There were a total of 13 maternal deaths, or 4.3 percent, approximately 200 times higher than occurs in general obstetrical practice.

Prediabetes and diabetes are disorders that are also known to affect the fetus unfavorably. Prediabetes is the name given to the condition in which a latent disorder of carbohydrate metabolism is present without the subject's being aware of it. By giving a woman a glucose-tolerance test, which is a test for determining the amount of sugar circulating in her blood after she has ingested 100 grams of glucose, it is possible to determine whether or not she is prediabetic. If after two hours of fasting the test shows more than 120 milligrams of sugar per 100 cubic centimeters of blood, the condition is almost certainly prediabetic. Prediabetic women generally have babies whose weight exceeds nine pounds, another indication of this condition. A large baby in a woman not known to be suffering from diabetes is cause for the suspicion that unless preventive measures are taken, she will probably develop full diabetes some time in the not-too-distant future.

Children of prediabetic mothers have a birth weight, on the average, 400 grams more than normal. When the father alone is prediabetic, birth weight is approximately 150 grams more than normal.

Diabetes is a disorder of carbohydrate metabolism as a consequence of which the sugar taken into the body is not burned by the body and converted into energy. The disorder is due to a deficiency of insulin. Insulin is normally secreted by the pancreas. It is known that in diabetic mothers the fetus grows very rapidly, often reaching the birth weight of the newborn long before it reaches term, and so may present considerable obstetrical difficulty. The mortality rates of fetuses and newborn babies of diabetic mothers are extremely high. For example, in one series, reported by Gaspar, of 49 deliveries in 45 pregnant diabetics, there were 19 stillbirths and 6 newborn deaths, a mortality rate of 51 percent! But in a comparable series reported by Stephens, Holcomb, and Page, the mortality rate was reduced to 8 percent by careful attention to the requirements of the diabetic mother, including preterm delivery of the child by Cesarean section. There is evidence that Cesarean section in such cases is an unnecessary practice.

In women with prediabetes in whom a latent disorder in carbohydrate metabolism may become accentuated during pregnancy, the adverse effects upon the fetus are the same as those associated with maternal diabetes. In mothers with diabetes the stillbirth and newborn death rates are increased for many years prior to the development of diabetes proper, and during the five years immediately preceding its onset the fetal loss may exceed 30 percent. Doctors Carrington, Reardon, and Shuman have shown, however, that by the early recognition and treatment of the prediabetic mother, the fetal loss may be reduced to zero with nothing but beneficial effects to the mother and her offspring. Of 111 cases, 46 were detected early enough in pregnancy to permit adequate treatment, and in these there were no fetal losses. Among the rest there was much neonatal morbidity, with a 25.6 percent infant-mortality rate among the 39 cases unrecognized before delivery.

Hyperthyroidism, excessive secretion of the thyroid hormone, thyroxine, from the thyroid gland, is another condition that adversely affects the development of the fetus. In a recent report by Battarino and Capodacqua on a series of 46 pregnant Italian women with hyperthyroidism, approximately 30 percent of the fetuses died. In 6 women, who had had 15 pregnancies and who had undergone operation for their hyperthyroidism, the mortality rate of the fetuses was 40 percent.

Undersecretion of the thyroid hormone in the pregnant mother is responsible in most cases for goiter and cretinism in the infant. There is some evidence, too, that maternal toxemias (general poisoning of the blood) and bleedings during pregnancy may in some cases produce defects in the brain of the fetus, and that a number of the children will become behavior problems or will be mentally retarded.

## 4. Maternal Sensitization

In those cases where the genes of mother and fetus differ for the substances borne on the surfaces of the red blood corpuscles, the mother may become sensitized and produce antibodies that unfavorably affect the development of the fetus. This usually results in causing fetal anemia at a relatively late fetal age. The Rh incompatibilities constitute a well-known example of this. As Landsteiner and Wiener discovered, when the blood of the rhesus monkey is injected into rabbits or guinea pigs, a special serum is obtained. The serum will "clump" the blood of about 85 percent of all white persons. The factor in the blood that makes it clump in response to the serum is the *Rh factor* ("Rh" from *rh*esus). Persons who have this type of blood are said to be *Rh positive*. Persons who do not are *Rh negative*. The exact way in which the Rh blood types are inherited is somewhat complicated. Three principal Rh factors, as well as several contrasting Hr factors, are known, and at least ten major genes are involved. These result in fifty-five combinations of genotypes. Several of the genes recently discovered increase the number of genotypes still further.

Understanding how the Rh factor operates and how it is inherited is extremely important in biology and medicine. As we shall see, it has meant the saving of the lives of a great many babies and the elimination of many family tragedies that not so long ago seemed an inseparable part of their destiny.

When a woman who is Rh negative marries a man who is Rh positive, the first-born child of such a marriage is usually normal. However, during the following pregnancies the fetus may be lost by stillbirth. Or it may be born in such an anemic or jaundiced state that it lives only a few hours or days after birth. The infant dies from a blood disorder called erythroblastosis (Greek *erythros*, red; *blastos*, germ; *osis*, increase). The name indicates that the baby's bloodstream contains numerous primitive red cells (erythroblasts), produced in response to destruction of the baby's red cells by the maternal Rh antibodies. The baby's normal red blood corpuscles, the erythrocytes (Greek *erythros*, red; *cyton*, cell), are subjected to wholesale destruction. Another name for the same disorder is hemolytic disease of the newborn.

The disorder may result when the fetus has inherited an Rh-positive gene from its father and the mother is Rh negative. The fetus produces Rh-positive substances, called antigens, in its blood. If fetal red corpuscles pass through the placenta into the mother's blood, the antigens stimulate the production of Rh antibodies.* These Rh antibodies in turn pass through the placenta into the bloodstream of the fetus. There they combine with the fetus's Rh-positive red blood corpuscles, which may then break down. The result of such destruction of the fetus's red blood cells is that it becomes anemic, and there is too little blood for its needs. The usually associated jaundice, which yellows the skin, is due to the infiltration of the skin by the breakdown products of the cells. In severe cases the skin may take on a greenish color.

Fortunately this disorder does not occur as often as the facts of heredity might lead one to expect. Erythroblastosis fetalis occurs in only about 1 out of every 200 pregnancies. Actually, about 1 in 12 pregnancies involves an Rh-negative mother carrying an Rh-positive fetus. Therefore, 1 out of every 12 pregnancies might be expected to yield an erythroblastic child instead of the 1 out of about every 200 that is in fact born. This fortunate discrepancy may be caused by the fact that in many cases the antigens from the fetus do not pass through the placenta. In other cases the mothers may not respond to the foreign antigens from the fetus. However, if we omit first-born children, who are seldom affected, the number of children actually affected would rise sharply to 1 in 17 or 18 pregnancies.†

While in most cases erythroblastosis is caused by Rh sensitization, maternal sensitization to other blood factors is occasionally responsible. In fact, if the mother belongs to group O and the fetus to group A or B, the baby may be affected, in which case it is said to have A-B-O hemolytic disease.

---

* An *antigen* is a substance that can cause the formation of another substance, called an antibody, with which it reacts specifically. An *antibody* is a protein produced in the body in response to contact of the body with an antigen, having the specific capacity of neutralizing or reacting with the antigen. Antibodies may or may not exert a protective function in this way, that is, by neutralizing infectious agents or reacting with proteins foreign to the organism.

† Why the first-born child is seldom affected is not definitely known, but there is good reason to suppose that it is due to the fact that the Rh-negative mother has to be adequately sensitized by the RH-positive antigens of her first-born before her own body can respond with the appropriate antibodies. This sometimes happens with her first pregnancy, but rarely.

## EFFECTS OF ENVIRONMENT IN THE WOMB

Erythroblastic babies may suffer serious damage of the liver and brain. The damage to the brain results from the deep jaundice, which is associated with the reddish bile pigment, *bilirubin*, a substance very harmful to brain cells. Since the red blood corpuscles carry the oxygen, which is circulated to the body on their surfaces, destruction of these corpuscles also serves to reduce the amount of oxygen available to the brain and other organs. We know that oxygen reduction in adults can have serious effects on the brain. There is no doubt that the oxygen reduction created by the destruction of red blood corpuscles in the fetus can be damaging and even cause death. If the fetus is deprived of an adequate amount of oxygen during any stage of its development, the brain may be permanently damaged. This may explain how mentally deficient children sometimes occur in families in which there is no previous record of mental deficiency.

The method of treating erythroblastic infants is to give them an exchange transfusion of Rh-negative blood. It is important to remove the baby's damaged red cells as well as to transfuse it with fresh red cells. In an exchange transfusion the baby's blood is removed from an artery at the wrist at the same time that an equal volume of Rh-negative blood is injected into a vein at the ankle. The reason for using Rh-negative blood rather than Rh-positive is because the baby already has maternal antibodies in its blood, and so, if Rh-positive blood were transfused to the baby, it would be destroyed too rapidly by the antibodies circulating in the baby from the mother. The Rh-negative blood helps the baby over the difficult period during which the antibodies circulating in its blood become exhausted.

The Rh antibody does not occur naturally in the blood; therefore its presence in the blood must be due to immunization. This may be brought about by a previous transfusion with Rh-positive blood in Rh-negative females or by previous pregnancies involving an Rh-positive child.

Understanding the meaning of the Rh factor is of great practical importance. Every woman planning marriage should consult her physician in order to learn the Rh types both of herself and of her prospective husband. There are various ways in which the undesirable effects of clashing Rh factors may be averted if doctors know about them beforehand.

Thus, once more, we perceive how a knowledge of the manner in which genes interact with one another to produce their effects upon the developing organism helps us to control those effects, and to master them.

There is much evidence that the fetus is affected by various protein bodies that pass to it from the mother through the placenta, and that excessive indulgence in certain proteins by the mother during pregnancy may produce all sorts of allergic reactions in the infant through sensitization of the predisposed fetus *in utero*. Ratner has shown that many cases of infantile allergy are probably due to this cause. He quotes the case of a mother who took seven to eight eggs a day throughout her pregnancy, mostly in the form of eggnogs. Her infant developed a severe eczema three months after birth. At fourteen months the child was brought to Dr. Ratner, who was then able to bring about a rapid improvement by the "absolute elimination of egg-containing foods."

Ratner quotes a dozen such cases, all with the same outcome and all involving the excessive consumption of eggs in one form or another by the mother during pregnancy. In other cases nuts were involved, and in still others the excessive consumption of milk by the mother during pregnancy was the responsible factor.

## 5. Nutritional Effects

Possibly the most important factor in influencing the development of the fetus is the nutrition it receives from its mother. The foods consumed by the mother are reduced to molecules, which are able to pass directly through the placenta into the fetal bloodstream. But there are many other factors in addition to maternal nutrition that affect the nutrition of the fetus. Among them are the mother's occupation during pregnancy, her health, general hygiene, and the sanitary conditions of her environment. These conditions generally reflect the socioeconomic status of the mother.

Fetuses and infants of mothers of low socioeconomic status are smaller and have a higher mortality rate than those of mothers of higher socioeconomic status. In itself small size is not necessarily a handicap, but in many cases it constitutes a symptom of basic organic deficiencies that are destined to play an important role in the later developmental history of the organism. By the end of the first postnatal month children who may otherwise appear to be normal will sometimes exhibit evidence of a deficient intrauterine environment in the form of lines of condensed bone, which show up in X-ray films as thick white lines in the ankle, or tarsal, bones. These lines, corresponding to the lines of retarded growth seen in the long bones of older children and adults, and caused by periods of prolonged illness, indicate that disturbances in nutrition, from whatever cause, arising during prenatal life are capable of inscribing their effects very substantially upon the structure of the developing organism.

Numerous experiments carried out in recent years on lower animals suggest that maternal nutrition in the early stages of fetal growth is a decisive factor in the production of certain physical abnormalities. It is not at present clear, in most cases, how the damage is done, but the evidence strongly indicates that it is mostly because of the lack of certain vitamins or proteins and also because of some toxic disturbance occasioned by the mother's state of malnutrition. It has been found that maternal diets deficient in vitamins B, C, and D were common among women with a high frequency of malformed fetuses. Fetal rickets as a consequence of the depletion of the mineral reserves of starved mothers is a well-known phenomenon. It is also known that maternal diet deficiency in vitamin D predisposes the child to early rickets.

It is generally agreed that when nutrition during pregnancy is inadequate, the fetus suffers more than the mother. If the mother's diet is good during pregnancy, then the infant is usually in good condition at birth. A number of different investigators have demonstrated the substantial importance of an adequate maternal diet during pregnancy for the health of the infant and the adult it grows to be —*if* it survives to be an adult.

In a Canadian study, carried out by Ebbs and his group, on 120 pregnant women who were on a poor diet compared with 90 pregnant women of the same socioeconomic status whose diet had been made good, it was found that not one of the 120 women on a poor diet showed any recognizable evidences of any deficiency diseases. But what they did find was that in every way the mothers and their offspring who were on a good diet did better than the mothers and their offspring who were on a poor diet. These facts, clearly brought out in Table II, are striking. They show that a diet that was inadequate, although adequate enough not to produce any recognizable deficiency symptoms in the mother, seriously interfered with her reproductive efficiency and affected the fetus much more than it did her.

These findings were abundantly confirmed by Burke and her coworkers at Harvard. In the Harvard study, carried out on 216 mothers and their infants during 1930–1941, it was found that every stillborn, with one exception, every infant dying during the first few days after birth, with one exception, the majority with malformations of various kinds, again with one exception, all prematures, and all functionally immature infants were born to mothers who had had inadequate diets during pregnancy.

TABLE II

COMPARISON OF 120 PREGNANT WOMEN ON A POOR DIET WITH 90 PREGNANT WOMEN OF THE SAME SOCIOECONOMIC STATUS WHOSE DIET HAD BEEN MADE GOOD

|  |  | Diet |  |
|---|---|---|---|
|  |  | Poor | Good |
| Prenatal maternal record | Poor-Bad | 36.0% | 9.0% |
| Condition during labor | Poor-Bad | 24.0% | 3.0% |
| Duration of first stage of labor | Primipara | 20.3 hours | 11.1 hours |
|  | Multipara | 15.2 hours | 9.5 hours |
| Convalescence | Poor-Bad | 11.5% | 3.5% |
| Record of babies during first two weeks | Poor-Bad | 14.0% | 0.0% |
| *Illnesses of Babies During First Six Months* |  |  |  |
| Frequent colds |  | 21.0% | 4.7% |
| Bronchitis |  | 4.2% | 1.5% |
| Pneumonia |  | 5.5% | 1.5% |
| Rickets |  | 5.5% | 0.0% |
| Tetany |  | 4.2% | 0.0% |
| Dystrophy |  | 7.0% | 1.5% |
| Anemia |  | 25.0% | 9.4% |
| Deaths |  | 3.0% | 0.0% |
| *Miscarriages and Infant Deaths* |  |  |  |
| Miscarriages |  | 5.8% | 0.0% |
| Stillbirths |  | 3.3% | 0.0% |
| Deaths: |  |  |  |
| Pneumonia |  | 1.7% | 0.0% |
| Prematurity |  | 0.8% | 0.0% |
| Prematures |  | 7.5% | 2.2% |

SOURCE: Ebbs, J. H., and others, "The Influence of Improved Prenatal Nutrition upon the Infant."

The effects of malnutrition upon the fetus and infant were very evident in babies born in Europe during World War II. Birth weight and length of infant decreased and premature births and stillbirths increased, as did severe rickets, severe anemias, and tuberculosis. A German study is of significance here. This study, by Dr. D. Klebanow, reports on 1,430 newborns who had been born during the years 1946–1948 to Jewish women who had been in German concentration camps and suffered great hardships previous to their pregnancies. All these women conceived *after* their concentration-camp experiences were over, in most cases several years later. Four percent of the newborn of these women showed congenital defects of various kinds. There were 12 with birthmarks (nevi and hemangiomas), 12 were cases of Down's syndrome, 8 had clubfeet, 5 had hydrocephalus, 4 had supernumerary fingers or toes, 4 were born without brains (anencephaly), and so on, in proportions greatly exceeding the normal expectancy. Normally one would expect to find 2 or 3 cases of Down's syndrome in such a number of births; in this series the number is increased 4 to 6 times! Dr. Klebanow has interpreted his findings as indicating that the consequences to the organism of malnutrition over a prolonged period, even after that period of deprivation is ended, may be seriously damaging to the ova, and otherwise so affect the maternal organism as to make the uterine environment she provides in her subsequent pregnancies in one way or another inadequate. Most of the mothers were under 30 years of age.

In those countries during World War II in which the food intake was rigidly and scientifically controlled, as in England, the health of children actually improved. Unfortunately, in other countries food shortages were not so ably handled.

TABLE III
NUTRITION OF MOTHERS DURING PREGNANCY AND BIRTH WEIGHT OF THEIR BABIES

| Nutrition of Mothers | Number | Average Birth Weight of Babies in Grams |
|---|---|---|
| Well nourished | 63 | 3,098.3 |
| Fairly well nourished | 272 | 2,991.9 |
| Undernourished | 29 | 2,897.7 |

SOURCE: Acosta-Sison, H., "Relation Between the State of Nutrition of the Mother and the Birth Weight of the Fetus."

The fact that well-nourished mothers tend to have well-nourished babies and poorly nourished mothers tend to have poorly nourished babies indicates that the well-being of the child before and after birth is significantly influenced by the nutrition of the mother both before and at the time of conception, as well as throughout the duration of pregnancy.

A recent extensive study by Harrell, Woodyard, and Gates has shown that the three-year-old children of mothers who had received a polynutrient vitamin supplement to their diets during pregnancy had an IQ that was 5 points higher than the children of mothers of the same socioeconomic group who had not received such a supplement. The IQ was 103.4 for the first group and 98.4 for the children of the non-supplemented group.

In another study by Acosta-Sison it was found that mothers who had been well nourished during pregnancy had babies whose birth weight was more than 100 grams* greater than the birth weight of babies of fairly well-nourished mothers and more than 200 grams greater than the birth weight of babies of undernourished mothers (see Table III).

It has been shown by Burke and her coworkers that babies born of mothers who consumed 85 grams of protein daily during pregnancy averaged, at birth, as much as 3 pounds more in weight and were 2¾ inches taller than infants whose mothers consumed 45 grams or less of protein daily during pregnancy.

Normally ingested food proteins are broken down into amino acids before they are absorbed into the circulation, but it is now known that unsplit proteins may enter the bloodstream directly from the gastrointestinal tract. Such antigens entering the fetus, as well as those antibodies it receives by filtration through the placenta, will attach themselves to various tissue cells. When similar antigens are introduced into the baby's body after birth, an allergic reaction may ensue, resulting from interaction of the protein and fixed tissue antibodies. The fixation of the antibodies may be to the skin, gastrointestinal tract, nervous system, respiratory tract, and so on, and it is through one or other of these systems that the sensitized individual will react. Such allergic reactions may persist for many years.

The evidence is now fairly substantial that mothers who have an excessive craving for particular foods during pregnancy may by their ingestion produce various allergic sensitivities that show up in their children after birth.

But enough has been said on nutrition. Its fundamental importance in development is evident.

* A gram is about 1/28 of an ounce.

## 6. Infections

All viruses and bacteria are capable of passing from the mother to the developing organism within the womb. Viruses reaching the fetus during the first trimester (first three months) of pregnancy often produce serious interferences with development, resulting in malformations of various sorts. Bacteria do not produce damage of this sort, but frequently result in infections of the fetus that may be more or less damaging.

When viruses are prepared for immunological purposes and are injected into the mother, the maternal antibodies will enter the fetus and confer passive immunity upon it for at least several months after its birth. We have only recently learned this, for example, in connection with the widespread immunization of the population against poliomyelitis. However, similar knowledge in connection with smallpox was already in existence as early as 1702, when Düttel recognized, from the striking effects on the fetal skin, that smallpox in the mother could be transmitted to the fetus in the womb. Edward Jenner, who introduced vaccination into England at the end of the eighteenth century, was aware of the fact that immunization of the mother against smallpox seemed to confer passive immunity to the disease upon her infant. This soon wears off and the baby must then be vaccinated in order to become actively immune.

Infections with the viruses producing smallpox, chicken pox, measles, mumps, scarlet fever, erysipelas, and recurrent fever have long been known to be transmissible from mother to fetus. There is also some evidence that the virus of influenza A can produce serious deformities in the developing embryo. German measles (rubella) is an example of a virus disease that if contracted by the mother during the first twelve weeks of pregnancy is capable of producing gross developmental defects as well as death in the embryo or fetus; a high proportion of surviving infants suffer from such conditions as cataract and deafness with mental defect. Between 12 and 20 percent of fetuses will be adversely affected. If the mother contracts German measles during the first twelve weeks of pregnancy, she has about an 8-to-1 chance of giving birth to a developmentally defective child.

Since the organ-forming period of the human embryo extends from approximately the fourth week to the end of the tenth week, this is the period during which invading organisms may be expected to do most damage. Since the virus of German measles is capable of doing so much damage during this period, it has been reasonably suggested that it would be a good idea to have girls deliberately infected with the virus, and thus immunized, before they reach childbearing age.

There is evidence that poliomyelitis contracted by the mother during the first three months of pregnancy may have serious effects upon the development of the fetus. The evidence is unclear for the ill effects of measles, though the fetus may be born with a clear case of measles, including the typical skin rash and other symptoms.

The bacterium causing syphilis (*Treponema pallidum*) can actually enter the embryo. If this happens miscarriage occurs. If the bacterium enters at a later fetal age, the child is born with signs of congenital syphilis—blindness, deafness, defects of the heart—or the disease may not show itself till later, in the form of severe disorder of the nervous system.

Tuberculosis is also transmissible to the fetus from the mother by means of the *Mycobacterium tuberculosis*. The fetal death rate from tuberculosis is high, and infants who are born with the disease usually die within the first year.

Malarial parasites are known to be transmissible from mother to fetus. Protozoan parasites, such as *Toxoplasma*, can be transmitted from mother to fetus and produce severe disorders of the nervous system and eyes: meningoencephalomyelitis (inflammation of the brain and spinal cord and their membranes), microphthalmia (reduction in the size of the eyes), chorioretinitis (inflammation of the retinal and choroid layers of the eyes), microcephaly (reduction in the size of the head), and hydrocephalus, convulsions, and idiocy or mental retardation.* This disease, toxoplasmosis, occurs in adults in such a modified form that it is rarely recognized.

Various forms of cancer (malignant melanoma, choriocarcinoma) have been transmitted by the mother to her fetus.

Since congenital malformations constitute the second highest leading cause of death of newborn babies, and it is estimated that about 23 percent of the human race dies before birth or shortly thereafter from developmental defects, it will be appreciated how necessary it is for us to become aware of the importance of providing an adequate maternal environment for the developing human being.

---

* For further material on these disorders see Appendix A.

## 7. Drugs

Since 1962, following the announcements of the birth of thousands of malformed children in West Germany and elsewhere in Europe as a result of their mothers' having taken the drug thalidomide, many people who were hitherto altogether unaware of the fact have become sensitive to the dangers of drugs to the unborn child when taken by his mother. It has now been established that the critical phase of toxic thalidomide action is between the twenty-seventh and fortieth day after conception. This does not mean that some developmental damage cannot be done before and after that phase during the first ninety days after conception. There is evidence that it can.

It is now believed that most if not all drugs are detrimental to the uterine organism. Among these are some commonly taken drugs. For example, many cases of congenital deafness have been traced to the mother's use of quinine for malaria during pregnancy. Morphinism has been reported in the infants of mothers who were morphine addicts. Inhalation of amyl nitrite by the mother for a few seconds has induced an increase in the fetal heartbeat, beginning during the third minute following the mother's inhalation. Subsequent inhalations produced a diphase (excitor-depressor) response.

The recent obstetrical practice of dosing the pregnant mother with barbiturates and similar drugs prior to delivery may so overload the fetal bloodstream as to produce asphyxiation in the fetus at birth, with either permanent brain damage or subtle damage that could lead to mental impairment. Fortunately, the trend today is away from heavy sedation.

It is known that when a barbiturate derivative such as secobarbital, usually prescribed as a sedative, is given to the pregnant mother, it will pass into the bloodstream of the fetus and cause a depression in the waves measuring the electrical activity of the cells in the gray matter of its brain, and that this depression persists for some time after birth.

The effects of maternal cigarette smoking upon the fetus may be considered under the present heading. The smoke, which does not have to pass into the lungs to get into the bloodstream, is mostly absorbed directly into the bloodstream of the mother through the mucous membranes of her mouth, nose, and throat. Tobacco smoke contains such noxious substances as nicotine, arsenic, carbon monoxide, furfural, pyridine, collidine, hydrocyanic acid, methanol alkali, and various tar products. It is known that the nicotine absorbed by the smoker acts upon the aggregations of nerve cells (ganglia) of the autonomic nervous system and upon the respiratory center, that it increases the acidity of the gastric tract, produces acidity of the whole gastrointestinal tract, constricts the arteries, increases blood pressure, causes the heart to pump harder—to beat more rapidly—and produces irritation and congestion of all organ membranes.

Is there any evidence that the pregnant mother's smoking affects the fetus? There is. It has been found by Sontag and his associates that the smoking of one cigarette generally produces an increase, and sometimes a decrease, in the heart rate of the fetus. At seven months' fetal age, for example, when the fetal heart beats at the rate of 140 beats per minute, a few puffs by the mother on a cigarette is sufficient in many cases to send the heartbeat up by 40 additional beats, that is, to 180 beats a minute, and sometimes to decrease the heartbeat by 17 beats per minute. The maximum effect is observed between the eighth and twelfth minute after the cigarette, and the response of the heart and blood vessels is more marked after the eighth month of pregnancy. It is quite possible that the products of tobacco entering the embryonic and fetal circulation adversely affect not only the heart and circulatory system but also many other organs. The increase in cardiac and circulatory disorders in recent years may not be unconnected, in part at least, with smoking by pregnant mothers.

In a recent study by Simpson of 7,499 mothers whose smoking habits during pregnancy were examined in relation to the frequency with which premature babies were born to them, some interesting findings were made. In this study a premature baby was defined as one weighing 5½ pounds or less. It was found that there was a significant correlation between the number of cigarettes smoked each day during pregnancy by these mothers and the frequency with which they were delivered of premature babies. Heavy smokers (more than ten cigarettes a day) had the highest prematurity rate; light smokers (between one and ten cigarettes a day) had a lower rate; the nonsmokers had the lowest prematurity rates. The prematurity rate, in fact, was approximately twice as high for smokers as it was for nonsmokers, being 11.4 percent for the smokers and 6.4 percent for the nonsmokers (see Fig. 13). Simpson's findings have been fully corroborated

EFFECTS OF ENVIRONMENT IN THE WOMB 109

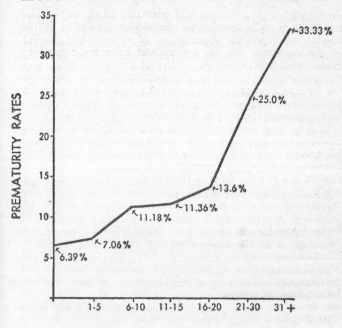

FIG. 13. Relationship of Number of Cigarettes Smoked a Day to Prematurity Rates. (After Simpson.)

by the work of Frazier in the United States, and Lowe, Zabriskier, and others in England.

## 8. Physical Agents

Differences in pressure *in utero,* whether induced through internal or external forces, differences in position, umbilical cord entanglements, and similar factors may more or less adversely affect the development of the organism. Deformity may be caused by faulty position, immobilization, mechanical shaking, temperature changes; asymmetry of the head may be produced by pressure of the head downward upon the thorax. Torticollis (wryneck) has also been observed, and pressure atrophy of the skin indicates the kind of continuous stimulation to which the fetus may be exposed.

Exposure to massive doses of X rays within the first two months of pregnancy will in many cases produce abortion of the embryo. Where abortion does not follow, serious injury has been found to result in a percentage of cases. In a series reported by Professor Douglas Murphy, 75 children whose mothers' pelvic organs had received therapeutic doses of X rays during pregnancy were studied. Only 37 of these children were normal; 38 were malformed. There were 18 microcephalic idiots, 1 case of hydrocephalus; 4 suffered from other serious defects of the nervous system; 6 were premature, weak, and diseased; 2 were mentally defective; 2 were stillbirths; and 5 were otherwise deformed. In other words, 50.7 percent of these children were abnormal.

In April, 1959, a report was released by the New York State Department of Health that indicated that high natural radiation areas in that state were associated with a high congenital malformation rate of more than 20 per 1,000 live births for the period 1948 through 1955. The statewide average is 13.2 malformations per 1,000 births.

In recent years we have learned that the fetus may respond in a convulsive or startled manner to sudden loud noises or vibrations, and that sound reaching it in any form may elicit increase in its activity.

It has, of course, long been known that the birth process itself may seriously affect the fetus, and that at this time damage to the brain or other organs may be produced by a difficult labor and attempts to deliver the child with instruments.

## 9. *Emotional Factors*

There is some evidence that emotional disturbances in the mother may affect both the structural and psychological development of the fetus. Emotional changes in the mother insofar as they produce chemical changes within the body of the mother may by this means cause the passage of excessive quantities of certain chemical substances into the developing embryo that produce disturbances in growth and development. Some evidence has recently been turned up suggesting that emotional disturbances during the first ten weeks of pregnancy may be responsible for the production of a large number of children with cleft palate. The bones that enter into the formation of the palate develop between the seventh and tenth week of embryonic age, and any disturbance at this time might well affect the development of these

# EFFECTS OF ENVIRONMENT IN THE WOMB

bones adversely. The means by which this might be accomplished is as follows: Maternal stress is known to cause hyperactivity of the cortex, or outer layer, of her adrenal glands (situated on the upper poles of the kidneys). Hyperactivity of the adrenals causes the secretion of the hormone hydrocortisone. Hydrocortisone can pass through the placenta into the fetus. When hydrocortisone is injected into mice or rats at the time when their palates are being formed, almost 90 percent of them are later born with cleft palates. The probabilities are high that a similar mechanism is at work in man.

Children born of mothers who have been seriously disturbed emotionally at some time during their pregnancy frequently exhibit psychological and psychosomatic irritability after birth. There is much evidence that strongly suggests that a human being's nervous system may even be permanently sensitized by the action upon it of excessive quantities of chemicals released as a consequence of the pregnant mother's emotional disturbances.

Emotional factors are known to play an important role in habitual miscarriage and sterility.

## 10. Other Environmental Factors

The direct and indirect effects upon the fetus of radiation from the fallout of atomic bombs is a matter important enough to deserve a separate chapter. The subject is therefore dealt with in Chapter 18. Here we may mention an environmental factor affecting the fetus *in utero* that may malform the human being for the rest of his life—iodine deficiency leading to endemic cretinism.

Endemic cretinism is due to a congenital deficiency of the thyroid gland, which causes it to produce an inadequate amount of its normal secretion, thyroxine, which is high in iodine salts. In areas of the world in which the soil is deficient in iodine all foodstuffs derived from that soil are likely to be poor in iodine, thereby causing iodine deficiency in the individuals living in such areas (see Plate 5). Iodine deficiency will almost always result in a goiter. It is highly likely that mothers suffering from iodine deficiency will give birth to children who are cretins. Cretins suffer from an almost complete lack of the thyroid hormone and are stunted both physically and mentally, are often more or less deaf, and, in addition to a characteristic slouching gait, are frequently affected by other abnormalities. All the children of a goitrous woman need not, however, be cretinous, and if the mother moves to an environment where her children are brought up on foods having an adequate iodine content, they need never become cretinous. If she stays where she is, her children need not be cretinous if she is treated throughout pregnancy with iodine. Healthy persons moving into an iodine-deficient area may develop goiter.

Cretinism is not always due to iodine deficiency in the mother; in some cases it is caused by a deficient gene.

Only a small fraction of the evidence has been cited bearing on the manner in which the development of the human being may be affected while in the womb *, but this should be more than sufficient to show the importance of environmental factors in the development of every human being. The evidence we have considered in this chapter constitutes something of a record of unhappiness, but as a consequence of our increased understanding of the causes of such unhappiness, it becomes possible for us to conclude this chapter on a happy note. We know that heredity, or constitution, is not a mystery tantamount to fate, or predestination, that on the other hand there is much that can be done to make the gestation period, the 266½ days of pregnancy of the mother, a thoroughly successful one in the vast majority of cases. The health of our children is largely in our own hands. We can do something about it. If we eat sour grapes, our teeth may be set on edge, but there is no reason why our children's teeth should. We need no longer be as doubtful or anxious about whether or not we are doing the right thing as parents. Abundant scientific knowledge is now reasonably available. All we need do is acquire that knowledge and act upon it. The health of the infant and child begins at conception. This being so, care for the quality of the genes that enter into that conception, by the intelligent selection of mates, should be at least as important a consideration as care of its prenatal and postnatal environment. Once the conception has been brought about, the care of the infant and child begins with caring for the human being developing in the womb. In this respect nothing is more important than the health and well-being of the mother who nourishes it.

---

* The subject is fully dealt with in my book *Prenatal Influences* (Springfield, Illinois: Charles C. Thomas, Publisher, 1962). A book for the general reader on this subject is my *Life Before Birth* (New York: The New American Library of World Literature, Inc., 1964).

## 7. ENVIRONMENT AFTER BIRTH

WHEN A BABY IS BORN he enters what is for him an entirely new environment, the postnatal environment, the environment of the world outside the womb. Having lived an entirely "aquatic" life in the womb, rocked, as it were, in the cradle of the deep of his mother's amniotic fluid and living off the oxygen, among other things, transmitted to him across the placenta from his mother, he now has to adjust to breathing atmospheric air, and instead of his food being supplied him in his blood, he now has to take it in through his digestive system. In a relatively short period of time—the period of birth—he is called upon to make literally scores of new adjustments. But however well he automatically makes these adjustments, if the environment fails him, the newborn will not do well. The most important of all the elements in the human infant's environment is other human beings, at least one. An infant must have oxygen, but this it normally obtains without any more effort than is involved in breathing. It must rest, sleep, and eliminate. All these things it is able to do without effort. It must also take in food and liquid. These things it cannot do without the provision of the necessary food and liquid by another human being. Usually it is the biological mother who provides the infant with its food and attends to its other needs. From the moment of its birth the quality of the mothering the infant receives, whether from its own mother or from someone else, is fundamentally related to the quality of its development physically and psychologically.

### The Importance of Mothering

If the infant has enjoyed optimum conditions for development in the womb, his postnatal development may nevertheless follow a seriously disordered course as a consequence of inadequate mothering. Furthermore, we now have ample evidence indicating that as a result of inadequate mothering some children will fail to grow and develop, and will die, especially during their first and second years. Such children present a pathetic picture of abandonment and hopelessness. They exhibit all sorts of regressive changes: pallor and wrinkling of the skin, dullness of the eye; they will lie motionless and quiet for hours or else cry for hours; they will regurgitate their food, suffer from diarrhea; they will virtually cease to grow and suffer a marked mental deterioration. Dr. René Spitz, in a classic study, reported his investigation of 239 children who had been institutionalized for one year or more, one group of whom were nursed by their mothers in an institution that he called the "Nursery." In the second institution, "Foundlinghome," the children were raised from the third month by overworked personnel, one nurse caring for from eight to twelve children. "Nursery" did not lose a single child through death, whereas in the "Foundlinghome" 37 percent of the children died during the two-year observation period.

"While the children in 'Nursery,'" writes Spitz, "developed into normal healthy toddlers, a two-year observation of 'Foundlinghome' showed that the emotionally starved children never learned to speak, to walk, to feed themselves. With one or two exceptions in a total of 91 children, those who survived were human wrecks who behaved either in the manner of agitated or of apathetic idiots."

The World Health Organization report entitled *Maternal Care and Mental Health*, written by Dr. John Bowlby and published in 1951, discusses all the material up to the year 1951 on this relationship between mothering and development, abundantly confirming the importance of good mothering for healthy development. It is now clear that a child who has not been adequately loved for any significant period during its first half-dozen years of life may suffer more or less severely not only during those first six years but also for the rest of his life.

Just as there are critical developmental periods in the organ-forming stage of the embryo's development, so there are critical developmental periods in the personalization stage of development. Personalization is the process of personality organization, the way in which the individual comes to be organized as a behaving organism in relation to others and to himself. There are at least three or four such criticial developmental periods, which may be described as follows:

1. The period during which the infant is in the process of establishing an explicit cooperative relationship with

a clearly defined person—the mother. This is normally achieved by five or six months.

2. The period during which the child needs the mother as an ever-present support and companion. This normally continues to about the end of the third year.

3. The period during which the child is in the process of learning to maintain a relationship with its mother during her absence. During the fourth and fifth years, under favorable conditions, such a relationship can be maintained for a few days or even weeks.

Inadequacies in the mothering experience during these periods are likely to have different damaging effects upon the development of the child. These relationships are under intensive study at the present time, and it will be some years before we shall be able to correlate the behavioral changes with the critical developmental periods. They have been studied carefully in dogs, and it has been shown in them quite clearly that there does exist a significant relationship between critical developmental periods for social behavior and susceptibility to defective behavior in response to certain kinds of environmental stimuli. For example, it has been found that when puppies are weaned too early (at about three weeks), they tend to show the canine equivalent of thumb-sucking; they will suck tails or ears or toes. When puppies are allowed a week or two more of nursing, they do not show such behavior. If one allows a puppy to be raised by its mother without the experience of human beings, say for the first twelve to fourteen weeks of its life, it then becomes extremely difficult to get the animal to relate to human beings.

If greylag goslings are exposed as soon as they hatch from the egg to a human being instead of to the mother goose, they will thereafter attach themselves to the human being as if he were the mother. They will form a similar irreversible attachment to any moving object. If one takes a newborn guinea pig away from its mother before it has nursed and keeps it separated from the mother for approximately three days, it is found that the guinea pig cannot learn to nurse when it is returned to its mother.

These examples are but a few of the evidences for the existence of critical developmental periods in lower animals. These critical periods are undoubtedly genetically based, and from them we have learned not only that certain environmental stimuli must be received within a circumscribed period of time if normal development is to occur but also that if other forms of stimuli are substituted, permanently abnormal forms of behavior will follow. Insofar as it has been possible to do so, these conclusions have in every way been confirmed in human beings, often in the most dramatic and spectacular manner.

No one, for example, would have suspected that lack of a warm emotional environment could possibly interfere with the bone growth of children. But it does. And not only the bones but also every aspect of physical growth may be affected. Studies were made of a group of children in a Cleveland institution for dependent and neglected children by doctors Ralph Fried and M. F. Mayer. They summarize their findings in the following words: "Socio-emotional adjustment plays not merely an important but a crucial role among all the factors that determine individual health and well-being . . . it has become clear that socio-emotional disturbance tends to affect physical growth adversely, and that growth failure so caused is much more frequent and more extensive than [is] generally recognized." These investigators found that there was only one treatment that could successfully overcome the retardation of growth suffered by the institution children they investigated, and that was removal to a warm, loving environment. A similar finding was made by the Medical Director of Schools, at Saskatoon, Canada, Dr. Griffith Binning. Dr. Binning, in a remarkable study of the effects of emotional tensions on the development and growth of 800 Saskatoon children, found "that events in the child's life that caused separation from one or both parents—death, divorce, enlistment of a parent—and a mental environment which gave the child a feeling that normal love and affection were lacking did far more damage to growth than did disease," that such an environment, indeed, "was more serious than all other factors combined."

The same children with the same genes in happier environments would respond to those environments very much more happily than they were able to do in the environments in which they found themselves.

What is the heredity of such children? Clearly, children inherit postnatal environments as well as genes and prenatal environments, and the interactive consequences of these last two with the postnatal environments they inherit can be just as lethal and retardative or stimulating and encouraging to the realization of the organism's potentialities as the prenatal environment. It is extremely important to emphasize this point, for it is widely believed that heredity is "what you are born with," and that "what you are born with" pretty much determines what you are going to become. The two beliefs taken together are both too limited and too comprehensive. Heredity is what you are born with, in the sense that you are born with palm prints and sole prints, but it is also true that heredity includes what you are born *into*. The potentialities of each individual are dependent on genetic endowment, but these potentialities have a wider range than is generally observed, so that the environment, *which is part of an individual's inheritance, is also the means by which the heredity of an individual can be changed, modified, enlarged, contracted,* and so on.

The child's potentialities at birth have by far the greater part of their development to realize in the years *after* birth, and how those potentialities will be realized will again depend upon the nature of those potentialities in interaction with the environment to which they are exposed. It may not be possible to make a silk purse out of a sow's ear, but it is most certainly possible to depress the environment to such an extent as to make sows' ears out of materials that would otherwise have become silk purses. But that is a metaphor, the first part of it too frequently used by those who believe that genes determine the fate of men. The facts, however, enable us to make a more correct generalization: the genes, or potentialities, being what they are at birth, what a human being subsequently becomes is then, within the limits set by the genes, largely determined by the experiences to which he is exposed. Let us consider the facts.

We are all born with potentialities for speech, but whether or not we will speak is dependent entirely upon the nature of the exposure to speaking human beings. As children, if we hear only a word now and again and are not involved in the necessity of learning the meaning of things, the probabilities are high that we will never speak. Some years ago in Ohio a little girl was found locked up in a closet, where she had been kept for five years with her deaf-mute mother. At the age of five years, when she was liberated, this little girl could not utter a word, but after approximately six months of intensive work with a speech expert she eventually learned to speak perfectly. She had, incidentally, received a great deal of love from her deaf-mute mother during their incarceration.

None of us would ever speak unless we heard other people speaking. We possess the general potentialities for speech, but what language we will speak will depend entirely upon the language we hear spoken. And so it is with all our behavioral potentialities. How we will behave about them depends largely upon the manner in which we have been taught by the environment around us. Our very senses have to be taught in this way. Our tactile sense, how we see, hear—all these we have to learn.

Compared to some nonliterate people we are very "thick-skinned." They often laugh at the length of time it takes a white man to discover that some insect has alighted upon his body, for he is generally unaware of it until after it has bitten or stung him. They can usually feel an insect the moment it has alighted. Their sense of visual discrimination is markedly more developed than ours simply because in their way of life visual acuity is vastly more important than in ours. But all human beings have to learn to see three-dimensionally and to evaluate the conditions of perspective.

The apparent simplicity of seeing is the result of a long learning process in which we eventually bring the whole of our past experience to bear on *how* and therefore on *what* we see. It is possible to demonstrate this in a number of ways, but perhaps in none more convincingly than in the "distorted room" experiments of Ames and Cantril. In this room the floor, ceiling, and back wall are all slanting. When one looks into this room with one eye through a hole one sees a normal rectangular room—that is, the *assumption* is made that it is a normal rectangular room, whereas there are actually no right angles in the room. When we enter the room and move about in it, we understand that the eye did not in fact see what was to be seen. We prejudiced what we saw with assumptions based on our former experience. Hence we understand that our ability to see things is not entirely a matter of having healthy eyes, but depends, too, on the conditioning of our past experience. All seeing is interpretation—and we interpret what we see according to the history of our experience.

How human beings will see, hear, and talk, provided they have the normal potentialities for these abilities, is, then, largely a matter of the experiences they undergo in the process of learning. With the exception of identical twins no two human beings are alike in their potentialities, and within the range of the normal the variation in the quality of potentialities is enormous among human beings regardless of ethnic affiliation, class, or caste. Thus we can be quite certain that because of the great variation in genotypes, even if we could make the environment identical for everyone and no matter how favorable we made that environment, there would still be an enormous difference in the range of abilities exhibited. In fact the variation in those abilities would be even greater than it is anywhere on earth today, for the more we maximize the opportunities for potentialities to express themselves, the more we increase the chances for individual differences to express themselves. In a society in which no one can write, all its members are alike in their incapacity to write. Teach them to write and very soon marked differences will be discovered in the abilities of different individuals to write. By maximizing opportunities we do not reduce inequality in ability, we maximize it. This refers to physical as well as to mental development. It should be clear that by maximizing opportunities we increase the possibilities not only for increasing likenesses but also for increasing the differences that exist between individuals.

## Physical Development

One of the best measures of physical development is increase in size. It has been known for many years that an unfavorable environment is capable of retarding growth. The kind of environment most generally involved is the socioeconomic environment, that is, the social environment and the kind of conditions that different amounts of money make possible. As might be expected, impoverished or low socioeconomic environments are associated with impoverished growth, and favorable socioeconomic environments are associated with favorable growth, and again this applies to mental as well as physical growth.

It has been shown that the rate of growth of children with two unemployed parents is slower than that of children with only one unemployed parent, and that the children of employed parents have the highest rate of growth. Children in private schools on the average have a higher rate of growth than those in the public schools. The upper socioeconomic classes have less weight for height than the lower ones. Japanese who migrate as children to California grow generally larger than those who stay at home. Growth rates of children of middle-class homes are significantly higher than those of institutional children. Rate of growth is higher for rural than for urban infants in the first eighteen months of life. Efficiency of the mother, as measured by the tidiness of the home and children, has been shown to be significantly related to the growth of infants; the more efficient the mother, the better the growth of the children.

Sickness and mortality rates are highly correlated with socioeconomic status. These rates are similar during the first week of life, becoming steadily dissimilar during the first and second months in favor of the higher socioeconomic classes, so that the rates for the infants of the poorest parents become 3 to 4 times greater. By the end of the year mortality rates of the infants of the poorest families are 8 to 10 times greater than those for the infants of middle-class families.

The higher morbidity (sickness) and mortality rates for infants of the poorer clases as compared with those of the more comfortably situated classes are almost certainly not due to genetic differences in the infants of these classes, to any inherent weaknesses in the infants born into poor families. Rather, the evidence indicates that the higher rates are due to the inadequacies—nutritional, sanitary, and the like —of the poorer infant's environment. The incidence and the mortality rates for such diseases as measles are not significantly different in the different socioeconomic classes. During epidemics the sickness and death rates for infants of higher socioeconomic classes are similar or tend to exceed those of the poorer classes. This suggests that there is no genuine difference in genetic susceptibility to disease in the infants of different socioeconomic classes.

The facts discovered about the growth of school children, and their illnesses and mortality rates, in relation to socioeconomic environment are much the same as those for infant and preschool children. These facts were brought out sharply in one of the earliest studies made of the relation of socioeconomic factors to growth. This investigation, reported by Elderton, was conducted during 1905–1906 in Glasgow, Scotland, and involved the heights and weights of approximately seventy thousand children between five and eighteen years of age. The results obtained were then related to the types of schools that the children attended. This yielded four district school groups of different socioeconomic ratings, from A to D, where A represented schools attended by children from the poorest districts and D those attended by children from the most prosperous. The results of this investigation are set out in Table IV.

From Table IV it will be observed that at practically every age, and for both sexes, an appreciable and regular difference was exhibited between the children of the four graded socio-

### TABLE IV
HEIGHT (INCHES) AND WEIGHT (POUNDS) IN 70,000 GLASGOW (SCOTLAND) SCHOOL CHILDREN, 1905–1906, BY ASCENDING ORDER (FROM A TO D) OF SOCIOECONOMIC STATUS OF SCHOOL

| Age: | 5 | 6 | 7 | 8 | 9 | 10 | 11 | 12 | 13 |
|---|---|---|---|---|---|---|---|---|---|
| *Boys* | | | | | *Height (inches)* | | | | |
| Group A | 41.3 | 43.0 | 45.1 | 47.0 | 48.8 | 50.6 | 52.3 | 53.8 | 55.2 |
| Group B | 42.1 | 44.0 | 45.9 | 47.7 | 49.5 | 51.1 | 52.8 | 54.3 | 55.5 |
| Group C | 42.1 | 44.0 | 46.2 | 48.1 | 49.9 | 51.5 | 53.5 | 55.0 | 57.2 |
| Group D | 43.0 | 44.8 | 46.9 | 49.0 | 50.9 | 52.6 | 54.2 | 55.9 | 57.7 |
| *Girls* | | | | | | | | | |
| Group A | 41.0 | 42.9 | 44.6 | 46.6 | 48.5 | 50.3 | 52.4 | 54.4 | 55.8 |
| Group B | 42.0 | 43.7 | 45.6 | 47.4 | 49.2 | 51.1 | 53.0 | 55.2 | 57.1 |
| Group C | 41.9 | 43.7 | 45.6 | 47.6 | 49.4 | 51.2 | 53.3 | 55.4 | 57.0 |
| Group D | 42.7 | 44.8 | 46.4 | 48.6 | 50.4 | 52.2 | 54.1 | 56.5 | 58.7 |
| *Boys* | | | | | *Weight (pounds)* | | | | |
| Group A | 40.9 | 44.2 | 48.0 | 52.3 | 56.7 | 61.6 | 66.4 | 71.7 | 75.6 |
| Group B | 42.0 | 45.6 | 49.6 | 53.9 | 58.4 | 62.7 | 67.8 | 72.9 | 77.3 |
| Group C | 42.5 | 45.9 | 50.1 | 54.4 | 59.5 | 63.9 | 69.1 | 75.6 | 82.2 |
| Group D | 43.3 | 46.6 | 51.2 | 56.3 | 61.2 | 66.3 | 70.8 | 76.9 | 83.2 |
| *Girls* | | | | | | | | | |
| Group A | 39.9 | 43.0 | 46.4 | 50.5 | 54.7 | 59.5 | 65.3 | 72.4 | 76.8 |
| Group B | 40.6 | 43.9 | 47.7 | 51.8 | 55.8 | 60.8 | 66.8 | 74.3 | 81.3 |
| Group C | 41.3 | 44.7 | 48.1 | 52.7 | 56.9 | 61.9 | 68.4 | 76.1 | 83.0 |
| Group D | 41.8 | 45.6 | 49.3 | 54.3 | 58.8 | 64.4 | 70.5 | 78.8 | 89.0 |

SOURCE: E. M. Elderton, "Height and Weight of School Children in Glasgow."

economic districts in both height and weight, and this always in favor of the higher socioeconomic groups. For example, at age nine the average height of boys in district A was 48.8 inches, in B 49.5 inches, in C 49.9 inches, and in D 50.9 inches. At every age a steady increase is observed from Group A to Group D. Boys from district D at age five are 1.7 inches taller and at age thirteen are 2.5 inches taller than those of district A.

Later studies carried out in many different parts of the world have fully confirmed the capital importance of the environment and its effects upon the processes of growth. One of the most striking of these studies, made by Craven and Jokl, evaluates the growth records of 1,067 physically substandard adolescent boys at the Physical Training Battalion in Pretoria, Republic of South Africa. It was found that within the first nine months these boys spent in the training station they grew in bulk, on the average, at a rate five times as great as they would have grown in their unsatisfactory home environment. The factors that produced this remarkable acceleration in growth would appear to have been mainly nutritional.

TABLE V

PERCENTAGE OF CHILDREN, AGE 5 TO 15 YEARS, WHO ARE UNDER AVERAGE HEIGHT, CLASSIFIED BY COUNTRY AND SOCIOECONOMIC STATUS

| Location of School | United States | Canada | Scotland | England | Ireland |
|---|---|---|---|---|---|
| Prosperous District | 7.5 | 11.0 | 11.2 | 18.5 | 22.7 |
| Average District | 18.3 | 23.4 | 24.6 | 24.4 | 24.3 |
| Poor District | 36.4 | 31.6 | 27.3 | 35.1 | 30.7 |

Adapted from Cudmore and Neal, *A Height and Weight Survey of Toronto Elementary School Children, 1939.*

In Table V are set out the figures for five different English-speaking countries with somewhat different dietary and other habits, showing the percentages of children under average height, between the ages of five and fifteen years, divided by socioeconomic status. These figures tell a remarkable, an almost dramatic, story. For one thing, they show that the least number of children under average height in schools situated in prosperous districts were, in 1939, to be found in the United States, and this was true also of the average districts, but in the poorer districts the United States had a higher frequency of children under average height than the poor-district children of the other countries. The evidence indicates that in the year 1959 English children of all districts, because of the great care that has been spent on the nutrition of children since the beginning of World War II (1939), are seldom under average height.

## Changes in Body Form with Changes in Geographic Environment

Beginning with Franz Boas' demonstration in 1912 that American-born descendants of immigrants undergo certain bodily changes, particularly in head form, a good number of similar studies on other groups have demonstrated even more fully how modifiable the form of the body is.

Table VI shows the kind and degree of differences between the measurements of immigrants' children born in the United States and those born in Europe. The results set out in this table prove that the form of the head may undergo certain changes with change in environment without change in descent. In other words, the pattern of the genotype may remain unaltered, but its physiological expression undergoes

TABLE VI
INCREASE (+) OR DECREASE (−) IN MEASUREMENTS OF CHILDREN OF IMMIGRANTS BORN IN THE UNITED STATES COMPARED WITH THOSE OF IMMIGRANTS BORN IN EUROPE

| Nationality and Sex | Length of Head (mm.) | Width of Head (mm.) | Cephalic Index | Width of Face (mm.) | Stature |
|---|---|---|---|---|---|
| Bohemians: | | | | | |
| Male | −0.7 | −2.3 | −1.0 | −2.1 | +2.9 |
| Female | −0.6 | −1.5 | −0.6 | −1.7 | +2.2 |
| Hebrews: | | | | | |
| Male | +2.2 | −1.8 | −2.0 | −1.1 | +1.7 |
| Female | +1.9 | −2.0 | −2.0 | −1.3 | +1.5 |
| Sicilians: | | | | | |
| Male | −2.4 | +0.7 | +1.3 | −1.2 | −0.1 |
| Female | −3.0 | +0.8 | +1.8 | −2.0 | −0.5 |
| Neapolitans: | | | | | |
| Male | −0.9 | +0.9 | +0.9 | −1.2 | +0.6 |
| Female | −1.7 | +1.0 | +1.4 | −0.6 | −1.8 |

SOURCE: Boas, F., "Changes in the Bodily Form of Descendants of Immigrants."

modification as a consequence of the effects exercised by a new environment. Furthermore, Boas showed that the influence of the environment makes itself felt with increasing intensity according to the time elapsed between the arrival of the mother and the birth of the child. This is well brought out in Fig. 14.

The American-born descendants differ in head form from their parents. The differences develop in early childhood and persist throughout life. The head index—that is, the head breadth taken as a percent of head length (cephalic index) —of the foreign-born remains practically the same no matter how old the individual was at the time of immigration. This might be expected when immigrants are adult or nearly mature, but even children who come to the United States when one year or a few years old develop the head-index characteristic of the foreign-born. For Jews this index ranges around 83; that of the American-born Jews changes suddenly. The value drops to approximately 82 for those born immediately after the immigration of their parents and reaches 79 in the second generation, that is, among the children of the American-born offspring of immigrants. The effect of the American environment makes itself felt immediately and increases slowly with the increase of time between the immigration of the parent and the birth of the child. Observations made by Boas in 1909 and 1937 yielded the same results, save that there is an appreciable increase in all measurements in the 1937 series.

Similar confirmatory observations have been made of the descendants of Puerto Ricans in the United States, of Russian Jews, Hawaiian-born children of Japanese, American-born Chinese, American-born children of Japanese, and American-born children of Mexicans.

What is the cause of such changes in the descendants of immigrants to the United States? We do not know. Is it nutrition? We cannot be sure. All that we can say is that the interaction between the developing genotype and the new environment has resulted in modifications of bodily form in the offspring of the immigrants. Once again, the importance of the environment is underscored.

## Environment and Longevity

While there cannot be the least doubt that duration of life, or longevity, has a genetic basis, there can equally be no doubt that the environment plays a major role in influencing the development of genetic potentialities. Environment, in interaction with the genes, determines the lifespan. An individual may start life with genes that would have been capable of maintaining him till he is eighty-five in a good environment, but a socioeconomically depressed environment, together with the psychosomatic (mind-body) changes that can afflict individuals in such environments, often makes it difficult for such an individual to attain half that age. On the other hand, an optimum environment enables many individuals to achieve a much longer life than would have been possible in a less favorable environment.

Perhaps the most spectacular effect of improved environmental conditions has been the increase in longevity during the last century. In 1858 the average duration of life in the United States was about 40 years. By 1900 this had jumped to 50 years; in 1920 this was 55 years, in 1930 a little over 60 years, and by 1959 the expectation of life for the average white American male was 67.3 years and for the average white female 73.9 years.* In Table VII the latest data are

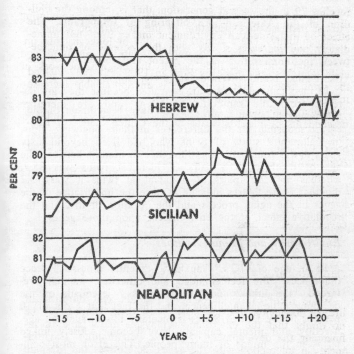

FIG. 14. Length-Breadth (Cephalic) Index of Immigrants and Their Descendants. (After Boas.)

presented on longevity in forty-seven countries, comparing the lastest available figures with those for the years around 1940. It will be seen from this table that the expectation of life at birth has increased appreciably throughout the world, the gains generally being largest in the countries that formerly had the least favorable record. In Puerto Rico, for example, the average length of life has increased more than twenty-two years in a fifteen-year period, from 46.0 years in 1939–1941 to 68.3 years in 1955. Even more striking is the record of Ceylon, where the expectation of life at birth rose from 42.8 years in 1946 to 59.9 years in 1954, an increase of seventeen years in the short period of eight years. Mexico,

---

* The expectation of life for nonwhites in the United States by 1959 was 60.9 for males and 66.2 for females.

Brazil, and Thailand have also made great gains, amounting to an increase in expectation of life at birth of about one year annually. These remarkable gains in longevity are undoubtedly due entirely to improvements in nutrition and sanitation, and in no little part to the magnificent work of the United Nations World Health Organization.

In identical twins, who are so called because they are derived from the same fertilized egg and are therefore genetically identical, environmental differences are undoubtedly responsible for the differences in their longevity. Thus, in Kallmann's study of twins who had lived beyond sixty years it was found that the difference in months between the age of death of 513 pairs of identical twins was 35.7, whereas in 1,226 pairs of nonidentical twins it was 73.7 months. These figures indicate at once the importance of the genes in the aging process and the duration of life, and the regulative effect of the environment upon those genes.

## Environment and Mental Abilities

In an age of "IQ's," "anxiety," "mental illnesses," "race differences," the feebleminded and otherwise mentally retarded, "the Jukes" and "the Kallikaks," jeremiads on the declining national intelligence, and the fact that 1 out of 10 persons in the United States is mentally ill—in such an age, heredity and mental capacity have become closely linked terms in the minds of many people. The error most commonly committed is to identify the level of mental ability achieved by the individual with his genetic endowment. Everyone agrees that Mozart was a musical genius by virtue of his extraordinary genetic endowment and the training received in his home. No one doubts that had Mozart not been possessed of an extraordinary genetic endowment for music, no matter how thorough the training he received at home he would not and could not have become the prodigy that he was. The genes had to be there first. Leonardo, Newton, Darwin, Einstein—all these men of outstanding genius were unquestionably genetically most extraordinarily well endowed. All they required was a minimal stimulus from the environment and their genius seemed to unfold itself almost in spite of themselves. Darwin regarded himself as a man of quite ordinary intelligence. Einstein never ceased to wonder what all the fuss was about. Such responses on the part of men of genius to their own accomplishments confirm most people in the belief that ability is a matter of

## TABLE VII
### EXPECTATION OF LIFE AT BIRTH, SELECTED COUNTRIES

| Country and Period | Total Persons | Male | Female |
|---|---|---|---|
| **North America** | | | |
| United States | | | |
| 1955 | 69.5 | 66.7 | 72.9 |
| 1939-41 | 63.6 | 61.6 | 65.9 |
| 1955 | | | |
| White | 70.2 | 67.3 | 73.6 |
| Nonwhite | 63.2 | 61.2 | 65.9 |
| Canada | | | |
| 1955 | 70.1 | 67.6 | 73.0 |
| 1940-42 | 64.6* | 63.0 | 66.3 |
| Costa Rica | | | |
| 1949-51 | 55.7 | 54.6 | 57.1 |
| El Salvador | | | |
| 1949-51 | 51.2* | 49.9 | 52.4 |
| Greenland | | | |
| 1946-51 | 34.9* | 32.2 | 37.5 |
| Guatemala | | | |
| 1949-51 | 43.7* | 43.8 | 43.5 |
| 1939-41 | 36.5* | 36.0 | 37.1 |
| Mexico | | | |
| 1949-51 | 48.7 | 46.7 | 49.9 |
| 1940 | 38.9* | 37.9 | 39.8 |
| Puerto Rico | | | |
| 1955 | 68.3 | 66.7 | 70.0 |
| 1939-41 | 46.0 | 45.1 | 46.9 |
| **South America** | | | |
| Argentina | | | |
| 1947 | 59.2* | 56.9 | 61.4 |
| Brazil (Federal District) | | | |
| 1949-51 | 52.9 | 49.8 | 56.0 |
| 1939-41 | 42.5* | 39.7 | 45.2 |
| Chile | | | |
| 1952 | 51.4 | 49.8 | 53.9 |
| 1939-42 | 41.8 | 40.7 | 43.1 |
| Ecuador (Quito) | | | |
| 1949-51 | 52.0* | 50.4 | 53.7 |
| Venezuela | | | |
| 1946 | 47.0* | 45.9 | 48.1 |
| 1941-42 | 46.7 | 45.8 | 47.6 |
| **Europe** | | | |
| Austria | | | |
| 1949-51 | 64.4* | 61.9 | 67.0 |
| 1930-33 | 56.5* | 54.5 | 58.5 |
| Belgium | | | |
| 1946-49 | 64.6 | 62.0 | 67.3 |
| 1928-32 | 57.9* | 56.0 | 59.8 |
| Czechoslovakia | | | |
| 1949-51 | 63.2 | 60.9 | 65.5 |
| 1929-32 | 53.6* | 51.9 | 55.2 |
| Denmark | | | |
| 1946-50 | 68.9* | 67.8 | 70.1 |
| 1936-40 | 64.7* | 63.5 | 65.8 |
| England and Wales | | | |
| 1956 | 70.5* | 67.8 | 73.3 |
| 1937 | 62.3* | 60.2 | 64.4 |
| Finland | | | |
| 1951-55 | 66.6* | 63.4 | 69.8 |
| 1931-40 | 57.0* | 54.5 | 59.6 |
| France | | | |
| 1954-55 | 68.2 | 65.1 | 71.4 |
| 1933-38 | 58.7 | 55.9 | 61.6 |
| Germany | | | |
| 1952-53 (East) | 67.1* | 65.1 | 69.1 |
| 1949-51 (West) | 66.5* | 64.6 | 68.5 |
| 1932-34 | 61.3* | 59.9 | 62.8 |
| Hungary | | | |
| 1955 | 66.7* | 64.7 | 68.7 |
| 1941 | 56.6* | 54.9 | 58.2 |

* Average of male and female.
SOURCE: Largely from United Nations Demographic Yearbooks and reports of various countries.

heredity, meaning genes.

Now, there is no doubt that genes and ability are fundamentally connected, but they are by no means the same things. Genes provide the potentialities, and it is the stimulus of the environment that brings those potentialities out. Even genius may fail to declare itself in the absence of the proper environmental stimuli. The French painter Paul Gauguin did not begin to paint until he was a mature man, and almost certainly would not have done so had he not

## TABLE VII
### EXPECTATION OF LIFE AT BIRTH, SELECTED COUNTRIES

| Country and Period | Total Persons | Male | Female |
|---|---|---|---|
| *Europe (cont.)* | | | |
| Ireland | | | |
| 1950-52 | 65.8* | 64.5 | 67.1 |
| 1940-42 | 60.0* | 59.0 | 61.0 |
| Luxembourg | | | |
| 1946-48 | 63.7* | 61.7 | 65.8 |
| Netherlands | | | |
| 1953-55 | 72.5* | 71.0 | 73.9 |
| 1931-40 | 66.5* | 65.7 | 67.2 |
| Northern Ireland | | | |
| 1950-52 | 67.1* | 65.4 | 68.8 |
| 1936-38 | 58.5* | 57.8 | 59.2 |
| Norway | | | |
| 1946-50 | 71.0* | 69.3 | 72.7 |
| 1931-40 | 65.8* | 64.1 | 67.6 |
| Poland | | | |
| 1952-53 | 61.4* | 58.6 | 64.2 |
| 1931-32 | 49.8 | 48.2 | 51.4 |
| Portugal | | | |
| 1949-52 | 58.0* | 55.5 | 60.5 |
| 1939-42 | 50.7* | 48.6 | 52.8 |
| Scotland | | | |
| 1956 | 68.6* | 66.0 | 71.2 |
| 1930-32 | 57.8* | 56.0 | 59.5 |
| Sweden | | | |
| 1951-55 | 72.0* | 70.5 | 73.4 |
| 1936-40 | 65.6* | 64.3 | 66.9 |
| Switzerland | | | |
| 1948-53 | 68.6* | 66.4 | 70.9 |
| 1933-37 | 62.7* | 60.7 | 64.6 |
| U.S.S.R. | | | |
| 1954-55 | 64 | 61 | 67 |
| 1926-27‡ | 44.4 | 41.9 | 46.8 |
| Yugoslavia | | | |
| 1950 | 56.3* | 54.5 | 58.2 |
| *Africa and Asia* | | | |
| Belgian Congo | | | |
| 1950-52 | 38.8* | 37.6 | 40.0 |
| Ceylon | | | |
| 1954 | 59.9* | 60.3 | 59.4 |
| 1946 | 42.8* | 43.9 | 41.6 |
| Cyprus | | | |
| 1948-50 | 66.2* | 63.6 | 68.8 |
| India | | | |
| 1941-50 | 32.1* | 32.5 | 31.7 |
| 1921-31 | 26.7* | 26.9 | 26.6 |
| Israel (Jewish) | | | |
| 1955 | 70.8* | 69.4 | 72.1 |
| 1939-41 | 63.5* | 62.3 | 64.6 |
| Japan | | | |
| 1955 | 66.1* | 63.9 | 68.4 |
| 1935-36 | 48.3* | 46.9 | 49.6 |
| Philippines | | | |
| 1948 | 51.2 | 48.8 | 53.4 |
| 1938 | 46.0* | 44.8 | 47.2 |
| Thailand | | | |
| 1947-48 | 50.3* | 48.7 | 51.9 |
| 1937-38† | 40.0* | 36.7 | 43.3 |
| Union of South Africa‡ | | | |
| 1945-47 | 66.0* | 63.8 | 68.3 |
| 1935-37 | 61.0* | 59.0 | 63.1 |
| *Oceania* | | | |
| Australia | | | |
| 1946-48 | 68.4* | 66.1 | 70.6 |
| 1932-34 | 65.3* | 63.5 | 67.1 |
| Hawaii | | | |
| 1949-51 | 69.5* | 67.8 | 71.3 |
| 1939-41 | 61.0* | 59.5 | 62.6 |
| New Zealand‡ | | | |
| 1950-52 | 70.4* | 68.3 | 72.4 |
| 1934-38 | 67.0* | 65.5 | 68.5 |

* Average of male and female. †Bangkok municipal area. ‡European populations.

lived in the center of the painter's world. Grandma Moses was an ordinary rural homemaker for the greater part of her life, until in her seventies she began to paint. Had Gauguin and Grandma Moses lived in environments that failed to stimulate their desire to paint or failed to make it possible for them to do so, Gauguin might have remained a banker and Grandma Moses a pleasant old lady, and the world would have heard of neither of them.

> Full many a gem of purest ray serene.
> The dark unfathom'd caves of ocean bear:
> Full many a flower is born to blush unseen,
> And waste its sweetness on the desert air.

Hackneyed as these lines may have become, we must still acknowledge their truth. Very well, it may be said, but this truth does not alter the fact that genes and abilities are connected. This is true enough, but this is a very different thing from saying that genes *alone* determine abilities, for this they certainly do not. Genes did not determine Mozart's musical abilities. What determined Mozart's abilities was a combination of two things: (1) his extraordinary genetic endowment and (2) the interaction of that genetic endowment with the appropriate environment.

But genius is an extreme manifestation of ability that is of rare occurrence in any generation of millions of human beings. Abilities at levels of achievement lower than genius appear to be even more dependent upon environmental stimulus. Genius seems to require only the very slightest stimulus. But at the nongenius levels of ability, human beings usually require considerable assistance from their environment and tend to realize their abilities in proportion to the amount of that assistance. Where human beings tend to receive little or no assistance from their environment in the development of their potentialities, there will be little or no development of them. Where individuals tend to receive a mediocre or moderate kind of assistance from their environment, their potentialities will tend to be correspondingly developed. Where human beings receive a high degree of assistance from their environment, their potentialities will tend to be highly realized.

It is perhaps necessary once more to recall that environment means anything, apart from the genes, that can act upon the individual. Bearing this in mind we will remember that the environment of human beings tends to be quite complex in the number and quality of the different variables it involves. A child may receive the best schooling available in the land, but if its home conditions are emotionally disturbing, in spite of a perfectly adequate genetic endowment that child may fail lamentably in schoolwork. On intelligence tests such a child may do very poorly. The damage he has suffered in his home environment may make it extremely difficult for him to recover his ability to learn for some years following an improvement in his domestic life. Such children are often puzzling. They do badly on intelligence tests; then on the tests of intelligence in everyday life they do well enough, but fail again very badly in their schoolwork. Often there is a reading disability, and should they tend to write backward and/or to read backward, this is often taken to constitute evidence that a genetic fault is somewhere involved. This may be so, and in some cases it is in fact so, but it by no means necessarily follows that these disabilities are of a genetic nature. Nor does it necessarily follow that if the domestic environment of a child has been disturbing to him he will fail at his schoolwork, or that the cause of failure if he does is definitely environmental.

Again, it must be emphasized that where human achievement is concerned causes are never exclusively genetic or exclusively environmental, but are the result of the interaction of both. Some children will respond differently from others under the same kinds of environmental stimulation; unfavorable environments affect some children much more unfavorably than they do others; and, similarly, favorable environments affect some children much more favorably than others. The differences in response are not necessarily principally due to differences in genotype—though in many cases this may be so—but may be due to a large extent to the environmental history of the individual since his birth—not to mention the substantial relevance of the prenatal history in many cases. Always, however, the resultant behavior is an expression of the interaction of the genotype and the environment. Hence we must rid ourselves of the erroneous notion that ability is an exclusive expression of genetic endowment. It is *not*. Ability is the expression of the *interaction* of the genetic endowment of potentialities and the environments that the individual experiences. Let us now proceed to consider the relevant materials.

Since this is an issue that has been much clouded by studies on "hereditary ability" or, at least, by the misinterpretation of the meaning of such studies, as well as befogged by certain famous studies on "hereditary disability," it is desirable to clear the atmosphere before proceeding further. In most such studies the common error is repeated of omitting to take into consideration the role played by the environment, and often what is due mainly to environment is credited to genes. The "goodness" of the genes that are transmitted in such families as the Russells, the Cecils, the Darwins, and the Wedgwoods cannot be doubted, but what can be questioned is whether any of the great men these families have given us would have achieved fame had they been born and raised in the slums. Favorable environments combined with their favorable genes enabled them to develop and express their abilities.

But what of those who occupy the other extreme of this spectrum of abilities, those who have practically no abilities or have mediocre abilities? Are they, in most cases, genetically deficient? That such persons *are* genetically deficient is a view that is widely held and supported by a number of famous works on the subject, the two most famous being those on the "Jukes" and the "Kallikaks." The subjects of these studies, the first published in 1875 and the second in 1912, have long been quoted as the horrid examples of the social and individual costs of "bad" heredity.

The "Jukes" and the "Kallikaks" are a euphonious combination of names that have become a synonym for depravity and degeneracy. Eugenists, criminologists, sociologists, and others have repeated the sorry story of these depraved families so often that they have come to enjoy not only a national but also an international reputation. Yet the first family is largely the creation of an unscientific, untrained, amateur criminologist, and the second has been saddled with a genetic history that, to say the least, is highly questionable.

The "Jukes" (the name is a pseudonym) were an extended family studied by Richard L. Dugdale in New York State in 1874. Dugdale seems to have had no preparation for such a study except that of interest, having recently become a New York Prison Association inspector. Dugdale did not quite invent the Jukes, but he certainly can be said to have compiled them, and compiled them after a method peculiarly selective, quite evidently calculated to give support to his belief that there is a significant relation between crime and heredity. In fact, his naïve assumption that crime is caused by heredity—"bad" heredity—is made quite explicit.

The difficulty with untrained investigators (and even with some trained ones) is that when they become enamored of a theory, they tend to become insensible to the facts that are in opposition to it. To bolster his theory of the hereditary causation of crime, Dugdale fell back upon his imagination when the facts failed him. In his report, *The Jukes: A Study in Crime, Pauperism, Disease, and Heredity* (1875), Dugdale covers 7 generations, 540 "blood" relatives, and 169 related by marriage or cohabitation, harking back to the middle of the eighteenth century. Since many of the individuals involved had been dead for many years and information on them was hard to come by, Dugdale was forced to fall back upon such characterizations as "supposed to have attempted rape," "reputed sheep-stealer, but never caught," "cruelty to animals," "hardened character," and the like.

Altogether apart from Dugdale's naïve assumption that crime, pauperism, disease, insanity, and other inadequacies are due to heredity, the inadequacies of his methods of inquiry and recording were so great that even in an age that was less deficient in reliable records than his own, his work would not be acceptable as throwing any light upon the causation of those conditions. Socially inadequate lineages undoubtedly exist, but even as a representative of such a lineage Dugdale's pedigree of the Jukes family is so full of assumptions and prejudices that it cannot be used for any purpose other than to provide an example of the manner in which such studies should *not* be made.

Were reliable pedigrees of socially inadequate lineages available, the problem presented by them would be not to prove them as being due either to heredity or to environment but to discover, if possible, what roles each of these factors actually played in the production of the social inadequacy in each individual.

Forty years after the original study of the Jukes, they were the subject of a follow-up study by another investigator wedded to the "heredity" theory of the causation of crime—Arthur H. Estabrook, whose study was published in 1916. He found that there were still a good many "inadequates" among the Jukes, but surprisingly enough he also found that there were a fair number of decent, respectable citizens, some even "superior," and others who had done quite well in the world. This was attributed to the fact that they had married outside the Jukes clan and received new infusions of "better" genes.

"The Kallikak Family" (a fictitious name formed from the combination of the Greek words meaning "good" and "bad") is the other famous family. The Kallikaks were studied by Dr. Henry H. Goddard, at the time the director of the Vineland Training School for the mentally retarded at Vineland, New Jersey. There were two clans of Kallikaks, the "good" and the "bad." The "good" and the "bad" ones were both descended from the same Revolutionary War soldier, Martin Kallikak, but from different women with whom he had formed unions. His first union, it is alleged, was with a feebleminded girl, who gave birth to Martin Kallikak, Jr., who was so bad that he became generally known as "Old Horror" and in turn became the father of no less than ten other "horrors," from whom all the other hundreds of horrible Kallikaks traced by Dr. Goddard were descended. The hundreds of "good" Kallikaks were, of course, descended from Martin's marriage with an estimable Quaker woman.

As Scheinfeld and others have pointed out, if "Old Horror" was a degenerate largely or entirely because of his genes, then it is clear that no single dominant gene could have been involved. So complex a condition as "social inadequacy" or "degeneracy" could not be due to a single gene and must, therefore, in genetic terms involve at the very least a pair of recessive genes. This being the case, then the estimable Revolutionary soldier Martin Kallikak, Sr., must have carried duplicates of the defective genes situated in the chromosomes of the feebleminded girl who gave birth to their alleged son who subsequently earned notoriety if not fame as "Old Horror." In which event it follows that something passing strange happened in the lineage of the "good" Kallikaks, for by some extraordinary piece of good fortune not a one of them seems to have inherited a pair of "bad" recessive genes for "degeneracy." If genes are involved, this is what should have occurred in some of them, for if Kallikak, Sr., was the father of the feebleminded girl's child, then he proved himself a carrier of defective recessive genes. In which case these should have been transmitted to some of his descendants, and some of these descendants would almost certainly have married individuals with similar genes, and a certain number of their offspring should, then, have shown "degenerate" traits. But they didn't. What, then, are we to conclude? That the "good" environment of the "good" Kallikaks was sufficiently strong to overcome the expression of their defective genes? Or that Kallikak, Sr., has all these years innocently borne this blot upon his escutcheon of the paternity of a child he never sired? Or did the "good" Kallikaks just by chance manage to avoid picking up any defective recessive genes? Or are they possibly a case of paramutation? The reader may be left to decide for himself.

The Jukes family and the Kallikaks are often quoted in books and articles as examples of what a "good" and a "bad" heredity can do to human beings and society. But no reputable scientist who has any acquaintance with the facts of genetics and its methods considers that such studies belong to anything but the category of the recklessly quaint, anecdotal method of investigation. The complete disregard by these early investigators of the possibility that the effects they described might have been due to environment rather than to genes renders their work valueless except for the fact that it demonstrates that extreme poverty and undesirable social traits can be *inherited through the environment* by successive generations. The Nobel Prize winner and great pioneer in the study of heredity, Thomas Hunt Morgan, categorized such studies very properly when he said, "The pedigrees that have been published showing a long history of social misconduct, crime, alcoholism, debauchery, and venereal diseases are open to the same criticism from a genetic point of view; for it is obvious that these groups of individuals have lived under demoralizing social conditions that might swamp a family of average persons. It is not surprising that, once begun from whatever cause, the effects may be to a large extent communicated rather than inherited."

And that is the point. It is quite possible that a genetic defect may have been involved in some of these pedigrees, but that was not demonstrated. The fact that a family line may for generations live in poverty and exhibit in many of its members undesirable traits of various kinds tells us nothing about the genetic endowment of those individuals. All that we know is that they have lived for generations in a "bad" environment, and we know that a "bad" environment has a way of making it difficult for people to extricate themselves from it.

The particular social environment of the individual, his *cultural environment*, adds a fourth dimension to what the individual inherits, a new zone of adaptation. Since the individual acquires most of his mental traits from his cultural conditioning, and since this is solidly accomplished by the time he reaches his seventh or eighth year, he thereafter carries his cultural, i.e., behavioral, environment around with him wherever he goes. His cultural conditioning virtually obscures whatever role the genes may have played in the development of his mental traits. It is this that makes it so difficult to discover what may have been due to genes and what to cultural conditioning. Hence, studies such as those of the Jukes and the Kallikaks are much worse than worthless, for they confuse the issues, make appear easy what is difficult to study, and with a solemn facility assert conclusions that cannot be substantiated by the evidence upon which they are supposed to be based. Finally, by attributing to "heredity," to genes, what may have been either wholly or, certainly, in part due to environment, such studies have served to misplace the emphasis and to orient our attention in the wrong direction. It should always be remembered that what is stated solemnly does not for that reason have to be taken seriously.

In the passages that follow, the intention is to show that no matter how well or ill endowed the individual is, he is likely to realize his potentialities, his abilities, only to the extent permitted by the environment with which he interacts. And this is, for the most part, the *cultural environment*. We had better describe more fully what the cultural environment is.

By *culture* the anthropologist means the man-made part of the environment, everything that the individual learns—the way of life of a people, its pots, pans, institutions, mores, beliefs, religion, social organization, educational practices, ways of bringing up children, occupations, and the things that are expected of human beings at their different ages and grades and in their different statuses and roles.

Numerous studies have shown that culture and ability are closely connected with each other. For example, persons living in different cultures often differ as a group in the development of certain abilities from the members of other cultural groups. Similarly, the different social classes within the same culture will often exhibit differences in ability. In the first instance such differences are often attributed to "racial" factors. In the second instance the differences are quite as frequently attributed to "hereditary" factors. This is, of course, an easy way to deal with such differences. All one need do is say that they are due either to "race" or to "heredity," and there we are, absolved from all responsibility of really doing the work of discovering what the true answer may be. Let us see what happens when we try to discover the truth.

It has long been well known that American Indians do not do well on intelligence tests. It has sometimes been said that this might be due to the fact that the intelligence tests we give them are foreign to their ways of thinking, to their cultural conditioning, that therefore we should not expect them to do as well on the tests as persons who belong to the culture for which those tests were devised. This is what the experts think, but others have preferred to think that American Indians simply are not as intelligent as whites. Let us take a case in point.

The average American Indian obtains an IQ score of about 80. This is when he is living on his own reservation among his own people and has adopted as little as possible of the ways of the whites. Some years ago oil was found on the reservation owned by the Osage Indians of Oklahoma. This enabled the Osages to improve their economic and social conditions very considerably and to bring them more in line with those existing among the whites. In a study by Rohrer, it was found that when given the Goodenough "Draw-a-Man" test, Osage children scored an average IQ of 104, whereas the white children scored 103. On a verbal IQ test the Osage children obtained an average score of 100, whereas the white children obtained a score of 98—the difference between the Indian children's score and that of the white children was insignificant. It showed that when the environmental opportunities are more or less equalized for Indian and white children, the Indian children can do at least as well as the whites.

American Indian children placed in white foster homes were found to have an average IQ of 102. Brothers and sisters of these children still living on the reservation were found to have an average IQ of only 87.5. The difference in favor of the foster-home children was undoubtedly due to the improvement in their environmental conditions.

It is widely believed that the IQ test measures "inborn intelligence." This is an error. What the IQ test measures for the most part is the response the individual is able to make to the tests as a consequence of his experiences. Though his genetic potentialities may be involved to some extent, often the experience of the tested individual has been of such a nature as to make it difficult for him to get anywhere near his potentialities. Children in our own culture may be so confused and upset by their life experiences as to make it extremely difficult for them to do well by their potentialities. Cross-culturally, that is to say, as we cross from one culture to another, ways of thinking about and perceiving the world become so different that tests designed in one culture are hardly applicable to the members of another. If the natives of New Guinea or of any other nonliterate people were to apply *their* IQ tests (supposing they had any), we should not, I am afraid, do very well.

When Negro children migrate to New York City or to Philadelphia after some years of residence in the South, their IQ scores undergo a gradual elevation until they are only slightly below those of the white children in those cities. But they are way above those of the Negro children of their own age who have remained in the South. The improvement in the conditions of the environment in the North, as compared with those under which the Southern Negro children live, is almost certainly responsible for the improvement in the IQ scores of the Negro children who have migrated from the South.

Negro and white babies tested during World War II at New Haven were found not to exhibit any significant differences in physical or psychological development. With good wages and improved working conditions the Negro parents were able to supply their infants with all that was required to equalize their conditions with those of the white babies.

Similarly, children who move from rural to urban environments achieve a higher IQ on the average than those who remain in the rural environment. On the whole urban dwellers do better on intelligence tests than rural dwellers, though the difference in their scores has become increasingly reduced in recent times. In earlier days rural areas were not as well equipped with schools as they are today; television and other modern communications media were not as widespread as they are today. Hence, with the improvement in the environment, rural dwellers have undergone a corresponding improvement in IQ. Such tests, however, are still devised by urban dwellers and contain more items with which the urban dweller is more familiar than the rural dweller. When tests were devised that were based on rural experience and administered to city and rural children, the children from the city did not do as well on the tests as the children from the rural areas. The opinion of Dr. Myra Shimberg, who applied this test, is that it was no less fair to the city children than the standard test is to the rural children, and that neither test sanctions any conclusion as to innate intelligence.

As an illustration of the manner in which response to intelligence tests is conditioned by environmental experience, we may cite the experience, reported by Pressey, of an intelligence tester among "poor white" children of the Kentucky mountains. The question was asked: "If you went to the store and bought six cents' worth of candy and gave the clerk ten cents, what change would you receive?" One child immediately responded, "I never had ten cents, and if I had I wouldn't spend it for candy, and anyway candy is what your mother makes." The examiner tried again: "If you had taken ten cows to pasture for your father and six of them strayed away, how many would you have left to drive home?" Whereupon the reply came forth, "We don't have ten cows, but if we did and I lost six, I wouldn't dare go home." The examiner made one final desperate effort: "If there were ten children in a school and six of them were out with measles, how many would there be in school?" Without hesitation the answer came back, "None, because the rest of them would be afraid of catching it too."

These were highly intelligent answers, but they were conditioned by the child's experience and not by that of the examiner. The same point is amusingly illustrated by the answers given an intelligence tester who asked a little boy who was accustomed to wearing a sombrero, "What would happen if I pulled your hat over your eyes?" "I couldn't see," replied the boy. "Now," said the examiner, "what would happen if I cut off your ears?" "I couldn't see," was the prompt reply. "You couldn't see?" exclaimed the astonished examiner. "Yes," said the boy. "My hat would fall over my eyes." This reply was, of course, entirely accurate, and a great deal more so than the reply the examiner was expecting, namely, that the boy wouldn't be able to hear—which would have been wrong, though the examiner would have thought it the right answer to the question.

TABLE VIII

MEAN TEST SCORES OBTAINED BY 16-YEAR-OLDS BY ASCENDING ORDER (FROM A TO D & E) OF DIFFERENT SOCIAL-STATUS GROUPS

| Social Status | No. | Stanford-Binet I.Q. | Wechsler-Bellevue Perf. I.Q. | Iowa Silent Reading Score | Paper Form Board | Minn. Mech. Assem. T-Score (Boys) | Chicago Mech. Assem. T-Score (Girls) |
|---|---|---|---|---|---|---|---|
| A | 13 | 98 | 103 | 45.6 | 31 | 53.0 | 45.3 |
| B | 49 | 104 | 102 | 48.9 | 31 | 48.8 | 48.5 |
| C | 44 | 112 | 109 | 51.0 | 40 | 51.6 | 52.0 |
| D & E | 9 | 128 | 118 | 58.0 | 44 | 46.8 | 62.1 |

Adapted from Janke, L. L., and Havighurst, R. J., "Relations Between Ability and Social Status in a Midwestern Community, II. Sixteen-Year-Old Boys and Girls."

The IQ test is a test of an actuality, not of a potentiality. One can get significant meaning out of an IQ test only when the environments of the subjects have been reasonably equal. Between socioeconomic classes a correction should always be made for differences in environmental experience.

Class differences in intelligence are as striking for children as they are for adolescents and adults. Class differences in intelligence express themselves as significantly in three-year-olds as they do in eighteen-year-olds. This has sometimes been taken to indicate that hereditary differences must exist between the classes, that those who have what it takes tend to rise in the scale, and that naturally such genetically superior individuals tend to have genetically superior children. This does not necessarily follow, however. Parents who have achieved a middle- or upper-class status need not have done so because of superior genes, but because of superior opportunities. The fact that gifted children come proportionately more often from families in which the father is a professional man than from families in which the father is a member of the skilled-labor class does not necessarily mean that significant genetic differences exist between the professional classes and the skilled-labor classes. Such differences may exist, but we have no definite evidence of this. On the other hand, it could just as well be true that there are no significant differences in the distribution of genes for ability between these classes, but that there have been very significant differences in the kinds of opportunities that the members of these classes have had to develop their abilities.

That environmental experience plays a major role in test results is demonstrated by the fact that achievement on these tests varies with the nature of the test, although on nearly all tests the tendency has been for the score to rise with the rise in social class. For example, in tests of sixteen-year-old boys and girls it was found that the boys of the lowest classes on the whole did better on a mechanical assembly test than the boys of the upper classes. The explanation offered for this is that the boys of the lowest classes had probably had more experience with mechanical things than the upper-class boys. Interestingly enough, the girls of the upper classes achieved a significantly higher mean score on this same test than the boys of any of the social classes!

In tests of primary mental abilities administered to thirteen-year-olds, the scores vary even more markedly for the classes with the function tested. Number, verbal comprehension, and word fluency show bigger differences between the classes than do the tests for space, reasoning, and memory. The strong suggestion is that there is a significant relationship between the test results on each of these items and social or cultural environment. As might have been expected, it has been found that children of the higher classes are much better at tests involving a knowledge of academic or bookish words than are children of the lower classes. Thus, tests may themselves be, and indeed often are, culturally biased. When the tests are put in simpler language, the responses markedly improve among those for whom the more bookish words have no meaning.

Twins reared apart are extremely interesting with respect to environment in relation to genes; for since identical twins are of the same genetic structure, we have in those who have been reared apart an excellent means of discovering to what extent different environments are capable of influencing the expression of the genes. But this is a subject that deserves a separate chapter.

# 8. TWINS, GENES, AND ENVIRONMENT

MULTIPLE BIRTHS, throughout the world, adhere more or less closely to the theoretical expectations. They are as follows:

Twins occur in 1 out of approximately 87 births.
Triplets occur in 1 out of approximately 7,569 births.
Quadruplets occur in 1 out of approximately 658,507 births.
Quintuplets occur in 1 out of approximately 57,289,761 births.
Sextuplets occur in 1 out of approximately 4,984,209,207 births.

There is an apparent mathematical relationship between the different kinds of multiple births. This is by no means exact but is of crude predictive value. Known as Hellin's rule, it states that triplets occur as the square (or second power) of the number of twin births; quadruplets occur as the cube (or third power) of the number of twin births; quintuplets occur as the fourth power of the number of twin births (or the square of the number of triplets); and sextuplets occur as the fifth power of the number of twin births.

There are only about fifty authentic records of quintuplet births, and of sextuplets there are no more than six authentic records. In the United States, for example, during the seven-year period 1951–1957, twins occurred in 1 out of 95 births, triplets in 1 out of approximately every 11,000 births, and quadruplets in 1 out of approximately every 900,000 births. No quintuplets or sextuplets were born during this period.

The chances of a multiple birth occurring varies with the age of the mother. The chances are least for adolescent mothers, being 6 per 1,000, and increase with maternal age to a maximum of 16 per 1,000 at ages thirty-five to thirty-nine. At ages forty to forty-four the chances are 13 per 1,000 and at age forty-five and over the chances are approximately 8 per 1,000.

The relative frequency of multiple births is greater among

Negroes than among whites, being 14 per 1,000 for American Negroes as compared with 10 per 1,000 for American whites.

For about every 95 deliveries in the United States one set of twins is born. This means that about forty thousand sets of twins are born each year in this country. About thirteen thousand of these sets are identical twins. There are two kinds of twins: *identical* (monozygotic = one-egg) and *fraternal* (or dizygotic = two-egg) twins. Identical twins develop from the same fertilized egg and are always of the same sex and contain the same set of genes, and so closely resemble each other that they are hardly distinguishable. Fraternal twins develop from two fertilized eggs and may be of the same or the opposite sex, and resemble each other no more closely than do brothers and sisters born at different times. A little more than one-fourth of all sets of twins born are identical; the other three-fourths are fraternal.

The frequency of monozygotic twins is fairly constant for all the populations of the world, namely, 3-4 per 1,000. There are, however, marked differences in the dizygotic rates for Negroes and whites. The average dizygotic rate for whites is about 7 per 1,000, whereas a rate of 20 per 1,000 is common throughout Negro Africa, reaching 40 per 1,000 at Ibadan, Nigeria, among a mainly Yoruba population. The dizygotic rate among Negroes outside Africa is 12-13 per 1,000—a difference that may be due to admixture with whites.

Because identical twins are genetically identical, they afford us an opportunity to study the relative effects of genes and environment in interaction with each other. Everyone knows of at least one set of twins who have been raised in the same family, dressed alike, treated alike, and who so closely resemble each other that even those closest to them sometimes experience difficulty in distinguishing one from the other. Such cases tell us that individuals with the same genotype when raised in the same environment will make much the same responses to it. But they do not tell us what would happen if these identical but separate individuals were raised in different environments. This is something about which we can learn by studying those identical twins who have been separated from each other since early infancy and have been raised in different environments. In their classic study, *Twins,* published in 1937, H. H. Newman, F. N. Freeman, and K. J. Holzinger give a detailed account of nineteen pairs of twins who were so separated. Since 1937, a fair number of other studies have been published on smaller series of identical twins, and all agree that there are both remarkable likenesses and remarkable differences between identical twins reared apart.

In spite of differences in their separate places of residence, the twins of these studies, who lived in America, were on the whole in much the same physical environment and experienced much the same nutritional history; hence, as might have been expected, in physical appearance, in height, and in weight, they maintained the closest resemblance to each other. The exceptions to this were cases in which one twin had developed a more or less severe illness and the other had not. There have been several cases recorded in which one identical twin was markedly taller and heavier than the other in spite of the similarity of the environment in which they were raised (see Plate 9). Differences in identical twins have also been recorded for their ability to taste certain substances, susceptibility to disease, and fecundity, or the ability to bear children. But on the whole everyone is impressed by the remarkable physical and psychological likenesses that exist between identical twins, even between those who have been separated from infancy. However, the differences that *are* found to exist between identical twins must be due to environment because the twins are genetically identical. Hence we arrive once more at the general rule that identical genotypes in different environments will respond in accordance with the differences set by those environments. All investigators agree that some traits are more susceptible to environmental influences than others; such traits, for example, are height, size, shape, dimensions of the head, and scores on a test for neurosis (Woodworth-Mathews test).

In the series of nineteen sets of twins studied by Newman and his colleagues, except for six of the sets, there were no more significant differences between the separated pairs of twins than were found to exist between unseparated pairs of twins. This strongly suggests the power of the genes and the limitation of the effects of environment. But in this connection it must be remembered that although the identical twins of the separated groups lived in families far removed from each other, nevertheless the environments in those families were not, on the whole, substantially dissimilar. Usually the twins were separated as a result of the death of the mother at or shortly after birth, whereupon one twin would be adopted by a member of the family and the other twin would be adopted by another member of the family, or one

## 144   HUMAN HEREDITY

chromosomes in the nucleus by *riboneucleic acid* (RNA), or both twins might go to an institution for a time and then be adopted by different families in different parts of the country. In every case the effort would be made to put each child in a similar home of a class background similar to that of its own family. Therefore, it should not be surprising to find that nonseparated and separated sets of identical twins do not differ markedly from each other in the development of any of their traits. The genes in each set being the same, and the environments for most sets being similar, one would not expect to encounter any markedly dissimilar developments. But in those cases in which there has been a more substantive difference in the environments of the separated twins, the differences between them, especially involving those traits that are more susceptible to environmental influence, are more substantial.

Let us describe the case history of a set of separated identical twins that throws a rather astonishing light upon the "strength" of the genes.

Edwin and Fred were seen by Newman when they were twenty-six years of age. They were separated in very early infancy and were adopted by two different families, of essentially the same socioeconomic status, living in the same New England town. They were each brought up as only children. They went to the same school for a time but never knew they were twin brothers, even though they had noticed their remarkable resemblance. When they were about eight years old, both families moved away and became permanently separated. The boys did not meet again until they were twenty-five. Edwin lived most of his life in a large city in eastern Michigan and Fred in a medium-sized city in western Iowa. On the whole there seems to have been no marked differences in their social environments.

Edwin learned of his twin brother's existence when he was repeatedly mistaken for his brother. Having informed his foster parents of these incidents, he was told by them that he pair to meet, and when they did so, the following facts emerged about them.

Edwin and Fred looked as like each other as identical twins reared together. Edwin was a half inch taller than Fred and weighed one and a half pounds more. Eye color, hair color and form, hairline, beard, complexion, ears, and other features were nearly identical. The teeth were irregular in both in the same way. In both, the upper-middle incisors were turned inward in the midline. Each had a supernumerary upper eyetooth placed too high in the gum, and Edwin had had this extracted. The extra tooth was on the right side in Edwin and on the left in Fred—a case of mirror-imaging. The hair whorls, however, ran clockwise. Both were right-handed. General body build and carriage were the same in both.

Though they had been so long and thoroughly separated, not being aware, indeed, of each other's existence, Edwin and Fred had led remarkably parallel lives. Both had been reared as only children; both had had about the same amount of education. They had each developed an interest in electricity, and both had become expert repairmen in branches of the same telephone company in their respective cities. They married young women of about the same age and type, each in the same year. Each had a baby son, and each—believe it or not—owned a fox terrier which they called by the same name, Trixie!

Make what you will of this astonishing story, the facts are as stated. Given environments as close as those of Edwin and Fred, and given their identical genetic background, astonishingly similar as their independently lived lives were, the likenesses are perhaps understandable. Given the same genotype and a similar environmental experience in two separated individuals, one may reasonably expect similar responses from them.

In behavior and personality there was a striking basic similarity, though Edwin was the more flexible, emotional, and easily aroused. He was more vivacious and responsive than Fred, possibly due to some difference in their environment. Edwin appears to have had better and more continuous schooling than Fred, though both were poor at their studies. However, Fred was the better speller and his handwriting was firmer; but even so, there was a good deal of similarity in their handwritings.

Edwin and Fred illustrate what happens when identical twins though separated grow up in similar environments. Now let us cite a case illustrating what happens to identical twins when they are brought up in contrasting environments. Such a case is that of Gladys and Helen.

Gladys and Helen were first seen when they were thirty-five years old. They were born in a small Ohio town and were separated when approximately eighteen months of age. They did not meet again till they were twenty-eight years old. Helen had been twice adopted. In the first instance the foster father turned out to be an unstable person and the foster mother became mentally ill two years after Helen had been adopted. Helen was therefore taken back to the orphanage, and after several months was again adopted, this time by a farmer and his wife who lived in southeastern Michigan. This was her home for the next twenty-five years. Her second foster mother, though she had had few educational advantages herself, was determined that Helen should receive a good education. So Helen eventually graduated from college. She taught school for twelve years, at twenty-six married a cabinetmaker with a high-school education, and had a daughter. In the course of her education she had acquired a great deal of polish, showed much ease in social relationships, and was possessed of considerable feminine charm.

Gladys had been less fortunate. She was adopted by a Canadian railroad conductor and his wife. When she was in the third grade, her foster father was stricken with an illness that necessitated his removal to a rather isolated part of the Canadian Rockies. Since there were no schools available, Gladys's formal education came to an end and could hardly be said to have been resumed when the family returned to Ontario. So she stayed at home and did housework till she was seventeen and then went out to work in a knitting mill. At nineteen she went to Detroit and worked as a saleswoman in stores and did some clerical work, finally getting a job in a small publishing establishment, where she became assistant to the president. Gladys married when she was twenty-one, her husband being a mechanic with a high-school education.

Helen was healthy as a child and had no serious illness as an adult. Gladys was very delicate as a child, had several serious illnesses, including scarlet fever, and very nearly died of measles. Both had quarrelsome foster parents. For the rest, their environments had been very similar, except with respect to the educational experience.

Physically, Helen looked her age at thirty-five, but Gladys looked forty. As for the remainder of their traits, they were physically highly similar: Helen weighed 140¾ pounds and Gladys 139¼ pounds. Helen was 62.1 inches in height and Gladys 61 inches. Gladys had broader shoulders and was more mannish in figure and carriage than the more feminine Helen. Hair color and form were the same, except that Helen's hair was a little grayer. Eye color was the same. Teeth were white and regular—each had six fillings of cavities affecting the molars and premolars. Thus, physically they were very much alike. The differences that distinguished them, differences in carriage of body and facial expression, were obviously associated with the different social lives they had led.

Helen was confident, suave, graceful, made the most of her personal appearance, and was the more overtly aggressive. Gladys was diffident, ill at ease, staid, and stolid. She was without charm or grace of manner. She made no effort to make the best of her physique or to create a favorable impression. As Newman and his colleagues remark, "As an advertisement for a college education the contrast between these two twins should be quite effective."

On all mental ability and scholastic tests Helen showed marked superiority to Gladys. Helen scored an IQ of 116 and Gladys an IQ of 92, a difference of 24 points. We have already seen that there were marked differences in personality traits. The styles of handwriting were very different, Helen's being mature, Gladys's that of a fourteen- or fifteen-year-old child, although there was a marked similarity in which both wrote a rather peculiar letter *f*.

Considering the nature of the environmental experiences of these identical twins reared apart, one whose formal education virtually stopped at the third grade, the other who went through high school on to college, from which she was graduated with a B.A. degree, the differences observed in Helen and Gladys as phenotypes are not surprising. Their case again underscores the fact that what we can do is set by the genes, but that what we actually do is largely determined by the environment.

An interesting fact about identical twins is the relative infrequency of high intellectual achievement among them. On the average, twins are one-quarter of a standard deviation inferior to the singly born in intellectual achievement. Whether this difference is due to genetic or environmentally limiting factors is unknown.

Attempts have been made by various writers to show that in spite of considerable differences in the environment of twins who have been separated for many years, they have nevertheless in virtually all respects remained quite similar. When, however, one comes to examine the cases these writers cite, it is found that the separation occurred as late as adolescence or even later, and furthermore, that the differences in the twins' environments were not really as great as they superficially appeared to be. For example, Dr. Franz J. Kallmann cites the case of identical twin sisters whose life history "approximated the widest possible discrepancy imaginable in our culture. It would be difficult to find twin partners separated by a greater distance and under more varied social, cultural, and climatic circumstances, or exposed to greater differences in dietetic and personal living conditions than were these twin sisters, now 85 years old." Both were raised in a small rural community. One sister married a local farmer at age eighteen, the other entered a Bible school in her middle twenties, after which she went as a missionary to the Orient, where she remained until she rejoined her widowed sister at the age of sixty-five. After forty-seven years of separation it was difficult to tell the twins apart, they were so much alike. The married twin who had had six children was twenty-eight pounds heavier, arrived at menopause seven years after her sister, at the age of fifty. Her vision was slightly more impaired, and she did not score quite as high on the intelligence test though she was of superior intelligence.

These twins were separated when they were between eighteen and twenty-one years of age. Up to that time they had lived in the same home; their basic personalities had already been formed and their basic education completed. It is not surprising that they should have remained as much alike in personality traits and abilities as they did. As for the "widest possible discrepancy" in their life histories, this obviously refers to the period after their separation, a period that insofar as the development of psychological or behavioral traits is concerned is in no way to be compared with the importance of the first twenty years of life; anything that happens thereafter, alas, is, as most of us know, not nearly as important to the formation of our basic character. Furthermore, in spite of the differences in cultural milieu and diet, the missionary sister probably spent the greater part of her time living in a mission with her own kind, so that the environmental differences between the twins were probably by no means as great as might be supposed. In any event, it is well known that regardless of differences in environment following adolescence, that is, when physical and psychological growth has been virtually completed, twins tend to age in a remarkably similar manner.

Since psychological traits depend so much upon experience, it is to be expected that significant differences in experience, in life history, will be reflected in those traits. On the other hand, traits that are not as susceptible as psychological ones to variations in the environment, such as hair color and form, eye color, form of features, teeth, and the like, are more likely to show a high degree of similarity in identical twins. To show the extent to which such identity may be realized in twins I shall refer to a study I carried out on the dentition of a set of identical twins. This study is somewhat unusual since it involved a mathematically and geometrically exact analysis of the position of every cusp on each tooth in both the upper and lower jaws and the exact location of these cusps compared in both twins.

## The Dentition of a Set of Identical Twins

These identical twin boys were selected at random in order to discover how closely their teeth resembled each other. The twins, whom we shall call Daniel and George, were studied over the course of several years. They were white Americans. George was born first, Daniel a few minutes later. George has always been bigger than Daniel. George was born with an enlarged heart and has always had a left auricular systolic murmur. Daniel is left- and George right-handed. Physically and temperamentally, they were as alike as could be, except that, according to the father, George was more reasonable and tactful than Daniel, whereas Daniel was more truthful than George. First examined when they were 6 years and 1½ months of age, the detailed measurements of the bodies of these boys proved that they were astonishingly alike. Examination of the teeth revealed striking likeness in every respect, but what was most interesting was that a complex identical pathological condition, involving the same teeth, had developed in both boys.

This remarkable pathology consisted of a marked resorption of the farthest roots and a portion of the crown of the second molar of the milk teeth on the right side of the upper jaw. This resorption had, at least in part, been produced by the pressure of the descending first molar of the second, or permanent, set of teeth. This had grown downward and forward into the milk tooth, its nearest portion lodging in the excavated basin of the milk tooth. This may be clearly seen in the X rays of the teeth shown in Plate 10. Such a pathological condition would be remarkable as an occurrence in any individual; its exact duplication in two individuals is little short of amazing. As we can see from the X rays, the state of development of the teeth is virtually identical in both twins. Their genes being identical and their environment as nearly similar as possible, it is not surprising that even a complex exaggeration of the normal eruption of a permanent tooth in relation to a milk tooth should follow a similar course in each twin.

In this case we have an illustration of the fact that where the genotypes are the same and the environment is as nearly the same as possible, the responses made by the organism, as a result of interaction of genotype and environment, will be the same.

Quite as remarkable in these twins is the virtually identical manner in which the teeth have erupted, and are placed within the jaws (see Plate 10, d-g). The cusps, grooves, ridges, edges, and the centroids of each tooth are almost indistinguishable, except for the right lateral incisor of the upper jaw, which has undergone a slight rotation in George. At the time the map shown in d was made, Daniel had lost his two upper central incisor teeth. George lost his five weeks later.

When it is considered that there are twenty teeth in the milk dentition and thirty-two in the permanent dentition, and that each of these teeth can erupt in any number of different ways, not to mention the number of possible variations in form, the chances of any two series of dentitions bearing an identical resemblance to each other in any two individuals are astronomically remote. In fact, the chance of getting any two like combinations in man is determined, on a purely numerical basis, by the number of his chromosomes, so that in the mating of two individuals each with a haploid set of twenty-three chromosomes, the chance would be $2^{23} \times 2^{23}$, or 1 in 70 trillion. This is actually a gross underestimate because it omits the complicating factors of crossing-over, consanguinity, and many other factors.

I have given an account of this case for two chief reasons: first because it illustrates how identical identical twins can be in their development, and second because it is the kind of case that causes many people to make the sort of unjustifiable inferences that they do. This type of inference takes the form: if the genes act as powerfully as they do, as in the case illustrated above, then they must act as powerfully for all other traits, and therefore it is genes with which we have to reckon rather than the environment.

This kind of reasoning, common as it is, is unsound on several counts. It neglects to consider the nature of the environment. It neglects the evidence that shows that there are different degrees of susceptibility to the environment of different developmental processes leading to different traits.

Even in identical twins, the expression of genes is influenced, as we have seen, by the environment with which they interact. If the environment has been similar, the expression of the genes will be similar; if it has been different, the traits "underwritten" by the genes will be different to the extent of the differing environment's influence. The action of the same environmental influences at different times often produces different effects. The fact that identical twins may often be virtually identical in all respects does not mean that genes alone are responsible, for genes alone are never absolutely responsible for any trait; but what such an identity always means is that the same genotypes of the two individuals have undergone development in a very similar environment. When the environment of identical twins is varied, we observe significant differences developing between them, and this is particularly true of those traits that are most subject to the influence of the environment, namely, psychological traits. An interesting aspect of this is the area of mental disorder.

## Mental Disorder in Twins

It has been said by a leading worker in the field of heredity and mental health, Dr. Franz Kallmann, that "the capacities for health and adequate adjustment are fundamental biological properties with the common denominator of hereditary potentiality." This is, of course, true, and it is, of course, also true that adjustment is always conditioned by the interaction of the genotype with the environment. Normal, healthy growth and adjustment can take place only in a normal environment. What is normal, healthy growth and adjustment? It is the process of development without defect, disorder, or disease, the realization of a state of physical and mental well-being. On that subject alone a sizable book could be written, but there is no space for the further development of that definition here.

It is the custom to emphasize the likenesses between identical twins. It is at least as important to draw attention to the differences. Suicide is one of them. It is extremely rare for both twins to commit suicide. In fact, it may be said that the chances are good that if one twin commits suicide, the other will not. Suicide is generally the end result of a psychological disturbance of some sort. In most recorded cases the twins were similarly psychologically disturbed; nevertheless, in the majority of cases only one committed suicide. Of the total of eighteen pairs of identical twins recorded by Kallmann, he found only one set in which both had committed suicide. There is little point to speculating on the reasons for this difference; the fact is that we don't know what they may be in detail, but it seems a not-unreasonable conclusion to draw that in general it is to be accounted for on the basis of environmental differences in the experience of the twins.

The reverse of the above findings has been recorded for homosexuals in a series of forty-five pairs of identical twins, in which only one pair were discordant for homosexuality. Kallmann thinks that such findings raise serious questions concerning the psychodynamic theories about the cause of homosexuality and feels that they strengthen the hypothesis of a gene-controlled disarrangement in the balance between male and female maturation (hormonal) tendencies. However, studies on homosexuals do not show such hormonal imbalances (see Kinsey), but homosexuals do show a certain similarity in the environment of their early lives. Without entering into a detailed discussion of the matter, it would appear that the vast majority of homosexuals are produced in response to an environment in which one or the other of the parents was markedly inadequate in some way, causing the child to identify itself either with the mother if the father was inadequate or with the father if the mother had failed. When both parents are inadequate, the child, as an adult, may become bisexual (see the works of Henry and Westwood). Hence, without denying the possibility that in some cases homosexuality may have something to do with genes, it is generally agreed among experts that this is a condition that is produced mainly by psychosocial factors during childhood. In spite of the fact that many homosexual twins state that they developed these tendencies independently of each other, the possibility must be taken into consideration that they may have been conditioned to do so in early life by similar environmental influences.

A similar interpretation is to be made of many mental disorders often attributed to genetic deficiencies but that upon careful examination quite frequently turn out to be not so much genetic as psychological. The fact that a whole family shows a high frequency of mental disorder does not necessarily mean that the family is genetically disposed to develop mental disorder. In many cases it is the environment that is "enough to drive anyone crazy." This is not to say that there are not gene deficiencies underlying many mental disorders. There is very good reason to believe that there are. And these are of two general classes: (1) gene deficiencies related to mental disorder that will express themselves in all known environments and (2) gene deficiencies related to mental disorder that will express themselves only in certain types of environments. Rather extreme cases conforming to the first class are such disorders as amaurotic idiocy and phenylketonuria, caused by rare recessive genes that appear unexpectedly in families in which the condition may never previously have been known to have occurred. We know of very few other mental disorders that fall into this first class. There is no agreement on which mental disorders may fall into the second class, the reason being that it is extremely difficult to distinguish such disorders—which undoubtedly exist—from those disorders that are more or less purely functionally determined. It would appear that all possible human genotypes are capable of becoming functionally disordered as a consequence of environmental factors, and this applies to physical functions as well as to mental ones. An individual endowed with as "good" a genotype as could be desired can, as a consequence of a disordering environment, become mentally disordered.

When, then, we are asked to view the evidence of mental disorder in twins, we must always be on guard against the tendency to attribute to genes what may in fact be due to environment.

The principal mental disorders for which a genetic basis has been claimed are schizophrenia, manic-depressive psychosis, and involutional psychosis.

Schizophrenia is a disorder characterized by fundamental disturbances in reality relationships, feeling, intellect, and behavior. It has been described as a fragmentation or rupture of the ego. Manic-depressive psychosis is characterized by marked emotional oscillation from manic to depressive states and vice versa. Involutional psychosis is a mental disorder characterized chiefly by depression, often in association with symptoms of insomnia, worry, guilt, anxiety, and delusional ideas.

In Table IX are set out the rate of concordance of schizophrenia in fraternal and identical twins as found by different investigators in Germany, Sweden, England, and America. From this table it will be seen how very much more frequent the concordance (occurring in both twins) of the disorder is in identical than in fraternal twins. The environments are generally as similar for fraternal twins as they are for identical twins, so that if we are to attribute a large part of the role played in producing schizophrenia to the environment, there ought to be many more schizophrenic pairs among the fraternal twins than we actually find. It is estimated that slightly under 1 percent of the general population suffers from one form or another of schizophrenia. Penrose has estimated the frequency of the gene involved as occurring in 1 out of 200 individuals. On the basis of the twin studies it is difficult to resist the conclusion that a significant number of these have some sort of genetic susceptibility to the disorder—but always, and until evidence to the contrary is forthcoming, we must remember, under the environmental conditions provocative of the disorder. If some cases of schizophrenia have a genetic basis, it has been suggested that it may be in the form of a recessive gene or genes with variable penetrance and expressivity, depending upon the resistance that has been built up within the individual.

TABLE IX
EXPECTANCY OF SCHIZOPHRENIA IN CO-TWINS OF SCHIZOPHRENICS

| Investigator | | Number of Pairs Two-Egg | One-Egg | Rate of Concordance Two-Egg | One-Egg |
|---|---|---|---|---|---|
| Luxemberger | (1930) | 60 | 21 | 3.3 | 66.6 |
| Rosanoff | (1934) | 101 | 41 | 10.0 | 67.0 |
| Essen-Müller | (1941) | 24 | 7 | 16.7 | 71.4 |
| Slater | (1951) | 115 | 41 | 14.0 | 76.0 |
| Kallmann | (1952) | 685 | 268 | 14.5 | 86.2 |

It has been found, for example, that when one identical twin develops schizophrenia and the other does not, the twin with the disorder is usually weaker physically and of lower weight. Frequently, when the general health of the latter twin is improved and weight is gained, the schizophrenic symptoms disappear and the individual may resume a normal life.

Manic-depressive psychoses in the general population occur in approximately 1 out of 200 persons. Figures given for the concordant occurrence of the psychosis in identical twins run as high as 90 percent and more. The condition is thought to be due to a dominant gene with varyingly incomplete penetrance.

Involutional psychosis in the general population occurs in from 3 to 7 out of every 1,000 persons, and is recorded as occurring concordantly in about 61 percent of identical twins. It occurs with significantly marked frequency among relatives of schizophrenics. It is not considered to be due to a single gene—the genetic mechanism is obscure.

Epilepsy is a neurological disorder that appears to fall into a wide variety of classes with respect to cause. Gene deficiencies are undoubtedly involved in many cases. More than 66 percent of identical twins have been found to be concordant for the disorder, whereas only 3 percent of fraternal twins show such a concordance. From such twin studies we are able to deduce that in many cases if not in all epilepsy depends on a gene-specific type of vulnerability, in which a number of genes (polygenes) are involved and are most probably carried in the recessive condition. In other words, the genes involved must come together from each of the parents in order to express themselves in some of the offspring.

As the reader may have gathered, we know comparatively little concerning the biology of the nervous disorders discussed, and still less of their genetics, but some form of genetic deficiency must be involved in most if not in all of them.

Twins have been used in a large variety of studies in order to throw some light on the heredity-environment problem. The studies that we have dealt with here must suffice as typical of the results yielded, inconclusive as they are on many points. There is one area, however, in which twin studies have been made, namely, in behavior disorders, particularly with reference to criminal behavior, that we must consider—and this we may proceed to do in the next chapter.

# 9. CRIME—GENES OR ENVIRONMENT OR BOTH?

A CRIME is an act committed in violation of the law. There are serious students of the criminal who have asserted that criminals are born, not made. The late Professor Earnest Hooton of Harvard was one such investigator. His claims have been thoroughly demolished, as have been those of Cesare Lombroso, the Italian criminologist who initiated criminological studies during the last century, and whose views as to the physical "stigmata" by which criminals might be recognized were widely disseminated throughout the Western world. Ever since Lombroso, serious attempts have been made to throw some light on this question, but most of them have suffered from fatal errors, both of method and of interpretation, which have vitiated their conclusions.

The approach to the problem of the causation of crime through the study of twins is by far the most satisfactory that has thus far been made. Five investigators—three German, one Dutch, and one American—have independently reported on criminality in twins. Their findings are extremely interesting (they are set out in Table X). When both members of a twin pair were found to be similar with respect to the commission of one or more crimes, they were termed "concordant," when dissimilar—that is, when one was found to have committed a crime and the other not—they were termed "discordant." From Table X it will be seen that of 104 pairs of one-egg twins examined, 70 were concordant and 34 were discordant. The concordant were more than twice as numerous as the discordant pairs. On the other hand, the two-egg twins showed a discordance more than twice as great as the concordance shown in this group of 112 pairs. The proportions are exactly reversed for the one-egg and the two-egg twins.

These are interesting figures, but what do they mean? It has been held by several authorities to indicate that genetic factors play a large part in the causation of criminal behavior. This is the opinion of the five investigators whose twin studies we are here considering. The truth, however, is that such studies do not prove any connection whatever between genetic factors and criminal behavior. One-egg twins are likely to have the same companions and to be otherwise closely associated in social activities; they are therefore more likely to encounter together such social influences as may lead them to criminal activities. All our knowledge of twins tells us that two-egg twins do not experience environments as closely similar as those of one-egg twins. Hence it is quite reasonable to account for the greater concordance of criminal behavior among one-egg twins on the basis of the greater similarity of their environmental experience as compared with the two-egg twins.

Unfortunately, the tendency has been to underplay the significance of the environmental factor in the interpretation of the findings in such twin studies of criminality. Once more the danger is underscored of unwarrantedly assuming that a higher frequency of concordance of any behavioral trait in one-egg twins as compared with two-egg twins indicates the greater likelihood of the operation of purely or partly genetic factors in the one-egg twins than in the two-egg twins. Where the environment has been highly similar, this is a distinctly dangerous inference to make, even if in some cases it turned out to be true. On the evidence that is available to us most of the time, however, it is an erroneous inference, for what is attributed to genes may, in fact, be largely or wholly due to environment.

The proportion of two-egg twins who are both affected and the proportion of one-egg twins where only one is affected are virtually identical, being 33 percent for two-egg concordance and 32.7 percent for one-egg discordance. Discordant one-egg twins ought not to be so frequent and discordant two-egg twins ought to be more frequent, according to the genetic theory of the causation of crime.

TABLE X
CRIMINAL BEHAVIOR OF TWINS

| Investigator | | One-Egg Twins Concordant | Discordant | Two-Egg Twins Concordant | Discordant |
|---|---|---|---|---|---|
| Lange | (1929) | 10 | 3 | 2 | 15 |
| Legras | (1932) | 4 | 0 | 0 | 5 |
| Kranz | (1936) | 20 | 12 | 23 | 20 |
| Stumpfl | (1936) | 11 | 7 | 7 | 12 |
| Rosanoff | (1934) | 25 | 12 | 5 | 23 |
| Total | | 70 | 34 | 37 | 75 |
| Percent | | 67.3 | 32.7 | 33.0 | 67.0 |

The actual findings, however, reveal that one-third of the one-egg pairs of twins investigated were discordant. Why did not the genetic factor for crime declare itself in the other member of this one-third of single-egg twins? Perhaps it did, but he had the luck not to get caught. If, however, the answer is that an environmental factor was operative in these cases, a factor that was absent in the case of the criminal sibling—or, if you like, an environmental factor was operative in the case of the criminal twin and was absent in the case of the law-abiding one—then the theory of the genetic causation of criminal behavior collapses beyond repair, for it then becomes clear that it was the absence of some environmental factors or the presence of others that constituted the indispensable condition in the causation of criminal behavior. We surely cannot assume that the environmental conditions were the same for both twins, with their identical genotypes, when one becomes a criminal and the other does not. If the difference is not to be found in their genes, the only other place it can be looked for is in their environment.

There almost certainly exist differences among individuals of a generalized genetic kind that *under certain kinds of environmental conditions* may make some individuals more ready than others to commit acts designated by society as criminal, but such environmental conditions are not like triggers that set off or release inborn tendencies to commit criminal acts. There are no such things as inborn tendencies to speed in a nonspeeding zone, to fail to return borrowed books, to commit petty larceny, robbery, felonious assault, arson, rape, murder, or any other kind of social offense. But what do exist are pressures of the environment that, together with specific psychosocial histories, cause the individual to commit such offenses *regardless of the nature of his genotype*. As we know, the genotype can never be entirely disregarded, but for many practical purposes it certainly can be disregarded in the sense that its causal contribution to a particular social effect may have been negligible or nonexistent. If, for example, all human genotypes were identical but social environments varied as they do today, there would still be marked differences in crime rates, and it is greatly to be doubted whether they would appreciably differ from those of the present time.

It is not "criminal genes" that make criminals, but in most cases "criminal social conditions," regardless of whether such "criminal" conditions are created by society or the family.

## CRIME—GENES OR ENVIRONMENT? 159

The conditions that drive men to crime are the effects of their accumulated environmental experiences, whether those environmental experiences are accumulated in a criminal or delinquent environment where moral and ethical standards are virtually nonexistent, or whether they are accumulated in homes where there is no love but only hostility, frustration, or indifference. Delinquents are not born; they are made by a delinquent society, which makes delinquent parents, who make the kind of delinquent environments with which we are familiar in the Western world.

Why is it that delinquency and crime are virtually unknown among Chinese immigrants and their children living in the United States? How does it happen that drunkenness is a condition that occurs so rarely among Jews? Is it because these groups have few or no genes disposing them toward such behavior? The answer is positively in the negative. The infrequency of criminal behavior among American Chinese and of drunkenness among American Jews is no more genetically influenced than is the tendency of Chinese immigrants to become laundrymen and of Jews to attend synagogues. These are all forms of behavior that can be shown to be determined by cultural and not by biological factors. The Chinese are brought up in an environment in which gentleness, kindliness, and honesty are traditionally strongly emphasized. When they emigrate to America, it is usually with the support of a relative who is already engaged in laundry work. Jews do not get drunk as often as their neighbors do because in the home drinking is largely regarded as a ritual act and is usually limited to wine. Because of the high negative sanctions that are enjoined against such a condition, drunkards are held in contempt. When Chinese and Jews become thoroughly Americanized and lose the ancient virtues of their own cultures, some of them become as proficient in the vices of their adopted culture as any of their fellow Americans.

What shall we say of the millions of murders committed on helpless individuals by the Germans during World War II? Shall we say that in the pre-Hitler environment the genetically determined impulses of the Germans to murder were kept under control, but that with the advent of Hitler and his sanction of such murderous behavior the environment was changed and the murderous impulses were released? This could be persuasively argued, but surely it is nearer the truth to say that in every country there are millions of frustrated individuals with violent impulses suppressed inside them, produced by the frustrating conditions of their lives, who would welcome a socially sanctioned opportunity to release their hostilities in some way. It is no more true to say that it is genes that make nations murderous than it is true to say that it is genes that make men so.

Human beings are not born with genes for "violence." Nor are they born like blank tablets upon which one can inscribe whatever one desires, for the organism is characterized by developmental limits, which are largely determined by its genes. However, within those genetic limits what the organism as a behaving individual becomes will largely, if not entirely, be determined by environmental experience. The discrepancy between what we are and what we were capable of becoming constitutes for many of us the real tragedy of life.

It is amusing to record the reply of the great hereditarian among criminologists, Cesare Lombroso, when he was confronted with his own student's investigation, that of Enrico Ferri, who found that 63 percent of soldiers showed Lombroso's so-called stigmata of degeneration (lobeless ears, small ears, receding chins, and the like). Lombroso attempted to explain this startling discovery away with the statement that when the "stigmata are found in honest men and women, we may be dealing with criminal natures who have not yet committed the overt act because the circumstances in which they have lived protected them against temptation."

So we see that in the end what Lombroso stated was that individuals exhibiting stigmata will be likely to commit crimes under certain environmental conditions. Since almost all individuals exhibit one or another of these so-called stigmata, we may unreservedly agree. Under certain environmental conditions all of us are capable of committing crimes, and most of us have. But what Lombroso meant, and always insisted upon, was that in almost all cases it was not the unfavorable environment that led to the commission of a crime but the biological predisposition to commit it, which could be foretold by the presence of stigmata. The stigmata were taken by Lombroso to be marks of biological inferiority, "atavisms" * (Latin *atavus,* ancestor; hence, process of reversion to an ancestral state), proofs of the reversion to more primitive forms of biological organization,

---

* There are no such things as "atavisms" or "throwbacks." See Ashley Montagu, "The Concept of Atavism," *Science,* vol. 87, 1938, pp. 462-463.

which was reflected in primitive levels of response or behavior. This criminal behavior was inseparably connected with biological inferiority; the biological inferiority was held to be the cause of criminal behavior.

As we have already said, attached ear lobes, low foreheads, receding chins, malformed ears, crooked noses, the so-called stigmata of Lombroso, are traits that are widely distributed among the populations of every land, and it is now known that there is no genetic relationship between such physical traits and behavior. Malformed features may, with the assistance of an unsympathetic environment, cause the individual to develop maladjusted behavior, but, again, such responses have nothing whatever to do with genes.

Hooton, unlike Lombroso, did not set out with any preconceived notion as to what criminal stigmata might be; he allowed the sample of 4,212 white prisoners of old American stock who were the subject of his investigation to tell him what these marks were. Hooton took the marks of biological inferiority to be any of the characters that were distinctive of the criminal aggregate when compared with the civilian sample. "Thus," Hooton writes in his book *Crime and the Man* (1939), "if we find felons to manifest physical differences from civilians, we are justified in adjudging as undesirable biological characters those which are associated in the organism with antisocial behavior. . . . It is the organic complex which must be estimated inferior or superior on the basis of the type of behavior emanating from such a combination of parts functioning as a unit. . . ."

"Whatever the crime may be," Hooton concludes, "it ordinarily arises from a deteriorated organism. . . . You may say that this is tantamount to a declaration that the primary cause of crime is biological inferiority—and that is exactly what I mean."

When Hooton's "marks of inferiority" are examined, it is found that they consist mostly of measurements of the body, indexes expressing one dimension as a percentage of the other, and measures of form. For his marks of inferiority Lombroso took characters that were for the most part apelike. If we classify Hooton's marks of inferiority by standards that are generally accepted to be marks of primitiveness, advancement, or as neutral, then Hooton's results when thus analyzed are by no means surprising. By such standards we find that of Hooton's combined anthropometric, indicial, and body-form characters only 4 percent are primitive, 15.8 percent are neutral, and 49.5 percent are *advanced* characters, the last appearing more frequently among criminals than among the noncriminal population.

By biological standards, therefore, we see that Hooton's findings actually make his criminal series a considerably more advanced group biologically than his noncriminal series of 313 native-born white civilians.

Dr. W. H. Sheldon has in more recent years claimed to have found a relationship between body type and delinquent behavior. In a study of 200 delinquent youths in Boston he found that the 16 criminals in the series belonged to a rather broad-chested muscular type, called the endomorphic-mesomorphic type. Sheldon does not suggest that this type of body build "predisposes toward criminality, but," he says, "it might mean that to make a go of being a criminal requires a certain amount of guts that is usually found only" in this body type. But that, of course, is precisely the point. There is a certain amount of social selection at work in criminal as well as in many other types of social and antisocial activities. Clearly, the robust, big-chested, tough-looking male has a great advantage over the roly-poly fatty or the long, linear, stringlike type in any activities involving the boldness of muscularity and the necessity of violence. But Sheldon is inclined to attribute this selection more to what he calls "guts" than to "occupational" requirements. In this opinion he stands with Hooton, with whom he identifies himself. As Eleanor and Sheldon Glueck point out, body type may be significantly related to delinquency but can by no means be regarded as a cause of it. Boys with a long, linear body type (*ectomorphs*) may be teased a great deal, and so may roly-polys (*endomorphs*), whereas broad-chested muscular types (*mesomorphs*) may be selected by certain environments for delinquent roles by other routes.

We may conclude, then, that all biologistically biased attempts and other more dispassionate attempts to prove a relationship between genes or "heredity" or "constitution" and criminal behavior have failed. Crime appears to be the result of a complex of factors, and there is very good reason to believe it has little if any relation to genetic or constitutional factors. The significant relationship is to social factors. Everyone agrees that there is a high correlation between poverty and crime, although, of course, criminals are by no means drawn exclusively from the poorer classes. And poverty by itself is rarely a cause of crime, but within the group of conditions that constitute the cause of crime in any one instance it is of frequent occurrence. No kind of geno-

## CRIME—GENES OR ENVIRONMENT? 163

type of which we have any knowledge is associated with criminal behavior. The evidence indicates that all individuals, whatever their genotype, as a result of exposure to certain social conditions can be caused to function in a criminal manner. There is every reason to believe that with changes in the social environment in the right direction criminal behavior could be reduced to the vanishing point.

## 10. GENES AND CONSTITUTION

"IT'S CONSTITUTIONAL." That statement generally carries the weight and finality of a Supreme Court judgment. If it's constitutional, there is nothing one can do about it except by constitutional amendment. Perhaps we can achieve a change in our understanding of what we mean by the "constitution" of human beings. Somehow the word "constitution" as applied to human beings has come to mean the bodily and functional expression of the biological endowment, the genes, of the individual. This is the biologistic fallacy once again at work. The constitution of the individual may quite properly be defined in terms of his bodily makeup and functions, but it is quite another thing to attribute that makeup and those functions to the exclusive determination of genes. "Constitution" is defined as the physical makeup of the body, including the mode and performance of its functions, the activity of its metabolic processes, the manner and degree of its reactions to stimuli, and its power of resistance to the attack of pathogenic organisms. All the criteria or terms of this description of constitution are determined neither by genes nor by environment, but by the *interaction* of genes and environment. The magnitude of the role played by the one or the other varies with the kinds of conditions that make up the elements of constitution. In some conditions the genes are almost exclusively responsible; in others the environment seems to play the major determining role. But the process of interaction between the two is always involved.

Normally when we use the word "constitution" it is in application to the postnatal organism, and in most cases we tend to use the term with respect to adults, but whether we use the term in connection with children or adults, we use it in a kind of final sense, meaning, "This is what this individual is" or "This condition is the result of the biological structure of this individual," and in either case, the implicit assumption is that there is nothing one can do about it because it is the individual's biological fate. In no case is this pessimistic conclusion sound. If there is one thing that can be predicted with complete certainty, it is that we shall all die, and most of us will die before we reach the age of eighty. But even in those cases in which we can predict that the individual will not live to attain five years, as in the case of amaurotic idiocy, in which the infant usually dies before he reaches his third year, it is now possible to prolong the life of such an individual appreciably. In some cases it may be debatable whether this is at all desirable. But in any event, even in the worst of cases we do not have to sit back in resigned hopelessness. For if we are aware that constitution is not something fixed and final but is something about which, even in the worst of cases, we *can* do something, however little it may be in the present state of our knowledge, there is the hope that we may someday be able to do much more. And by citing the facts, this is what I want to make quite clear in the present chapter.

It should be clearly understood that there is little that is terminal about constitution, for constitution is a *process* rather than a fixed, final, and unchangeable entity. It is important to understand that constitution is not a biological *given,* a structural system predestined by its genotype to function in a predetermined manner. As, I hope, has been sufficiently emphasized in these pages by this time, the manner in which all genotypes function is determined by the interaction of the genotype with the environment in which it undergoes development. What, so to speak, the genotype—that complex of genetic potentialities with which the organism is endowed—asks is: "What kind of responses are going to be made to my chemically accelerating and developmental overtures, my tentative advances? How will I impress? How will I be impressed? For the outcome of all this will be my constitution."

The point that must be emphasized here is that every genotype is a unique physicochemical system comprising particular kinds of potentialities having definite limits. These limits vary from individual to individual, so that were the genotype to be exposed to identical environmental conditions, its interactive expression would nevertheless continue to vary from individual to individual. But in point of fact the environmental conditions are never the same for two individuals, not excluding one-egg twins. This fact renders it necessary for us to recognize that constitution is constituted not merely by the genotype but also by the genotype as modified by the environment in which it has developed, as the result of the dynamic interaction of the two. As the sum total of the structural, functional, and psychological characters of the organism, constitution is in large measure the integral expression of the genetic potentialities influenced in varying degrees by internal and external environmental factors.

Let us now proceed to examine the constitutional factor in relation to the conditions with which it is most frequently discussed, namely, disease. It is often implied that when a disease is stated to be constitutional that that is the long and the short of it—it will always be with the individual, and there is very little or nothing that can be done to alter the situation. The disorders and diseases discussed in what follows are all cited in order to show that this view is unsound and requires modification in the light of the facts.

## Diabetes

Diabetes has already been cited as a constitutional disease about which a great deal can be done (see page 81). True diabetes (diabetes mellitus) is caused in most cases by the failure of the pancreas to secrete a sufficient amount of its hormone, insulin. The consequent failure in the proper conversion and utilization of sugar results in its excessive accumulation in the blood, with seriously damaging results to the organism. There is no means of curing genetically conditioned diabetes at the present time, but the sufferer from the disease can be maintained on insulin for many years as an efficient and for all intents and purposes a comparatively healthy human being. In cases in which the disease is caused by an excessive secretion of hormones from the pituitary or adrenal glands, removal of one or another of these glands may alleviate the condition. One out of every four diabetics comes from a family in which one or more relatives are also diabetic. In the many cases that are of genetic origin the condition is due to a pair of recessive genes. In other cases it is suspected that the disorder may be produced by emotional conditions combined with overeating and similar disordering states. Thus in the genetic cases environmental means may alleviate the condition; in the other types of cases the stress of environment may bring on the condition, which by the appropriate means may be either alleviated or altogether cured.

By making simple glucose-tolerance tests it is possible to diagnose the prediabetic state and to prevent the development of full diabetes in these individuals. This is now possible on a mass basis.

## Tuberculosis

Tuberculosis is a disease produced by the tubercle bacillus. But susceptibility to the disease presents marked individual differences and even familial differences. Since the disease is highly infectious, it is to be expected that several members of a family will show it, and this is, indeed, the case, but there are many families in which only one member develops the disease, whereas the others, though they may have been infected with the tubercle bacillus, show not the slightest evidence of the disease. The fact is that millions of human beings have been infected with the tubercle bacillus but have not become diseased. Indeed, only a small proportion of those who do become infected become diseased. These facts strongly suggest that there exist genetically influential differences in resistance to tuberculosis. The breeding of different strains of rabbits that show marked differences in resistance to tuberculosis has actually been accomplished. This takes the form of differences in the virulence with which the disease attacks these different strains of rabbits. Kallmann and Reisner found that whereas only 25.6 two-egg twins were concordant for the disease (the same rate as for ordinary brothers and sisters), 87.3 one-egg twins showed concordance. The probability of a genetic factor connected with susceptibility to tuberculosis is therefore high, and it would appear likely that many genes are involved. However, it is clear from these same twin studies that in the discordant cases in which there is a difference in the health of the twins, it is invariably the weaker twin who is affected. So the resistance is not entirely genetically determined. And the truth is that the best protection against tuberculosis seems to be the development of a healthy body in a healthy environment. Resistance to tuberculosis, then, can be built up by building up the health of the individual. Essentially this means the provision of good conditions for prenatal and postnatal development and the maintenance of those conditions throughout the life of the individual. This also implies a general improvement in the conditions of the environment. As a result of the general betterment of the health of the average individual and the improvement of the environment during the last sixty years, there has been a spectacular decline in the tuberculosis disease and death rates. In 1900 there was an annual death rate from tuberculosis of 194 persons per 100,000 of population. By 1950 this was down to only 20 persons per 100,000. This means that in the year 1950 instead of there being 315,000 deaths at the 1900 rate, there were only 35,000 deaths, a saving in that year alone of 280,000 lives. This saving was undoubtedly due to an improvement in the constitution of many individuals as well as to a general improvement in environmental conditions and improved methods of detection and treatment. Vaccination with BCG (bacillus Calmette-Guérin) affords protection against TB infection. No matter what the constitutional predisposition of the individual may be to tuberculosis, if he contracts the disease its course can be modified by the proper environmental procedures, especially by treatment with antibiotics such as streptomycin.

The principles involved in the modification of the course of tuberculosis, its prevention, and cure are applicable to all diseases, whether they are only partly conditioned or wholly determined by genetic factors.

Being born with a genetically determined disorder does not condemn, as it always did in the past, the individual to the consequences that inevitably followed. We are now able to control the expression of some genetically determined disorders, and had we possessed the knowledge of these matters fifty years ago that we possess today, the whole course of world history might have been altered. I have in mind the famous case of hemophilia, or "bleeder's disease," as it worked out its tragic consequences in Russia.

## Hemophilia

Hemophilia, or "bleeder's disease," is a disorder characterized by a defect in the clotting power of the blood, and as a consequence of this, intractable and profuse bleeding following the slightest scratch, cut, or bruise may result in death. In former years very few hemophiliacs managed to survive into adulthood. But since there are varying degrees of severity with which the disorder manifests itself, there have always been a number of adults of quite advanced age who were hemophiliacs.

Hemophilia is caused by a sudden change in the chemical structure of a normal gene situated on an X chromosome. Because the effects of such a defective mutant gene are always linked with the X chromosome, and because it is carried in the recessive condition, the hemophilia gene is recognized as a sex-linked recessive. It has been calculated that the mutation from the normal "big $H$" to the hemophilia gene, "little $h$," occurs in every generation in 1 out of every 100,000 persons. This is a relatively high rate of mutation and means that in the English-speaking world at the present moment there are probably over 10,000 persons in whom the hemophilia gene has come into being quite spontaneously. To see how simple the genetic mechanism of hemophilia transmission is, let us take an actual case.

The normal gene for blood clotting undergoes mutation in a man. This occurs in his X chromosome, for his Y chromosome has no corresponding gene. This man marries a normal woman with two healthy X chromosomes. They have five sons and five daughters. Since the male transmits his X chromosome to his daughters only, they will all become carriers of the defective gene, but because they have inherited a healthy gene for clotting in the X chromosome they have received from the mother, the disorder will not express itself in them. That is why it is exceedingly rare for a female to be afflicted with this condition. What would be necessary for a female to suffer from hemophilia would be the inheritance of a defective gene on the mother's X chromosome and a defective gene on the X chromosome contributed by the father. Several such cases have been reported, though there is evidence indicating that in some of these cases the sex-chromosome constitution was XY. Since the male with the defective gene for blood clotting does *not* transmit his X chromosome to his male offspring, the sons of a hemophilic father never suffer from the disorder except in the rare event that the father married a carrier. His daughters, marrying normal men, will transmit the defective X chromosome to about half their sons and to half their daughters, but the sons will be hemophilic, whereas the daughters receiving the defective chromosome will not, for their "good" X chromosome protects them from the "bad" one, whereas the male has only a Y chromosome, which contains nothing that can afford him protection.

In September, 1959, Dr. Inga M. Nillson and her collaborators at the University of Lund, Sweden, reported a case of severe hemophilia in a child aged sixteen months. The child presented every appearance of being a little girl, except that her clitoris was a little larger than normal. Upon investigation, cells from this child's skin were seen to be sex-chromatin negative; relatively few of the white blood cells known as polymorphonuclear leucocytes (because they are characterized by irregularly shaped or more than one nucleus) exhibited "drumsticks" protruding from the nucleus, and those present were small—all characteristics suggesting

170 HUMAN HEREDITY

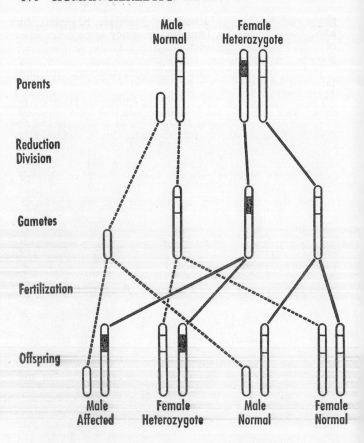

FIG. 15. The Transmission of the Hemophilia Gene.

a male chromosomal constitution. In the female, "drumsticks" are frequent and large. Further study revealed that the sex-chromosome constitution of this little "girl" was almost certainly XY—"almost certainly" because it is not yet possible to identify the sex chromosomes with complete certainty. However, when the chromosomes were arranged in their size and form relationships, there seemed to be no doubt that this was a case of the normal number of 44 autosomes and an X and a Y chromosome, so that this little phenotypic "girl" was actually a genotypic boy who, for some reason, had been feminized. Possibly some other cases of hemophilia in apparent females may be similarly explained.

Hemophilia is one of the most tragic disorders. The most famous family line affected by it is that of Queen Victoria (1819-1901). Victoria was the daughter of the Duke of Kent, the fourth son of George III. It is believed that the mutation from the normal gene for blood clotting to the abnormal gene occurred in the person of Queen Victoria's father, that he transmitted his defective X chromosome to her, with the consequences set out in Fig. 16. One of Victoria's four sons, Leopold, Duke of Albany, was a hemophiliac. He married and transmitted the defective gene to his desecendants. Two of Victoria's daughters transmitted the disorder to the royal families of Russia and Spain. The heirs to both thrones were hemophiliacs, and there can be little doubt that this fact played a role in leading to the revolutions that occurred in both countries. The czarevitch was the answer to his mother's devoted prayers. She had given birth to four girls when finally an heir to the throne was born. The tragic blow that befell her when the czarina learned that her son was a hemophiliac is one from which she never recovered. This caused her to resort to soothsayers, quacks, and mystics of every kind, and this is what led her to fall under the influence of Rasputin, the scheming monk who claimed to work remarkable cures and whose influence at court created that atmosphere of intrigue and duplicity that alienated millions of Russians and helped prepare the way for the Revolution. Neither the dethronement of the czar nor that of Alfonso XIII of Spain was the direct result of the fact that their sons and heirs were hemophiliacs, but there can be little question that the knowledge that the successors to the throne were invalids played a role in bringing about the fall of these monarchies. Today the discovery that there are different varieties of hemophilia and hemophiloid conditions and the discovery of some of the factors that enter into the causation of these conditions have made it possible to alleviate, at least in part, something of their severity. The expectation is that further intensive research will yield much more successful modes of treatment. In January, 1959, a new coagulant for treating hemophilia became available. This antihemophilic globulin, which consists of a white powder obtained from the blood of pigs, is nearly twenty times as potent in its antihemophilic properties as are human blood

# 172 HUMAN HEREDITY

transfusions. Antihemophilic globulin has been successfully used to control bleeding in surgical operations on hemophiliacs that formerly would almost certainly have ended fatally. In spite of great advances made in our knowledge of hemophilia, the average life expectancy of the hemophiliac is seventeen years (Ikkala).

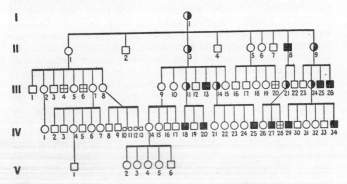

◐ Known to be Transmitter      ⊞ Died in Childhood or Infancy

FIG. 16. Hemophilia in Queen Victoria's Descendants.

I. (1) Victoria, 1819-1901, Queen of England.

II. (1) Victoria, 1840-1901 × Frederick III, Emperor of Germany. (2) Edward VII, 1841-1910, King of England. (3) Alice, 1843-1878, × Prince Louis of Hesse. (4) Alfred, 1844-1900, Duke of Edinburgh. (5) Helena, 1846-1923, × Prince Christian of Schleswig-Holstein. (6) Louise, 1848-1939, × Duke of Argyll. (7) Arthur, 1850-1942, Duke of Connaught. (8) Leopold, 1853-1884, × Princess Helena of Waldeck. (9) Beatrice, 1857-1944, × Prince Henry of Battenberg.

III. (1) William II, 1859-1941, Emperor of Germany. (2) Charlotte, 1860-1919, × Duke of Saxe-Meiningen. (3) Henry, 1862-1929, Prince of Prussia. (4) Sigismund, 1864-1866. (5) Frederica, 1866-1929. (6) Waldemar, 1869-1879. (7) Sophie, 1870-1932, × Constantine, King of Greece. (8) Margaret, 1872-1954, × Frederick, Duke of Hesse-Cassel. (9) Victoria, 1863-1950, × Prince Louis of Battenberg (Marquess of Milford Haven). (10) Elizabeth, 1864-1918, × Grand Duke Sergius of Russia. (11) Irene, 1866-1953, × Prince Henry of

Prussia [see III (3)]. (12) Ernest, 1868-1937, Grand Duke of Hesse. (13) Frederick William, 1870-1873. (14) Alexandra, 1872-1918, × Nicholas II, Czar of Russia. (15) Mary Victoria, 1874-1878. (16) Christian Victor, 1867-1900. (17) Albert, 1869-1931. (18) Victoria, 1870-1948. (19) Marie Louise, 1872-1956. (20) Harold, 1876-1876. (21) Alice, 1883       , × Earl of Athlone. (22) Charles Edward, 1884-1954, Duke of Albany. (23) Alexander, 1886-1960, Marquess of Carisbrooke. (24) Victoria Eugénie, 1887-       , × Alfonso XIII of Spain. (25) Leopold, 1889-1922, Lord Mountbatten. (26) Maurice, 1891-1914, Prince of Battenberg.

IV. (1) Feodora Maria, 1879-1898, × Henry XXX of Reuss. (2) George II, 1890-1947, King of Greece. (3) Alexander, 1893-1920, King of Greece. (4) Helena, 1896-       , × Carol II of Romania. (5) Paul, 1901-       , King of Greece. (6) Erene, 1904-       . (7) Catherine, 1913-       . (8) Frederick William, 1893-1916, Prince of Hesse. (9) Maximilian, 1894-1914. (10) Philipp, 1896-       . (11) Wolfgang, 1896-       . (12) Richard, 1901-       . (13) Christoph, 1901-       . (14) Alice, 1885-       , × Prince Andrew of Greece. (5)    Louisa, 1889-       , × King Gustav VI Adolf of Sweden. (16) George, 1892-1938, Marquess of Milford Haven. (17) Louis, Lord Mountbatten, 1900-       . (18) Waldemar, 1889-1945, Prince of Prussia. (19) Sigismund, 1896-1927. (20) Henry, 1900-1904. (21) Olga, 1895-1918, Grand Duchess. (22) Tatiana, 1897-1918, Grand Duchess. (23) Marie, 1899-1918, Grand Duchess. (24) Anastasia, 1901-1918, Grand Duchess. (25) Alexis, 1904-1918, Czarevitch. (26) May, 1906-       , × Sir Henry Abel-Smith. (27) Rupert, 1907-1928. (28) Maurice, 1910-1910. (29) Alfonso, 1907-1939, Prince of Asturias, later Count of Covadonga. (30) Jaime, 1908-       . (31) Beatrice, 1909-       . (32) Maria, 1911-       . (33) Juan, 1913-       . (34) Gonzalo, 1914-1934.

V. (1) Michael, 1921-    , King of Romania. (2) Margarita, 1905-       , × Godfrey of Hohenlohe-Langenburg. (3) Theodora, 1906-       , × Margrave of Baden. (4) Cecilie, 1911-1937, × Grand Duke of Hesse. (5) Sophie, 1914-       , × Christopher of Hesse. (6) Philip, 1921-       , Duke of Edinburgh.

---

Since October, 1959, when Dr. A. Fonio of Switzerland published his findings, it has become possible to identify female carriers of the hemophilia gene. Studying the thrombocytes (Greek *thrombos,* clot; *kytos,* cell) in thirty-one hemophiliac patients and in nineteen female carriers, Dr. Fonio found that the onset of decomposition and of formation of fibrin, necessary for clotting, occurs considerably later in the thrombocytes (platelets) of hemophiliacs than in those of normal individuals. For example, on the third day of observation and occasionally even on the fourth and fifth days, the thrombocytes of hemophiliacs, seen in the dark field of a microscope, are still sharply outlined, but are smaller, rounder, and brighter than normal thrombocytes. In other words, the hemophilic thrombocytes are more resistant to decomposition than their normal counterparts. The deficient blood coagulation of hemophiliacs is due to this resistance of the thrombocytes to decomposition and formation of thromboplastin. An additional defect of hemophilic thrombocytes is an abnormal retraction of the plasma coagulum.

Dr. Fonio found that the thrombocytes obtained from the nineteen female carriers were structurally identical with those obtained from hemophiliacs. Therefore, for the first time it became possible to identify female carriers of the hemophilic gene, and to counsel them accordingly. It was interesting that the thrombocytes obtained from sisters of hemophiliacs were in some cases identical with hemophilic thrombocytes and in some cases identical with normal platelets.

There are still other examples of constitutional conditions the course of which may be altered by altering the constitution of the individual.

## Phenylketonuria

Phenylketonuria, a disorder of metabolism inherited as a recessive, is characterized by a failure of the enzyme system involved in the conversion of phenylalanine into tyrosine. The enzyme phenylalanine hydroxylase, which converts phenylalanine into tyrosine in the liver, is absent from the liver of a phenylketonuric. The result is that phenylalanine, derived from proteins in the food, tends to accumulate in the blood and cerebrospinal fluid and to be excreted in large amounts in the urine. The associated intellectual impairment, thought to be due to the concentration of so much phenylalanine in the blood with the associated reduction of serotonin in the brain, may be very severe.

Most of the damage done to the nervous system occurs within the first six months of life. A characteristic of the disorder is the appearance of light hair. Apparently the presence of excessive amounts of phenylalanine inhibits the action of the tyrosine-tyrosinase system, which normally leads to the formation of the pigment melanin. When affected individuals are put on a phenylalanine-restricted diet or are fed massive doses of tyrosine over a period of months, there is a darkening of the new-grown hair.

About 1 out of every 25,000 individuals is born with phenylketonuria. But this disorder need no longer be the tragedy it once was, for as a result of the advances made in our knowledge of its biochemistry, it is now possible to prevent its development in individuals who have inherited it. This may be done by maintaining such individuals on a diet low in phenylalanine, a substance that occurs in milk, vegetables, legumes, etc. The restricted diet should start within a few days after birth. "Lofenalac," a powder that may be mixed with water to make a milk substitute, provides a balanced diet, which may be supplemented with fruits and vegetables. The diet may be terminated by the age of four. Sometime in the future it may be possible to supply the missing enzyme to infants in whom it is absent as routinely as insulin is now taken by diabetics.

Infants with phenylketonuria will usually begin to manifest evidences of the disorder by the presence of phenylalanine in their urine. The condition can be easily determined by the ferric-chloride test. A few drops of ferric-chloride solution are added to a fresh specimen of urine, or a test paper known as "Phenistix" may be used, which can be applied to wet diapers. If a green color results, the test is positive. All infants should routinely be given the test at three weeks of age and again at six and eight weeks. In this way the condition could be detected, and all affected infants could be enabled to develop as normal, intelligent human beings.

In some cities the newborn is now sent home from the hospital with a supply of gauze pads containing ferric chloride for its diapers. These are later returned at intervals from the end of the third week to the hospital laboratory, where they are examined for the presence of phenylalanine. In 1961 Dr. Robert Guthrie of Buffalo described a simple method of detecting elevated phenylalanine blood levels, thus overcoming some of the limitations of the ferric-chloride method. By the Guthrie method it is possible to determine the presence of phenylketonuria, in the hospital laboratory, a few days after the birth of the baby.

Since the parents of phenylketonuric children carry the defective gene in a single dose, and it is now known that they also exhibit, after fasting, a higher phenylalanine level in the blood than do noncarriers of the gene, it is possible by means of the phenylalanine-tolerance test to determine which individuals carry the defective gene. Heterozygotes exhibit a phenylalanine level about twice that of normal controls. The heterozygous carriers in every way appear to be otherwise normal, though it is clear from the test that their phenylalanine enzyme-conversion system is not as efficiently at work as it is in the normal individual.

Carriers of the defective gene can therefore be apprised of their condition and what it entails, and will thus be able to control its expression in their offspring.

## Hereditary Hemolytic Icterus

Hereditary hemolytic icterus, or hereditary spherocytosis, is a very serious form of jaundice combined with anemia. It is inherited as a dominant autosomal gene and is due to the enlargement of the spleen, which keeps too many red blood cells out of circulation and gradually destroys them. The result is jaundice and anemia and often very rapid death. It is possible not only to save such persons by a timely removal of the spleen but also possible to prevent the development of the disease in those individuals whom it would certainly kill, by the simple device of examining the offspring and immediate relatives for any signs of the precursors of the disease. Examination of their blood for changes in number, size, shape, variety of cells, and certain simple reactions makes it possible to determine whether the disease is likely to manifest itself in any of the examined individuals. Removal of their spleens, which in no way seems to have any untoward effects, will save them from much suffering. In this manner many lives have already been saved that were formerly doomed. We have cases on record in which several members of a family have willingly undergone the removal of their spleens and thus saved their lives, whereas other members of the same family have refused to undergo the operation and have died as a result.

Incipient cases of hypertension, pernicious anemia, certain types of cancer, and many other genetically based disorders have also been diagnosed and consequently prevented from developing in a similar way.

From the cases cited it will be observed that not only can the effects of a constitutional disorder be alleviated but also

## GENES AND CONSTITUTION 177

in many instances it can be prevented from developing. Hence, the simple conclusion can be drawn that one's constitution is not unalterable or unmodifiable, that it is not the equivalent of implacable fate, but that it is in many respects quite amenable to the changes induced by environmental means.

## 11. SEX

THE TERM "SEX" (Latin *sexus*) is probably derived from the Latin verb *secare*, meaning to cut or divide. Sexual species are divided into male and female. It is by means of the union of the sexes that new gene combinations are brought into being in an ever-varying but constant variety. By this means the species is reproduced and maintained from generation to generation. The great variety of genotypes thus produced endows the species with a high degree of adaptability, so that under gradual or sudden changes in the environment there are generally likely to be genotypes that will be able to respond to the challenges of the changed environment. The genotypes unable to make the necessary responses are unlikely to leave a sizable progeny behind them.

On the other hand, in organisms that usually reproduce asexually, with no division into male and female—like the one-celled amoeba or the slipper-shaped, one-celled paramecium—variability is reduced to a minimum. Among such organisms reproduction usually occurs by simple fission, or budding, an exact duplicate thus being formed, which varies ever so slightly from the maternal or sister organism. Different sets of genes are usually not brought together and recombined in the offspring, so that it will be understood why evolutionary change is likely to be arrested until sexually reproducing species make their appearance. Sexual reproduction is the great conduit through which most biological change is established within the species. By whatever agencies changes are produced in organisms, it is only through the reproductive process that they can be distributed within the population, and in some groups may even be made to cross over from one species and even from one genus to another.

In all sexual species it is the female who harbors the eggs, eggs with which, as in the human species, she is already endowed—to the number of about half a million—at birth. The male, on the other hand, is endowed with sperm-forming organs, the testes, which gradually develop the capacity to manufacture new sperm, of which he manufactures many billions in the course of a lifetime. The eggs of the human female are passed out on the average about every twenty-eight days from about (on the average) the age of sixteen to about the age of fifty, whereas the male normally continues to manufacture spermatozoa till he is beyond sixty and is not infrequently capable of producing fertilization in a young female at eighty years and beyond. Records of the offspring of such unions show that they are perfectly normal in every way.

Whereas the eggs age in the aging female, apparently the organs producing the sperm cells in the male preserve their integrity somewhat more efficiently till an advanced age. There is undoubtedly some adaptive advantage in this arrangement from the evolutionary standpoint. The explanation that suggests itself is that since there are always fewer males in any population than females, there will always be an excess of females. During the early evolution of man, it would have been of advantage to the group if an older man whose wife had ceased to be fertile were able to take an additional younger woman or two as members of his family. We still find this custom (polygyny) very widely distributed among nonliterate peoples. From the standpoint of the survival of the group with a small population, the custom has very obvious advantages.

The female has a much shorter reproductive life than the male, both (1) during her twenty-eight-day cycle (ovulation occurs from the fourteenth to the seventeenth day, counting from the first day of menstruation, with the ovum fertilizable for about two days at most during the period after it has entered the fallopian tube and (2) in the total duration of her capacity to reproduce, which lasts about thirty-five years. In addition, the human female takes, on the average, 266½ days from conception to delivery to gestate a child,* whereas it takes the male only a few minutes to initiate the process that eventually leads to the fertilization of an egg. With these facts the reader may begin to perceive that the maintenance of the species is rather more dependent upon the female than it is upon the male. In order for the species to be maintained, the female must be impregnated, and this can be achieved within a very narrow range of time. Once impregnated, she must be preserved for at least 266½ days if the species is to be perpetuated. The male who pro-

---

* The 280 days reckoned by the obstetrician refer not to *conceptional* age but to the gestation period reckoned from the first day of the last menstruation.

duced the pregnancy, however, could die immediately after conjugation, even before fertilization had occurred, and the continuity of the species would be quite unaffected.

For all these reasons (and a good many more) the female is the biologically more valuable part of the species capital—and this is true for similar reasons throughout the whole realm of animated nature. She has to be constitutionally stronger, genetically more resistant than the male, and, indeed, in every species characterized by sexual reproduction thus far investigated, the female has proved so. It is unlikely that any exceptions to this fundamental rule will be found, for it reflects a basic difference in the biological structure of the sexes, the very difference that determines sex itself.

As I have already explained, until April, 1959, it was believed that a female owes her sex to the fact that she has acquired one X chromosome from her mother and another X chromosome from her father; whereas a male owes his sex to the fact that he has acquired one X chromosome from his mother and a Y chromosome from his father. In 1958 doctors J. H. Tjio and T. T. Puck measured the X and Y chromosomes and found the X chromosome to be three times larger than the Y chromosome. Females have approximately 4 percent more DNA at their disposal than males. Each X chromosome contains a full complement of genes, whereas the Y chromosome probably does not. Until 1956 it was thought that the Y chromosome contained genes for overhairy ears, horny, scalelike, or barklike skin (ichthyosis hystrix), a nonpainful nodular affection of the hands and feet, and webbing between the second and third toes. But the first case is based on a single pedigree, the second has been shown to be unsound, and the two last conditions occur in both males and females, which renders it something less than certain that Y-chromosome inheritance is involved in any of the very few cases reported. Although we do not know precisely what genes the Y chromosome contains, we can be reasonably certain that it contains *some* genes, but the existence of these remains to be demonstrated. In fact, every condition of which we have positive and verifiable knowledge that it is linked in heredity with a sex chromosome is *always linked with the X chromosome,* and these conditions are known as *sex-linked* conditions. Hemophilia, which we have already described, is an example. We shall shortly return to a discussion of sex-linkage. What we have to emphasize here is that the female possesses two complete X chromosomes and the male possesses only one. It is this difference that endows the female with the biological superiority that, as we shall see, enables her to resist the assaults and insults of the environment so much more effectively than the male.

In April, 1959, publication of two papers, one by W. L. Russell, Liane Brauch Russell, and Josephine S. Gower, and the other by W. J. Welshons and Liane Brauch Russell, all of the Oak Ridge National Laboratory, Tennessee, showed that the Y chromosome is vitally important in sex determination in mammals. This is in contrast to Drosophila, in which sex is determined by the balance between X chromosomes and autosomes, with the Y playing a neutral role. By ingenious breeding and cytological studies involving certain exceptional mice, it was found that these animals, which were females of almost normal productivity, were of chromosome constitution XO. That is, they carried only one X and therefore had only thirty-nine chromosomes instead of the normal forty. This is clearly different from Drosophila, in which XO animals are males. The results in the mouse show that the Y chromosome is the bearer of the male-determining factors. They also indicate that all female-determining factors are situated on the autosomes or, as seems more likely, that the X chromosome carries at least some female-determining factors.

If the X chromosome carries some female determiners, then in those cases in which the organism has acquired an extra sex chromosome (as a result of nondisjunction at either mitosis or meiosis) and is of constitution XXY, we should expect either an intersexual or possibly even a female. There is some evidence suggestive of this in the literature, and in January, 1959, Jacobs and Strong, of the University of Edinburgh, reported such an apparently XXY condition in an intersexual young man of a nuclear-sex-chromatin type indicating a genetic female.

If we ask ourselves how it comes about that the female possesses two X chromosomes and the male only one, we may conjecture that the possession of a complete set of X chromosomes endows the female with a greater capacity for survival than the male. By biologically strengthening the female, the higher probability of the survival of the species is brought about. This interpretation of the facts the reader will be able to make for himself as he follows the evidence in the succeeding pages.

## 182 HUMAN HEREDITY

### Differential Survival Rates of the Sexes

Since 50 percent of spermatozoa carry an X chromosome in their heads and 50 percent carry a Y chromosome, it would seem that half the eggs fertilized should be females and half should be males. But this is not the case. Y-bearing sperm fertilize between 120 and 150 eggs for every 100 eggs fertilized by X-bearing sperm. The same ratio of difference has been found also in other mammals. The reason for this difference in fertilization is unknown. It has been suggested that because of their lesser mass Y-bearing sperm swim faster than X-bearing sperm. This is almost certainly not the explanation, for in birds, in which the chromosomal contents of the germ cells are reversed (that is, eggs are of two kinds,

TABLE XI
MORTALITY RATES FROM EACH CAUSE BY SEX
WHITE POPULATION OF CONTINENTAL UNITED STATES FOR THE YEAR 1957
RATES PER 100,000

| Cause of Death | Male | Female |
|---|---|---|
| Tuberculosis of respiratory system | 9.5 | 3.1 |
| Tuberculosis, other forms | 0.5 | 0.3 |
| Syphilis and its sequelae | 2.5 | 0.9 |
| Virus diseases | 1.6 | 1.5 |
| Malignant neoplasm of buccal cavity and pharynx | 5.6 | 1.5 |
| Malignant neoplasm of digestive organs and peritoneum | 59.1 | 48.3 |
| Malignant neoplasm of respiratory system | 36.1 | 6.3 |
| Malignant neoplasm of breast and genito-urinary organs | 28.4 | 58.2 |
| Malignant neoplasm of other and unspecified sites | 17.6 | 15.0 |
| Neoplasms of lymphatic and hematopoietic tissues | 17.2 | 11.9 |
| Benign neoplasm | 1.0 | 1.6 |
| Neoplasm of unspecified nature | 1.7 | 1.3 |
| Allergic disorders | 5.5 | 2.4 |
| Thyroid diseases | 0.2 | 1.0 |
| Diabetes mellitus | 12.9 | 19.0 |
| Other endocrine gland diseases | 0.5 | 0.5 |
| Avitaminoses and other metabolic diseases | 0.8 | 0.7 |
| Pernicious anemia | 0.4 | 0.5 |
| Hemophilia | 0.1 | 0.0 |
| Psychoses | 0.7 | 0.6 |
| Psychoneurotic disorders | 0.0 | 0.1 |
| Disorders of character, behavior, and intelligence | 2.1 | 0.5 |
| Cerebral hemorrhage | 63.6 | 67.9 |
| Meningitis | 1.1 | 0.8 |
| Multiple sclerosis | 0.8 | 1.0 |
| Other diseases of central nervous system | 5.4 | 4.2 |
| Epilepsy | 1.3 | 0.9 |
| Motor neuron disease and muscular atrophy | 0.8 | 0.5 |
| Diseases of ear and mastoid process | 0.3 | 0.2 |
| Rheumatic fever | 0.5 | 0.4 |
| Arteriosclerosis and degenerative heart disease | 292.2 | 242.1 |
| Other diseases of heart | 14.4 | 9.3 |
| Aortic aneurysm, nonsyphilitic | 6.4 | 2.2 |
| Peripheral vascular disease | 0.3 | 0.1 |
| Thrombo-angiitis obliterans | 0.2 | 0.1 |

Based on *Vital Statistics—Special Reports, National Summaries,* vol. 50, No. 2, February 4, 1959. U. S. Department of Health, Education and Welfare, Public Health Service, National Office of Vital Statistics, Washington, D.C.

X-bearing and Y-bearing, whereas male sperm contain only X-bearing chromosomes), males conceived still outnumber the females conceived. On the average at each successful coitus the male ejaculates approximately 250 million spermatozoa. Traveling at the rate of 1 inch in 20 minutes,

TABLE XI

MORTALITY RATES FROM EACH CAUSE BY SEX

WHITE POPULATION OF CONTINENTAL UNITED STATES FOR THE YEAR 1957

RATES PER 10,000

| Cause of Death | Male | Female |
|---|---|---|
| Arterial embolism and thrombosis | 0.3 | 0.2 |
| Varicose veins, lower extremities | 0.1 | 0.2 |
| Hemorrhoids | 0.1 | 0.0 |
| Varicose veins, other sites | 0.3 | 0.1 |
| Influenza | 4.2 | 3.6 |
| Lobar pneumonia | 8.8 | 5.2 |
| Bronchopneumonia | 18.0 | 13.5 |
| Primary atypical pneumonia | 2.9 | 2.2 |
| Pneumonia, other and unspecified | 4.0 | 2.9 |
| Bronchitis | 2.8 | 1.3 |
| Bronchiectasis | 1.9 | 0.9 |
| Stomach ulcer | 4.5 | 1.6 |
| Duodenal ulcer | 5.2 | 1.4 |
| Gastrojejunal ulcer | 0.3 | 0.1 |
| Gastritis and duodenitis | 0.2 | 0.1 |
| Appendicitis | 1.5 | 0.7 |
| Abdominal hernia | 2.3 | 1.9 |
| Gastro-enteritis and colitis | 2.1 | 2.0 |
| Chronic enteritis and ulcerative colitis | 1.7 | 1.7 |
| Peritonitis | 0.4 | 0.4 |
| Cirrhosis of liver | 15.8 | 7.5 |
| Cholelithiasis | 1.7 | 2.4 |
| Cholecystitis | 0.9 | 1.2 |
| Other diseases of gall bladder and biliary tracts | 0.5 | 0.8 |
| Diseases of pancreas | 1.6 | 1.1 |
| Chronic nephritis | 6.9 | 5.7 |
| Hydronephrosis | 0.2 | 0.1 |
| Cystitis | 0.2 | 0.1 |
| Rheumatoid arthritis | 0.4 | 0.7 |
| Osteitis deformans | 0.1 | 0.0 |
| Myasthenia gravis | 0.1 | 0.2 |
| Inborn defect of muscle | 0.5 | 0.2 |
| Curvature of spine | 0.1 | 0.1 |
| Congenital malformations | 13.8 | 11.5 |
| Intracranial and spinal injury at birth | 2.7 | 1.5 |
| Other birth injuries | 4.9 | 3.4 |
| Postnatal asphyxia and atelectasis | 11.1 | 7.7 |
| Pneumonia of newborn | 2.0 | 1.3 |
| Other diseases peculiar to infancy | 17.7 | 12.9 |

only one of these, if it reaches an egg, will produce fertilization. Whatever the physical reasons may be that result in more eggs being fertilized by Y-bearing spermatozoa, the evolutionary "reason" would appear to be that *since the male is the constitutionally weaker organism, he must be conceived in greater numbers than the female if a relatively harmonious numerical balance is to be achieved between the sexes during the reproductive life of the female.* We have already observed that the greater duration of reproductive capacity in the male compensates somewhat, insofar as the survival of the species is concerned, for his average lesser overall durability.

At birth for every 100 females that are born there are about 105 males born. At first glance this would seem to mean that out of the 120 to 150 that have been conceived, between 15 and 45 males have died *in utero*. This would be true only if every female conceived survived to birth, but this is not the case: some females also die *in utero*. The prenatal mortality rate is therefore higher than the figures given. We do not know how many. Evidently, fewer female conceptions fail to come viably to term than male ones. At every stage of prenatal and postnatal development the mortality rate is higher for the male than for the female. At every age during postnatal life more males are dying than females.

Fetal deaths are 50 percent higher among males than females. Within the first month following birth, the male death rate exceeds that of the female by 40 percent, and among prematures by 50 percent. Within the first year of life male mortality exceeds that of the female by 33 percent. For ages five to nine, male mortality exceeds that of the female by 44 percent, from ten to fourteen by 70 percent, and from fifteen to nineteen by 145 percent. The disparity in the rates increases steadily till twenty-one, when the death rate of the male exceeds that of the female by 130 percent, and then grows increasingly less, being almost equal at thirty to thirty-four. At ages sixty to sixty-four there are 23 percent more women than men, and at seventy-five and over, there are nearly twice as many women as men.

The statement is often made that men die earlier on the average than women because they work harder than women. This statement is a good example of the kind of mythology to which many of us subscribe even in the face of the facts to the contrary.

Newborn boys do not work harder than newborn girls; yet they die more frequently than girl babies. One-year-old boys do not work harder than one-year-old girls, but the boys die more frequently than the girls. And so one can go on for every age, with the difference in mortality being in favor of the female.

It is found that the married man lives longer on the average than the bachelor, and that the married woman lives longer than the spinster, so, if anything, the state of marriage helps both sexes to greater longevity. Men and women engaged in the same kind of work, and men and women who don't work, still show mortality rates that are in favor of the female. A study, published in 1957, of the longevity of Catholic Sisters and Brothers who for many years had been living the same kinds of lives showed the same sort of disparity in their mortality rates as the rest of the population. The data were obtained on nearly 30,000 Sisters and more than 10,000 Brothers. The investigator, Father Francis C. Madigan, working at the University of North Carolina, found that the expectation of life at the age of forty-five was 34 years more for the Sisters and only 28 years for the Brothers, the difference favoring the Sisters being 5½ years.

At the present time the average male child born in the United States may expect to attain an age of approximately 67.3 years, whereas the average female child may expect to attain an age of more than 73.9 years. Table VII, on pages 127–28, gives the expectation of life at birth for a number of countries. It seems most unlikely that we shall be able to close the gap that exists between male and female longevity in the near future, but as we learn more about the nature of the hereditary factors involved it is highly probable that we shall not only be able to lengthen the life of the male and reduce this gap but also lessen the prenatal and postnatal mortality rates that exist for both sexes. When we have succeeded in doing even part of this, new social problems will be facing us as a result of the unusual increase of males, unless, of course, we foresee the nature of those problems and do something to control the proportions of the sexes born so that they are harmoniously balanced. This brings us to the matter of the artificial determination of sex. We shall deal with this later on in this chapter, but now we must discuss the mechanisms involved in producing the sexual differences in the rates of functional capacity, sickness, and death.

As I have already explained in connection with the manner of inheritance of hemophilia, the possession of a double number of X chromosomes protects the female against the development of this deficiency disorder even though she has inherited the defective gene that is the cause of the condition in the male. It is this protection afforded her by the extra X chromosome that confers a basic biologically determined natural superiority upon the female as compared with the male, who possesses only one X chromosome. The genes for traits or characters that are associated in inheritance with

TABLE XII
SEXUAL DIFFERENCES IN SUSCEPTIBILITY TO DISEASE

| MALES | | FEMALES | |
|---|---|---|---|
| Diseases | Preponderance | Diseases | Preponderance |
| Acute pancreatitis | Large majority | Acromegaly | More often |
| Addison's disease | More often | Arthritis deformans | 4.4-1 |
| Amebic dysentery | 15-1 | Carcinoma of genitalia | 3-1 |
| Alcoholism | 6-1 | Carcinoma of gall bladder | 10-1 |
| Angina pectoris | 5-1 | Cataract | More often |
| Arteriosclerosis | 2.5-1 | Chlorosis (anemia) | 100% |
| Bronchial asthma | More often | Chorea | 3-1 |
| Cancer, buccal cavity | 2-1 | Chronic mitral endocarditis | 2-1 |
| Cancer, G.U. tract | 3-1 | Combined sclerosis | More often |
| Cancer, head of pancreas | 4.5-1 | Diabetes | Slight |
| Cancer, respiratory tract | 8-1 | Diphtheria | Slight |
| Cancer, skin | 3-1 | Gall stones | 4-1 |
| Cerebral hemorrhage | Greatly | Goiter, exophthalmic | 6 or 8-1 |
| C.S. meningitis | Slight | Hemorrhoids | Consid. |
| Childhood schizophrenia | 3-1 | Hyperthyroidism | 10-1 |
| Chronic glomerular nephritis | 2-1 | Influenza | 2-1 |
| Cirrhosis of liver | 3-1 | Migraine | 6-1 |
| Coronary insufficiency | 30-1 | Multiple sclerosis | More often |
| Coronary sclerosis | 25-1 | Myxedema | 6-1 |
| Duodenal ulcer | 7-1 | Obesity | Consid. |
| Erb's dystrophy | More often | Osteomalacia | 9-1 |
| Gastric ulcer | 6-1 | Pellagra | Slight |
| Gout | 49-1 | Purpura haemorrhagica | 4 or 5-1 |
| Heart disease | 2-1 | Raynaud's disease | 1.5-1 |
| Hemophilia | 100% | Rheumatoid arthritis | 3-1 |
| Hernia | 4-1 | Rheumatic fever | Consid. |
| Hodgkin's disease | 2-1 | Tonsilitis | Slight |
| Hysteria | 2-1 | Varicose veins | Consid. |
| Korsakoff's psychosis | 2-1 | Whooping cough | 2-1 |
| Leukemia | 2-1 | | |
| Mental deficiencies | 2-1 | | |
| Muscular dystrophy, Ps.h. | Almost exclusively | | |
| Myocardial degeneration | 2-1 | | |
| Myocardial infarction | 7-1 | | |
| Paralysis agitans | Greatly | | |
| Pericarditis | 2-1 | | |
| Pigmentary cirrhosis | 20-1 | | |
| Pleurisy | 3-1 | | |
| Pneumonia | 3-1 | | |
| Poliomyelitis | Slight | | |
| Progr. muscular paralysis | More often | | |
| Pseudohermaphroditism | 10-1 | | |
| Sciatica | Greatly | | |
| Scurvy | Greatly | | |
| Syringomyelia | 2.3-1 | | |
| Tabes | 10-1 | | |
| Thromboangiitis obliterans | 96-1 | | |

Source: Montagu, *Introduction to Physical Anthropology*.

sex are of three kinds: (1) *sex-linked,* (2) *sex-influenced,* and (3) *sex-limited.*

## Sex-Linked Traits

When a trait is conditioned by a gene lying in the X chromosome, it is said to be *sex-linked*. What we mean by sex-linkage is that the X chromosome carries genes for characteristics other than those that merely determine sex. This means that such characteristics will be linked with sex in heredity. It does *not* mean that such characteristics are linked to a particular sex but that such characteristics will follow the distribution of the X chromosome in both sexes. However, because the male carries only one X chromosome, any defective genes present in that chromosome are likely to express themselves considerably more frequently in him than in the female.

Let us consider the X chromosome. It contains genes that are lacking in the Y chromosome. These genes will always express themselves in the male offspring because there is nothing present in the Y chromosome that can possibly affect their expression. The effects of such genes appear in the female only if the gene is present in both X chromosomes. The genes passed on in this manner are said to be wholly sex-linked. Approximately sixty traits are known and approximately twenty others are thought to be due to sex-linked, or X-linked, genes. These traits are shown in Table XIII. We have already seen how sex-linkage works in the case of hemophilia. Let us deal with another sex-linked condition, namely, color blindness. There are many different kinds of color blindness. At one extreme are the persons who cannot distinguish any colors. At the other extreme are those persons who have some slight defect in their ability to distinguish red from green. Red-green blindness is the commonest of all inherited defects that show a sex-linkage. Among whites, 2 men out of every 25 suffer from some lack of ability to recognize red and green. Less than 1 woman out of 200 women suffers from a similar defect.

One way in which color blindness can be inherited is shown in Fig. 17. The dark *X* is an X chromosome that carries the gene for color blindness. If a man inherits such a gene, he will be color-blind. There is no corresponding gene in his Y chromosome that can counteract the effect of the defective gene in his X chromosome. If such a man marries a woman with two normal X chromosomes, all the children

## TABLE XIII
## Some Human Traits Reported as Dependent upon X-Linked Genes

Partial color blindness, deutan series
Partial color blindness, protan series
Total color blindness
Glucose-6-phosphate dehydrogenase deficiency
Xg blood-group system
Muscular dystrophy, Duchenne type
Muscular dystrophy, Becker type
Hemophilia A
Hemophilia B
Agammaglobulinemia
Hurler syndrome
Late spondylo-epiphyseal dysplasia
Aldrich syndrome
Hypophosphatemia
Hypoparathyroidism
Nephrogenic diabetes insipidus
Neurohypophyseal diabetes insipidus
Oculo-cerebro-renal syndrome of Lowe
Hypochromic anemia (Cooley-Rundles-Falls type)
Angiokeratoma diffusum corporis universale
Dyskeratosis congenita
Dystrophia bullosa hereditaria, typus maculatus
Keratosis follicularis spinulosa cum ophiasi
Ichthyosis vulgaris
Anhidrotic ectodermal dysplasia
Amelogenesis imperfecta, hypomaturation type
Absence of central incisors
Congenital deafness
Progressive deafness
Mental deficiency
Börjeson syndrome
Spinal ataxia
Cerebellar ataxia with extrapyramidal involvement
Spastic paraplegia
Progressive bulbar paralysis
Charcot-Marie-Tooth peroneal muscular atrophy
Diffuse cerebral sclerosis (Pelizaeus-Merzbacher)
Diffuse cerebral sclerosis (Scholz)
Hydrocephalus
Parkinsonism
Ocular albinism
External ophthalmoplegia and myopia
Microphthalmia
Microphthalmia, with digital anomalies
Nystagmus
Megalocornea
Hypoplasia of iris with glaucoma
Congenital total cataract
Congenital cataract with microcornea
Stationary night blindness with myopia
Choroideremia
Retinitis pigmentosa
Macular dystrophy
Retinoschisis
Pseudoglioma
Van den Bosch syndrome

*Conditions for which the evidence for X-linkage is considered inconclusive*

Inability to smell cyanide
Incontinentia pigmenti
Wildervanck's syndrome
Male hypogonadism
Male hypogonadism and ichthyosis
Testicular feminization syndrome
Male pseudohermaphroditism
Kallmann syndrome
Anosmia
Zonular cataract and nystagmus
Myoclonic nystagmus
Combined Charcot-Marie-Tooth disease and Friedreich's ataxia
Choroidal sclerosis
White occipital lock of hair
Diffuse cortico-meningeal angiomatosis
Albright's hereditary osteodystrophy
Aptitude for spatial visualization
Microcephaly with spastic diplegia
Familial obstructive jaundice
Paget's disease of bone
Phosphorylase-deficient glycogen storage

After V. A. McKusick, "On the X Chromosome of Man."

will receive a normal X chromosome from their mother. The daughters will also receive an X chromosome from their father, which carries the gene for color blindness. They will usually have normal vision. The gene for color blindness is recessive; that is to say, it is a gene that will not normally express itself in the presence of the normal gene, which is dominant. The recessive gene has no effect on the phenotype unless it is carried in the homozygous state or is carried on an X chromosome in the male. Thus, a person receiving both a defective gene and a normal gene will be normal for the trait; a person receiving only the defective gene will exhibit defective red-green vision. The gene for normal vision is said to dominate the defective gene for red-green vision in the female. In the male if the gene is normal, there will be normal vision; if it is defective, there will be defective vision. Since the male never transmits his X chromosome to his sons, but the female transmits one of her two X chromosomes to all her children, the sons of a color-blind mother, regardless of whether her husband has normal or defective red-green vision, will all be color-blind, but the daughters will all have normal color vision if the father is normal in this respect. But these daughters all become carriers of the recessive gene for color blindness, and if they marry normal men, they will transmit the defective gene to about half the sons, and these sons will be color-blind because they have no masking gene in their Y chromosome that compensates for the defective gene. Half the daughters will receive the defective gene from their mother. They will not be color-blind because the normal X chromosome inherited from their father will be dominant over the gene for color blindness. These daughters in turn will be able to transmit color blindness to half their sons. Color blindness in a daughter can be produced only if a color-blind man marries a color-blind woman or a woman who is carrying the defective gene in the recessive state.

Color blindness is an important problem in a society in which, to cite but one example, the ability to distinguish between red and green traffic signals may mean the difference between life and death. Many color-blind persons have learned to distinguish between the different traffic lights not by their color but by their standard position in the light. Even so, color blindness does present a small but definite social problem. What should be done with the color-blind? One solution would be to prevent them from having children. Another, and obviously far more sensible, would be to make

190 HUMAN HEREDITY

color filters available to them so that they could easily detect essential color differences.

In the male, sex-linked genes can be represented on only one chromosome, the X chromosome. Sex-linked genes can be represented on both chromosomes in the female. From these facts and the fact that almost all sex-linked mutant genes are recessive, it is possible to give specific advice to families known to be carrying such genes. If the men in such

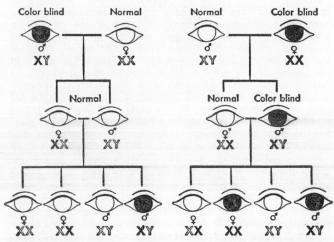

FIG. 17. The Inheritance of Color Blindness. 1. In the mating of a color-blind man with a normal woman, since the defective gene is sex-linked, it is transmitted through the daughters of this mating and appears in half of their sons. 2. In the mating of a color-blind woman with a normal man, the defect is transmitted to all the sons but to none of the daughters, who are, however, carriers. Such a carrier is mated to a color-blind man in this diagram in the second parental generation, and half the grandsons and half the granddaughters are color-blind. (After Dunn.)

a family do not exhibit the trait, then clearly the gene for that particular expression of the trait is not present on the male's X chromosome, and he cannot therefore transmit the gene for the condition to any of his offspring. He may safely have children. An affected father can transmit his defective X chromosome to his daughters, who will *all* be carriers. Since the sons receive only his unaffected Y chromosome from him, *none* of them will be affected. The sons may safely marry. When the carrier daughters marry a normal man, as they are most likely to do, they will transmit the defective X chromosome to half their daughters and to half their sons. The daughters will be unaffected but half the sons will be.

There are some conditions that appear to be due to *incompletely sex-linked genes*. The manner in which these conditions behave in inheritance leads to the conclusion that the human Y chromosome must possess some genes. If completely sex-linked genes are present in both chromosomes of both sexes, they will behave like autosomal genes in inheritance, with this difference: About half the families in which the fathers carry the gene will contain more affected sons and unaffected daughters than would be expected on the basis of autosomal inheritance, and the other half will contain more affected daughters and unaffected sons than would be expected on the basis of autosomal inheritance. The reason for this is that such genes can cross over from the X to the Y chromosome, and vice versa in the male, and thus will be differentially transmitted as to sex. If the male receives the defective gene from his mother, he transmits it to a majority of his daughters and a minority of his sons.

About ten or so traits have been attributed to incompletely sex-linked genes. These are: total color blindness (as opposed to red-green blindness), xeroderma pigmentosum (a childhood skin disease characterized by numerous pigmented spots, lesions, and a glossy-white thinning of the skin, often terminating fatally), Oguchi's disease ( a type of night blindness), spastic paraplegia (a neuromuscular defect), the recessive form of epidermolysis bullosa (malignant skin blisters), the dominant form of retinitis pigmentosa, hereditary hemorrhagic diathesis (a blood abnormality), and a type of cerebral sclerosis (a mental defect).

## Sex-Influenced Traits

Sex-influenced traits are conditioned by genes carried in the autosomes, and hence are inherited equally by both sexes and transmitted equally by both sexes. Sex, however, controls the dominance. The gene that is dominant in one sex is recessive or intermediate in the other, and vice versa. Several sex-influenced traits have been described in man, such as one form of white forelock, absence of the upper lateral incisor teeth, simple ichthyosis (scaling of skin), and Heberden's nodes (enlargement of terminal joints of fingers). Though baldness has customarily been attributed to sex influence, it is in fact an example of sex limitation and not of sex influence. We shall deal with baldness under sex-limited traits.

An example of a sex-influenced trait is provided by a form of white forelock described by Holmes and Schofield in 1917, which appeared in four generations in a family of 32 individuals. It was present in 14 males but not in 3 males; nor was it present in 15 females. It could be transmitted through either the males or the females and was dominant in males and recessive in females, and hence was sex-influenced.

## Sex-Limited Traits

When certain traits are expressed in one sex but not in the other, they are said to be *sex-limited*. Sex-limited characters are conditioned by genes that are carried either in the autosomes or in the sex chromosomes. There are, therefore, many different types of sex-limited heredity. The expression of sex-limited traits depends largely upon the presence or absence of one or more sex hormones or, to put it more accurately, the amount of such hormones present within the organism. Complete sex limitation, that is, complete development of the trait in one sex and complete absence of it in the other sex, is not frequent. Examples of complete sex limitation are milk production and menstruation in females but not in males. Another example of such complete sex limitation is the appearance of coarse hairs on the external ear of white men during the process of aging, and the absence of such hairs in women. I am not aware of any evidence indicating that the administration of male hormone to women at any age will induce the growth of such hairs on their ears.

Another example, but not quite so complete, is the presence of a developed beard and mustache in the male and their absence in the female. It has been shown that women have exactly the same number of hairs on the face as men, but the hairs simply do not develop and grow as they do in men. However, under the proper endocrine stimulation with male hormones women are capable of growing fairly respectable beards and mustachios.

A familiar example of sex limitation is baldness. That baldness is not due to a sex-linked gene is clear from the fact that bald fathers, depending upon whether they carry two baldness genes or only one, transmit their baldness to all or only half their sons, and it will be remembered that since sex-linked characters are conditioned by those carried in the X chromosomes, a man cannot transmit X-linked genes to his sons, but only to his daughters, in whom they remain recessive. Furthermore, hereditary baldness in the female is extremely rare. In the male, baldness in varying degrees (see Plate 15) occurs in more than 40 percent of those over the age of thirty-four. Baldness can show up in males when neither parent is bald and when one or both parents contribute a baldness gene to their sons. A bald father may have some sons who become bald and some who never become bald, or all the sons may become bald in varying degrees. Daughters, however, will experience thinning or partial baldness, but complete hereditary baldness of the masculine

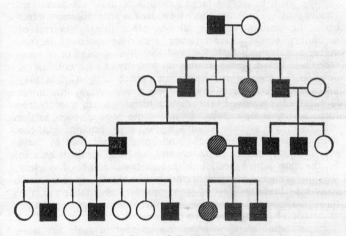

FIG. 18. A Family Pedigree of Balding. Solid symbols indicate balding in males. Cross-hatched symbols indicate females showing thinning of hair.

type is of excessive rarity in the female. When baldness of the masculine type affects a female, it is usually due to disease or disorder. As Obermayer and others have shown, baldness in varying degrees may be produced in both sexes by psychogenic factors, hence the statistics on baldness among women patients in mental hospitals that have been used in some books to interpret the genetic mechanism of baldness have served only to confuse matters.

In some cases baldness is due to disease of the hair follicles and in others to hormonal disorders, but these are the exceptional cases. In most cases baldness is due to the action not of genes that have been acquired from the parents but to the action of hormonal factors upon these genes. The genes for baldness derived from father and mother are identical, but their action—their expression—differs very markedly in each sex. Obviously the fact that the same genes express themselves differently in male and female is another way of saying that the same gene acts differently in each sex, owing to some difference in the constitution of the sexes. The nature of this constitutional difference is suggested when, as not infrequently happens, a woman develops a tumor of the adrenal gland or within an ovary. Often such a tumor will produce abnormal stimulation of certain cells, causing them to secrete large amounts of male sex hormones (androgens). These androgens produce a masculinizing effect upon the woman, causing, among other things, reversal of secondary sexual characteristics, including those of larynx, voice, and mental attitudes, and the development of a mustache and beard. Some women do not, under such conditions, develop thinning of the head hair, which suggests that they probably do not carry any baldness genes. When the tumor is surgically removed, the masculinized woman rapidly returns to her female traits, though some residua, such as the masculine voice and Adam's apple, may remain, but the tendency is for the mustache and beard unsustained by male sex hormones to return to normal, and this is also likely to be the case where head hair has thinned. Hence it is clear that the presence of male hormones has something to do with baldness. Investigation of this possible association in the male has, indeed, revealed that there is a very significant association between these two phenomena.

Experiment has shown that no amount of male sex hormones will produce baldness in males who have not inherited the necessary genes. It has long been known that men who have been castrated in youth, eunuchs, never develop baldness, often in spite of a family history of baldness. On the other hand, some males who for some reason have suffered a more or less extreme reduction in the quantity of these hormones circulating in their bodies will, when male sex hormones are administered to them, undergo a loss of hair. In the actual cases to which I am referring it was noticeable that the men came from families with a tendency to hereditary baldness. It was also found that in those men who came from families with no hereditary tendency to baldness, the administration of male sex hormones had no effect whatever upon the growth of their head hair—baldness could not be produced in them. Head hair has occasionally regrown in adult men following castration.

The sex-limited effect of baldness is, therefore, clear. The baldness gene must be present if baldness is to occur, but baldness will occur very much more frequently in the male than in the female—even in the attenuated form that it may take in her—principally because the male's sex hormone enables the gene to express itself more completely. Or put in another way, the male's sex hormone is more devastating to the hair follicle in the presence of the baldness gene than is the female sex hormone in relation to the baldness gene.

The expressive or dominance relations of the genes for scalp-hair growth seem, then, to be sex-controlled, or sex-limited. It is not that baldness is a dominant trait in the male and recessive in the female. If this were so, then in the case of two recessive baldness genes in the female, baldness would result, but we know that it does not in fact do so. The genes involved, we have already said, are the same in each sex; it is simply that their expression is differentially limited by the factor of sex. The hereditary mechanism is quite simple. If big *B* is the gene for baldness and little *b* the gene for nonbaldness, then the following genotypes and phenotypes appear:

| Genotype | Males | Females |
|---|---|---|
| BB | Bald | Thinning of hair or partial baldness |
| Bb | Bald | Nonbald |
| bb | Nonbald | Nonbald |

In other words a *B*aldness gene derived from one parent and a *B*aldness gene derived from the other parent will express themselves in double dose in the male, but will be limited in their expression in the female to the extent only of thinning or partial baldness. When the *B*aldness gene has been derived from one parent and the *b*, nonbaldness, gene from the other parent, the *B*aldness gene will express itself in baldness in the male, but the *B*aldness gene associated with the nonbald *b* allele in the female will remain unexpressed, and the female will therefore be nonbald. A nonbald gene *b* derived from one parent and a nonbald gene *b* derived from the other parent will always yield nonbaldness in each of the sexes. Figure 18 shows an actual pedigree of baldness that demonstrates the effects of sex limitation.

Baldness cannot be said to be entirely genetically determined because both the thickness of growth of head hair and the hairline are markedly modified by gonadal and adrenal hormones. Male baldness follows a genetic pattern both in time and in intensity, and is tempered by the individual hormonal milieu.

Since almost all cases of baldness are genetically and hormonally conditioned, and there is at present no known means of influencing this genetically based and hormonally influenced sex-limited trait, all claims promising to alleviate or "cure" the condition must be regarded as fraudulent. This does not mean that at some time in the future a way may not be found for controlling baldness. It probably will. But at the present time no such way exists.

Sulzberger and his colleagues have recently stated that there appears to have been a sharp increase in the incidence of balding among European and American women. The cause is unknown. Possibly one of the conditions entering into the cause—if not the sufficient cause—is that women are less inclined to be ashamed of acknowledging the condition than they were in earlier years, and hence appear more frequently in the dermatologist's consulting room.

Sexual-hair growth is largely under the influence of androgenic (maleness-influencing) hormones in both sexes. Failure of such sexual-hair growth is usually correctable by the administration of androgens. When the treatment is stopped, the sexual hair usually disappears. Reinstitution of therapy results in regrowth of the hair.

The difference between sex influence and sex limitation is that the same trait when it is *sex-influenced* (the genes being carried in the autosomes) can be equally frequently expressed in both sexes, but in one sex the trait will appear to be transmitted as a dominant, whereas in the other it will act as if its expression were due to a recessive. In *sex-limited inheritance* (genes carried on both autosomes and sex chromosomes) the trait is fully expressed in only one sex. It is not the genes that condition the expression of the trait but the sex of the individual.

Gout, for example, is an *incompletely sex-limited* trait because although the gene is dominant and is carried on the autosomes, it is expressed in about 95 percent of males and only in 5 percent of females. Gout does not occur equally frequently in both sexes, but is largely limited to males, occurring in females not only in a much smaller proportion of cases but also in a less exacerbating form. The female constitution exercises a strikingly modifying effect not only upon the expression of certain genes but even upon the damage that certain infective organisms will cause in her as compared with the male. Fo rexample, syphilis follows so benign a course in the female as compared with the male that it has been said that it manifests itself in the female "almost as if she were of another species." This is true of many other conditions.

There are disorders inherited as recessives and that should therefore be equally expressed in male and female that are, however, more frequently expressed in one sex than in the other. Such conditions more frequently expressed in the male than in the female are albinism (sometimes dominant), alkaptonuria (black-urine disorder), retinitis pigmentosa (progressive inflammation of the retina), and amaurotic idiocy.

Similarly, some conditions that are due to recessive genes are expressed more frequently in the female, such as diabetes (sometimes dominant), manic-depressive psychosis, Sydenham's chorea, and Niemann-Pick disease.

The secondary sex characters, such as mustache and beard, body-fat distribution, form of body, breast development, texture of hair, distribution and growth of body hair, as well as many differences in response of the nervous system, depend upon sex-limited factors. The degree of secondary sexual development in the sexes is obviously differently limited by the hormones of the glands of internal secretion (endocrine glands)—the factors of sex are in themselves genetically limited—but this limitation is to some extent under the control of the hormones.

This is clearly seen even in prenatal development in the case of the *free martin,* which is a sterile, unsexed female with a tendency toward the development of male characteristics. Similar cases occur in human beings, but not quite in the same manner in which they sometimes occur in cattle when twin calves of opposite sex are born. In such cases there is a connection between the blood vessels on the outer (chorionic) of the two membranes enveloping the fetuses, and since the male hormones develop earlier than do those of the female, they get through to the female fetus, inhibiting the normal development of the female organs and serving to masculinize them, so that the ovaries tend to form testicular tissue and are incapable of producing ova. In the human

## 198 HUMAN HEREDITY

species it is the action of freely circulating excessive amounts of androgens that produces masculinization of the female embryo, usually by the sixth week of uterine development.

## Modes of Inheritance

It is important to note that the modes of inheritance of the same trait may differ in different families. In one family a trait may be due to a sex-linked dominant gene; in another family the same trait may be due to an autosomal recessive gene; in still another family to an autosomal dominant gene. Usually this means that the chain of events leading to phenotype A can be switched at different steps to give B, the steps corresponding to different genes that may be anywhere on the chromosomes.

Mutant genes and linkage between such genes occur in autosomes, but their study is considerably more difficult than sex-linkage. Several possible linkages have, however, been described. Examples are the linkages between ear flare and finger length, finger length and eye color, ear size and ability to taste a substance known as phenylthiocarbamide (or PTC for short), hair whorl and cross-eyes, skin color and hair color, eye color and tongue curling, hair shade and hair color, ability to taste mercaptobenzoselenazole (MBS) and ear size, ability to taste MBS and tongue curling, sickle-shaped red blood corpuscles and the MN blood type, Lutheran blood group and the secretor gene, and several others.

A gene that affects two or more parts or characters of the body that have no obvious relationship is termed *pleiotropic* (Greek *pleion,* more; *trope,* a turning). Most genes are pleiotropic. In man this pleiotropism with respect to the pigmentary system, for example, is apparently exhibited in hair, eye, and skin color, dark-skinned people usually having dark hair and eyes, fair-haired people usually having light-colored skin and eyes. One seldom sees black-haired people with blue eyes, although this combination occurs among the Irish, thus indicating that separate genes are at work in influencing the expression of these two traits.

## Penetrance, Expressivity, and Viability of Genes

Under the conditions in which genes develop they show varying degrees of activity. They are influenced, as we have seen, by internal and external environmental conditions, and one nonallelic gene may act upon another to inhibit its action (*epistasis*), or may itself be inhibited by a nonallelic gene (*hypostasis*).

*Penetrance.* When a gene regularly produces the same effect in all the individuals exhibiting it, it is said to have *complete penetrance*. When the effect is not produced in some individuals, even though they carry the gene in either homozygous or heterozygous state, the gene is said to have *reduced penetrance*. Penetrance relates therefore to the *either-or* state of the gene—either it is expressed in the form of a definite condition or it is not. Dominant genes with low penetrance may be mistaken for recessives. The blood groups are an example of genes that show complete penetrance. Every individual inheriting a gene for a particular blood-group trait develops that trait. On the other hand, it has been shown that such a disease as diabetes insipidus (the rarer form of diabetes, not to be confused with the common form, diabetes mellitus) is probably due to an incompletely dominant gene that manifests itself in about 10 percent of those carrying it.

*Expressivity.* When the manifestations of a trait vary from individual to individual, the gene is said to have variable expressivity. When the manifestation of a trait is constant, the gene is said to have constant expressivity. The dominant gene for allergy shows variable expressivity and may take such forms as asthma, eczema, hay fever, angioneurotic edema (sudden appearance of urticarial, or hivelike, swellings on face or upper extremities).

*Viability.* Genes carried in the homozygous state that shorten the life of the individual are known as lethal genes. Lethal genes are incompatible with life at various stages during the development of the individual. This means that such genes may exercise their effect at fertilization and at almost any time thereafter. Most of these lethal genes are recessive. In Table XIV are listed approximately twenty conditions that have been reported as caused by lethal genes.

For thousands of years human beings have desired the power of determining the sex of their children. This is not the place to consider the strange practices that have been resorted to and the fancy theories that even the most serious persons have entertained on this subject. Let it be sufficient to say of all of them that they were very wide of the mark.

Next to the power to be able to determine sex, men have always wanted to be able to foretell the sex of their children after they have caused their wives to conceive. Since 1950, this has for the first time in the history of man become a scientific reality, and it is since 1955 that the determination of sex has become something more than a scientific possibility.

In 1949 Dr. Murray L. Barr, of the University of Western Ontario, and his coworkers, discovered that from the eighteenth day of development any cell of an organism destined to be a female could be distinguished from any cell of an organism destined to be a male. On the inside of the nuclear membrane of almost every female cell there is situated a structure called the *chromatin body,* which takes a deep stain. In the cells of the male this body is not present. Hence, in order to be able to predict the sex of any human being immediately after conception, all one has to do is to obtain some of its cells, stain them, and examine them under a microscope. If they contain nuclear sex chromatin bodies, then the developing embryo is going to be a female; if such bodies are absent, the organism is almost certainly going to be a male. It is as simple as that. (See Plate 19.)

How does one obtain the cells of a living embryo or fetus *in utero?* At present this is done by a procedure called transabdominal amniocentesis, which means that the cells are obtained from the amniotic fluid, into which they have dropped from the organism surrounded by the fluid, by the passage of a needle through the mother's abdomen directly into the amniotic sac. This procedure has been practiced for other purposes, such as withdrawing excessive accumulations of amniotic fluid where its pressure has caused painful symptoms (polyhydramnios). This is not a procedure to be recommended, for it has sometimes resulted in ill consequences; some more satisfactory means will have to be found of obtaining the cells of the developing uterine organism. This will undoubtedly be done, but in any case we are now in a position to predict the sex of the human being long before birth.

The discovery of the method of determining sex by means of the chromatin body enables us to do for the first time, with virtually complete accuracy, many more things than the mere prediction of sex. For example, it now becomes possible for scientists to study the development of an organism whose sex is known even before there is the appearance in the embryo of the least external evidence of sex, which in man occurs at about the seventh week of development. Already great advances have been made in the study of de-

## Table XIV
### Conditions Reported as Due to Lethal Genes

#### Recessive Lethals

Acute idiopathic xanthomatosis (Niemann-Pick disease). Great enlargement of spleen and liver with discoloration of skin

Amaurotic idiocy
  (a) Infantile type    Impairment of vision leading to total blindness,
  (b) Juvenile type            degeneration of nervous system, and idiocy

Degeneration of the cerebral white matter
  (a) Acute infantile type
  (b) Subacute juvenile type
  (c) Convulsive type

| | |
|---|---|
| Epidermolysis bullosa | A skin disease in which blisters form on the slightest pressure |
| Gargoylism | Multiple-growth derangement, gargoylelike face |
| Glioma retinae | Tumor of the retina |
| Ichthyosis fetalis | Scaling of the skin |
| Infantile muscular atrophy | Wasting of muscles with paralysis |
| Microphthalmia of the sex-linked type | Abnormally small eyes |
| Pseudohypertrophic muscular dystrophy | Muscular enlargement and paralysis |

#### Semidominant Lethals

| | |
|---|---|
| Minor brachydactyly | Shortness of fingers |
| Pelger's nuclear anomaly | Unsegmented leucocytes |
| Sebaceous cysts | Cystic tumors of sebum-secreting glands |
| Spina bifida | Congenital cleft of vertebral column |
| Telangiectasia | Dilatation of capillaries, particularly serious nose bleeding |

Adapted from L. H. Snyder, "The Mutant Gene in Man."

velopmental anatomy as a result of this discovery. The method of chromatin-sexing has already been used in the prevention of the birth of children with such sex-linked recessive disorders as hemophilia. Furthermore, in many cases of true hermaphroditism, in which the individual exhibits the organs of both sexes, it is for the first time possible to determine with a high degree of probability the individual's genetic sex by taking some of the cells from his skin or the inside of his cheeks or any other part of his body and finding whether or not the nuclear chromatin body is present. This has already been done in many cases, and constitutes a most important advance in what was hitherto a very difficult area. The technique has been used in medicolegal and forensic cases in the identification of fragmentary remains of bodies.

What is the chromatin body? Investigation indicates that it

## 202  HUMAN HEREDITY

is one of a pair of X chromosomes that curls up on itself and becomes condensed, so that under the microscope it appears as a dense, dark spot. The other member of the pair of X chromosomes in the female behaves like an autosome, and most of its genes have nothing to do with sex. What those genes are concerned with is the production of enzymes from the multiplying cells. For anything from a few days to a few weeks after fertilization, or embryogenesis, the enzyme genes of both X chromosomes pass on their chemical patterns to some of the body's multiplying cells. After this, one of the X chromosomes becomes deactivated and assumes the form of a chromatin body. The big remaining X chromosome continues to retain its form and to act like an autosome.

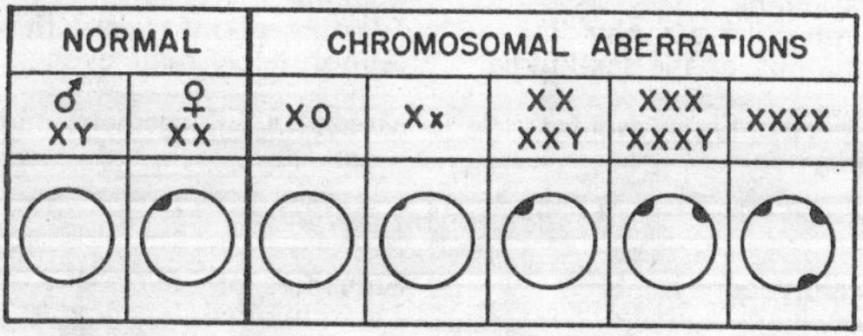

| NORMAL | | CHROMOSOMAL ABERRATIONS | | | | |
|---|---|---|---|---|---|---|
| ♂ XY | ♀ XX | XO | Xx | XX XXY | XXX XXXY | XXXX |
| ◯ | ◉ | ◯ | ◉ | ◉ | ◉◉ | ◉◉◉ |

FIG. 19. Diagram showing the relation between the number of X chromosomes (n) and the number of sex chromatin bodies in somatic cells. This relation is expressed by the formula $(n - 1)$. The first two columns show the relation in normal cells. In males, with an XY complement, the sex chromatin test is negative. In females, with an XX complement, the test is positive. In columns three-seven are shown the results of the sex chromatin test in various types of chromosomal aberrations. With the XO complex, the test is negative. With the Xx complex, the test is positive, but the sex chromatin body is smaller than normal, indicating deletion of one of the X chromosomes. With complexes containing more than two X chromosomes, the chromatin test always shows one chromatin body less than the number of X chromosomes.

SOURCE: Othmar Solnitzky, "The Human Chromosomes," *The Georgetown Medical Bulletin*, vol. 15, No. 1, August, 1961.

In females, then, enzymes derived from both parental X chromosomes are active in the earliest stages of development. This has been usefully demonstrated by Dr. Ernest Beutler, of the City of Hope Medical Center, Duarte, California. Dr. Beutler found that many women suffering from the inborn error of metabolism due to a deficiency of the enzyme 6-phosphate dehydrogenase (G-6-PD) carried on an

X chromosome, have approximately half the normal concentration of G-6-PD in their blood. These proved to be women who had one affected and one normal parent. Since the red-blood-cell lines are formed soon after conception, and it is the red blood cells that carry the enzyme, the indication was that these women had derived their blood cells from both parents. And this, indeed, is what Dr. Beutler found to be the case—females are genetic mosaics, which is a highly advantageous trait.

The chromatin body has already proved extremely helpful. Its presence or absence tells us whether or not the individual is a genetic female, and this simple test is invaluable in identifying as well as clarifying certain abnormal conditions that have hitherto been somewhat puzzling, such as Klinefelter's syndrome. Klinefelter's syndrome (testicular dysgenesis) is a condition in which "males," although they become masculinized, frequently develop small breasts as in the adolescent female (gynecomastia), and their testicles remain small and produce no spermatozoa, so that such "males" are sterile all their lives. It has been discovered that in all "males" characterized by Klinefelter's syndrome the sex-chromosome number is abnormal. Yet these individuals, except for the adolescent feminine appearance of their breasts, appear in every other way to be male.

It is a principal function of the sex chromosomes to direct the normal differentiation and development of the gonads (testes or ovaries). When for some reason an aberration in the number of the sex chromosomes occurs, there follows a disturbance in the development of the primary and secondary sex characteristics of the organism. This may lead to the phenotypic (the apparent) sex being somewhat confused, as in the case of Klinefelter's syndrome, in which the phenotypic appearance is that of a male and the chromosomal constitution is XXY, the fertilization of an XX egg by a Y sperm having produced a sterile male with a few female traits.

In Turner's syndrome, in which the chromosomal constitution is XO, there is a developmental failure of the ovaries with complete absence of ova, as well as associated physical and, sometimes, psychological abnormalities.

In the case of the so-called superfemale, with the chromosomal constitution XXX, there is an underdevelopment of the breasts, the external genitalia remain infantile, the vagina is small, and the menopause may occur very early.

Studies first published in April, 1959, have conclusively shown that chromatin-positive Klinefelter's syndrome is associated with forty-seven chromosomes, the additional chromosome being an X chromosome, the sex chromosomal constitution of Klinefelter's syndrome being XXY. Evidence is now available that some cases of Klinefelter's syndrome are chromosomal mosaics, some cells bearing the normal forty-six chromosomes and others the abnormal number of forty-seven.

Studies of the mental level of individuals with Klinefelter's syndrome generally reveal some mental retardation. For example, a psychological study of forty-seven such patients carried out by doctors J. Raboch and I. Šípová at the Charles University in Prague revealed that twenty of them were "poor" and twelve "subnormal"; only two had above-average results on the psychological tests. The results support the view, generally confirmed, that not only is the activity of the sex glands deranged in Klinefelter's syndrome but also that the nervous system is affected, as revealed by the mild mental deficiencies in this group of patients.

We have already seen that Down's syndrome is characterized by forty-seven chromosomes, the extra chromosome being an autosome. Down's individuals exhibiting Klinefelter's syndrome have been shown to possess forty-eight chromosomes, one additional autosome and one additional X chromosome.

C. E. Ford and his collaborators first showed in 1959 that there are only forty-five chromosomes present in Turner's syndrome, and that such cases are of sex-chromosomal constitution XO. These findings have since been fully confirmed by other workers.

The physical abnormalities seen in Turner's syndrome are shortness of stature, failure to menstruate, impalpable ovaries, absence of ovulation, undeveloped breasts, small uterus, negative sex chromatin, and a characteristically elevated urinary excretion of gonadotrophin, principally the follicle-stimulating hormone from the pituitary gland, which normally acts upon the ovaries. A reverse direction of the hair whorls is also sometimes present on the back of the neck, and occasionally there is a peculiar webbing of the skin of the sides of the neck, and there may be heart defects, deviation of the extended forearm to the inner side of the limb axis (cubitus valgus), and mental retardation.

The fourteen-year-old English girl exhibiting Turner's syndrome described by Ford et al. is anatomically and psychologically female, but nuclear-chromatin-negative, the latter being a condition usually associated with the male. Although she possesses only half the normal component of sex chromosomes, namely a single X chromosome, there is, however, no reason to assume that this girl is in any sense a male. Such an individual is not an example of sex reversal but of a female with an abnormal genotype, a female, however, who is an incomplete female—a female who would have developed as a completely normal female had she been endowed with another X chromosome, but who would have developed as a male had a Y chromosome been bestowed upon her. Thus, from cases such as this we learn that the hitherto enigmatic Y chromosome is the carrier of masculinizing factors.

In some cases of Turner's syndrome there is one normal and one abnormal X chromosome. In some of these cases part of one X chromosome is missing; in others, an "isochromosome" (a chromosome with two identical arms) is encountered that is formed of two long arms with no short arm. Mosaics also occur. Mosaicism refers to the constitution of the body by two or more chromosomally different lines of cells. The proportions of these cells usually vary in different tissues, though they intermingle throughout the body. In Turner's syndrome one line of cells is always XO; others may be XX, XY, XYY, or XXX.

The occurrence of mosaics tells us something about the origin of chromosomal anomalies. Anomalies affecting all the body cells must have originated either in egg or in sperm or, less often, in the fertilized ovum. For example, a mosaic individual with equal numbers of XO and XXX cells might arise if at first division of the fertilized ovum an X chromosome got into the wrong cell. Errors of this sort may be doubled or trebled during cell division, so that, for example, the loss of a Y chromosome during an early division would result in an XX/XXY mosaic, from an originally faulty XXY zygote—a partial correction of the original mistake. It is possible that similar autocorrections of chromosomal anomalies may result in the reestablishment of normal chromosome arrangements at very early stages of development.

Another disorder that has been shown to be associated with the loss of a sex chromosome is the Bonnevie-Ullrich syndrome, one of the characteristics of which is gonadal dysgenesis. It is unlikely that the other multiple anomalies of the body characteristic of this disorder are due to the chromosome loss alone.

In October, 1959, Jacobs and Keay reported a case of the

Bonnevie-Ullrich syndrome in a five-year-old girl with negative Barr body and an XO sex-chromosomal constitution. But this was probably a case of Turner's syndrome.

True reversal of sex occurs in the syndrome of testicular feminization. Here apparent females are, in fact, sex-reversed males. This syndrome is not due to either the loss or the addition of chromosomes, for the individuals involved, as Jacobs and her collaborators have shown, possess the normal number of diploid cells, are chromatin negative, and are of XY chromosomal sex. The condition is hereditary in certain families, being transmitted through the maternal line either as a sex-linked recessive or as a sex-linked dominant.

In testicular feminization the individual presents the appearance of a female: the external genitalia are female, and so is the external appearance of the body as a whole, except that pubic and axillary hair is likely to be either scanty or absent. There is primary amenorrhea, that is, complete absence of menstruation, and the vagina is incompletely developed. Testes can be located either in the abdomen or in the inguinal canals or the labia majora. An epididymus and vas deferens are commonly present on both sides, and there may be a rudimentary uterus and fallopian tubes.

The first case of a human XXX, or "triplo-X," female was reported in September, 1959, by Patricia Jacobs and her collaborators. This was a thirty-five-year-old woman, 5 feet 9 inches in height and weighing about 137 pounds. She was the youngest of three girls. The mother's age at conception was forty-one and the father's probably forty. Pregnancy was full term and normal. This female ceased menstruating with accompanying menopausal symptoms at age nineteen. Examination of her ovaries showed that follicle formation had been deficient, the ovaries themselves presenting a postmenopausal appearance.

The external appearance presented by this triplo-X female was in every way perfectly normal except that her breasts were slightly underdeveloped, but not more so than in many normal women. The external genitalia were infantile in appearance, and the vagina was small. Urinary gonadotrophin secretion was persistently high.

Study of bone marrow and skin cells showed a chromosomal constitution of forty-four autosomes and three X chromosomes, a total of forty-seven chromosomes. The mother showed the normal sex-chromosome constitution, XX. When somatic cells were studied for the chromatin body, 57 percent contained a single chromatin body and 14 percent two chromatin bodies. The proportion of cells containing chromatin bodies (71 percent) was greater than that found in buccal cells from normal women (48 percent), and more of the latter showed two chromatin bodies.

This triplo-X female probably never ovulated. The development of the breasts and genitalia was arrested at an early stage of adolescence, and the ovaries were prematurely aged. In Turner's syndrome, where the female has only one X chromosome, her ovaries fail to produce ova, and if a Y chromosome is added to an XX, the individual is masculinized even though the testes are incapable of producing sperm.

In June, 1960, Jacobs and her colleagues described another triplo-X female. This female, a high-grade mental defective, began to menstruate at nine years of age, and at the time of the examination, at the age of twenty-one, was continuing to do so regularly. No details of family history were available. There were no physical abnormalities. The possibility is suggested by this case that XXX females are sometimes fertile, in which case they would produce eggs with X and XX chromosomes, and their progeny would therefore include individuals with an XXX or an XXY constitution.

Another case described by the same investigators was a mosaic, that is, a female many of whose body cells were XXX while others were XO. This patient was born in 1908, her mother being thirty-two and her father thirty years of age at her conception. This woman's intelligence was low normal. Trunk, pelvis, breasts, upper extremities, and head hairline were of masculine type. The external genitalia were undeveloped, except for the clitoris. The vagina was absent and no uterus could be palpated. Apart from some deeply pigmented nevi on face and neck, there were no evidences of Turner's syndrome.

Observations such as these help us to understand more fully the roles played by the X and Y chromosomes in development. Recently Tanner and his associates have shown that the normal differences in the rate of maturation of males and females are probably due to different genes situated on the X and Y chromosomes. It has long been known that at all ages girls are physically more mature than boys. Tanner and his coworkers have shown that XXY individuals have a rate of skeletal maturation corresponding to that of XY individuals, and XO individuals a rate corresponding to that of XX individuals. As might have been expected, the

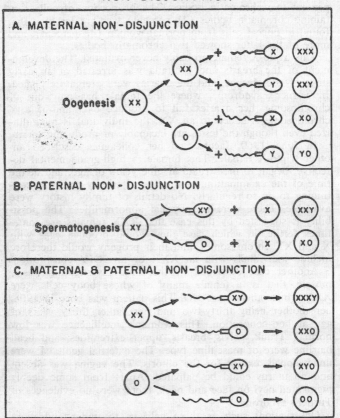

FIG. 20. Diagrammatic representation of the various types of chromosomal complexes resulting from maternal, paternal, and both maternal and paternal nondisjunction. Of these complexes, the YO and OO are nonviable. Practically all the other types of chromosomal complexes have been identified in human cells.

SOURCE: Othmar Solnitzky, "The Human Chromosomes," *The Georgetown Medical Bulletin*, vol. 15, No. 1, August, 1961.

presence of a Y chromosome is associated with a retardation of maturation in the presence of either one or two X chromosomes, and without a Y neither one nor two X chromosomes produce this effect. It appears evident, then, that the genes on the Y chromosome are responsible for the normal sex differences in the rate of development.

The abnormal chromosomal constitutions discussed above are best explained as due to failure of paired chromosomes to separate at mitosis or during the formation of the gametes at meiosis, a failure termed *nondisjunction*. Instead of the members of a pair of chromosomes passing to opposite sides of the spindle and into separate gametes, for some reason they remain together and pass as a pair into the same gamete, with the opposite gamete receiving no chromosomes at all. Thus, Turner's syndrome could result from nondisjunction in the formation of the egg, causing a loss of the X chromosome, or in the fertilizing sperm, causing a loss of either its X or Y chromosome.

It will be readily seen that nondisjunction in the sex chromosomes of the egg could lead, on fertilization, to individuals with either XXX, YYY, XO, or YO chromosomal constitution; the last is probably lethal at a very early stage shortly after fertilization. Nondisjunction in the sperm could lead to individuals with either XXY or XO chromosomal constitution. To date, cases of XXXX, XXYY, XXXY, XXXXY, and many forms of mosaics have been identified and described. The greater the number of X chromosomes beyond the normal number, the greater is the tendency to mental retardation. In the presence of a Y chromosome, the greater the number of X chromosomes, the greater the number of body anomalies.

In December, 1958, Holub and his collaborators reported a pair of seventeen-year-old identical twins with Klinefelter's syndrome. Since twinning occurs more often in older mothers and nondisjunction probably occurs in aging cells, this is an especially interesting case in that it tends to support the idea that chromosomal aberrations are frequently the result of an aging effect in the mechanism of cell division.

Mothers of Klinefelter individuals are older than the average, and in all known cases the syndrome has been proved to have originated by nondisjunction in the mother. Maternal nondisjunction is also the origin of Turner's syndrome, but in this syndrome maternal age is not increased.

Jacobs and her coworkers in June, 1960, reported the case of a thirty-seven-year-old female of low-normal intelligence who had never menstruated (primary amenorrhea), whose mother was twenty-two and father twenty-seven years of age at the time of her conception. The firstborn of four siblings, her two sisters have menstruated regularly, and her married brother has one child. The patient has scanty pubic and axillary hair, no breast development, infantile external genitalia, a rudimentary uterus, and only one scarcely identifiable gonad containing a few primitive follicles. Only 7 percent of cells contained a chromatin body, and the size of this was smaller than usual. There were forty-six chromosomes, but one of the pair of X chromosomes was only 60 percent of the size of the other X chromosome. Apparently the reduction in the length of the smaller chromosome was brought about by part of it having been deleted. This deletion of one X chromosome was sufficient to result in inadequate development of the primary and secondary sexual characters.

It is clear, then, that two normal sex chromosomes must always be present if normal development is to occur, at least one of which must always be an X chromosome. In those cases in which a single sex chromosome is present it must always be an X if the organism is to survive. If a Y chromosome is the only sex chromosome present, that is, if a Y sperm fertilizes an ovum empty of an X chromosome, such an ovum will fail to survive. The suggestion, therefore, is that the Y chromosome carries chiefly masculinizing factors, and very little, if anything, else.

There is considerable evidence that abnormal metabolism of maternal or placental hormones during pregnancy, when the gonads of the embryo are still undifferentiated, may produce sex reversal. Excessive amounts of progesterone or testosterone may produce masculinization of the female embryo so that it develops as a male or as a hermaphrodite or intersexual pseudohermaphrodite. In a recent study of 1,911 baby boys in Winnipeg, Canada, Moore discovered that the sex chromatin was that of a genetic female in five cases. This is a frequency of 0.26 percent, which, in effect, means that approximately 1 out of ever 400 phenotypic males may be a genotypic female! It is probable that in some cases male to female sex reversal is produced in a similar manner. Beregmann at the Gynecologic Clinic at the University of Bern found 4 (0.21 percent) out of 1,890 newborn boys to be chromatin-positive, and 1 out of 1,838 girls to be chromatin-negative. Thus, it may be that 1 out of every 2,000 females is actually a genotypic male, but before accepting such a figure we shall require further studies of sex-chromatin determinations in newborn infants.

Intersexuality has been experimentally produced in the female progeny of pregnant rats who were injected with male sex hormones, androgens (A. R. Greene et al.). Similarly, intersexuality appears to have been produced in some human genetic females by the administration to the mother during the first trimester of pregnancy of male sex hormones such as testosterone (Gold and Michael). Female to male sex reversal is often associated with mental defect. In a study of 663 prepubertal mentally handicapped English boys Ferguson-Smith found that eight of them, or 1.2 percent, were chromatin-positive females. Klinefelter's syndrome has recently been described in a set of seventeen-year-old identical twins. Holub, Grumbach, and Jailer consider that a genetic predisposition to develop this disorder probably exists.

In 1932 Dr. V. N. Schröder, of the U.S.S.R., was able by means of a simple electrical method to separate X-bearing from Y-bearing spermatozoa. In this method sperm are placed in a solution, the hydrogen-ion concentration of which can be controlled, contained in an electrophoretic apparatus. This apparatus has a positive electrode at one end and a negative electrode at the other. When an electric current is passed through the apparatus, some sperm will migrate toward the positive pole (anode) and others toward the negative pole (cathode). It was found that the majority of sperm that migrate toward the positive pole are X-bearing and those that migrate toward the negative pole are Y-bearing. Dr. Schröder was able to predict in 80 percent of the cases the sex of rabbits whose mothers had been artifically fertilized with such "positive" and "negative" sperm. In America, using similar methods, Gordon has been able to predict the sex of rabbits in 67.7 percent of cases. Sherry Lewin in England has reported similar behavior in human sperm. In May, 1960, Dr. L. B. Shettles, of Columbia University, reported his findings indicating that sperm with elongated heads are potential producers of females and that round-headed ones carry the genetic instructions for the production of males. But this work has been seriously questioned. There can, however, be little doubt that it will not be long before artificial determination of the sex of human beings will become a practical possibility.

Dairy farmers may soon be able to produce all the heifer calves they desire and raise only enough bulls for the purposes of reproduction.

212  HUMAN HEREDITY

In the case of man, what we need to learn is not so much how to determine sex as how to control the effects of sexual activity, more particularly, both the quality and the quantity of human populations.

## The Control of Population

In the year 1800 the total population of the world is estimated to have been approximately 906 million; in 1850 it was 1,171 million; in 1900 it was 1,608 million; in 1930 it was 1,987 million. Today it is approximately 3 billion. By the year 2000 it will reach 6,250 million. Every second of the day three babies are born in the world, 180 every minute, almost 11,000 every hour, approximately 260,000 every day, well over 1.5 million every week, and approximately 95 million every year. If we go on at this rate, then by the year 2100 we may expect to have a world population of 10 billion. This would give us a population of between 175 and 200 persons per square mile covering the whole face of the earth including the Antarctic Continent and the Sahara Desert. And after that, what then? Do we put on our space suits and migrate to other habitable planets? An idea the impracticability and absurdity of which have been effectively demonstrated by Dr. Garrett Hardin. This deluge of population has already seriously jeopardized the existence of the human species, for it has brought in its train poverty, famine, economic convulsions leading to revolution and war, the desire of nations for more room—*Lebensraum*—and finally the hydrogen bomb as a means by which powerful nations can implement their expansionist intentions or protect themselves against such intentions by others, fancied or real.

Not only this—certain religions actually encourage their members to have many children regardless of the consequences to the families and the societies involved, for by this means of increasing their numbers such organized religions increase their political and social power. But it is not numbers that human beings should aim for but quality. Beyond a certain point, number itself greatly increases the difficulty of a problem. When the size of cities and towns increases out of all proportion to society's capacity to handle the attendant consequences, the difficulties of dealing in a humane and intelligent manner with these human problems may become insurmountable.

We now have not only the hydrogen bomb to contend with but also the "population bomb."

This is not the place to consider whether or not the population will outrun the food supply—by technological means we may be able to meet the needs of even 10 billion people, but that is not the real point. There are numerous reasons other than mere subsistence that should give us pause, for man does not live by bread alone. Human life is sacred and must be preserved. Every living human being should enjoy as his birthright the fullest opportunities for the development of his potentialities; instead, by the indiscriminate breeding that characterizes the human species almost everywhere on the face of the earth today, we effectively deprive most human beings of their birthright—their sacred right to development—and we jeopardize the chances of all save a fortunate few to achieve that development.

We should not be so much concerned about the increase in the number of mentally inadequate persons, the otherwise deficient, and the hereditarily disordered as we should be about the total increase in the number of human beings. We have nothing to fear from the inadequate—it is the increase, beyond what is good for the survival of the species as a whole, of the humanely undeveloped that we have most to fear. Certainly we should always be concerned with the quality of human beings, and the best way to do so is to see to it that the sum of human beings is kept within controllable dimensions. First things should come first.

The birth rate has not risen appreciably in the last fifty years. What has served to produce the tremendous increase in population is the lowering of the death rate. In India the excess of births over deaths has trebled within the last thirty years. Medical and sanitary advances in the control of disease have been chiefly responsible. But as William Vogt has asked, "Is there any kindness in keeping people from dying of malaria so that they can die more slowly of starvation?" Professor Conway Zirkle has asked, "Have we who believe in birth control a moral right to send food to peoples who don't?" These are very real questions, but however they are answered, it is certain that the conditions that caused these questions to be asked at all should never have been allowed to arise. But they *have* arisen, and they are very much with us. What are we going to do about them? In other words, *what are we going to do about the population problem?* "The whole human race is rumbling on to destruction," declares Lord Boyd-Orr.

The present methods of limiting fertility, the Malthusian quartet—war, famine, disease, and voluntary control—are

## 214 HUMAN HEREDITY

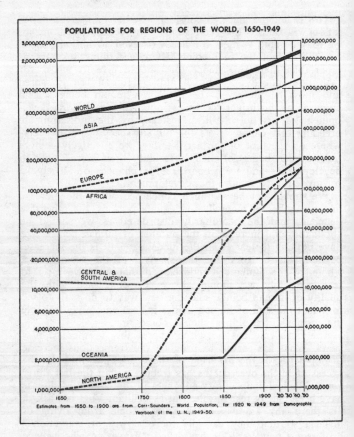

FIG. 21. World Population. (Courtesy of U.S. Public Health Service.)

not good enough. In improved methods of voluntary control, however, lies the only hope of our being able to do something about limiting the explosive rate of uncontrolled human multiplication. If we had only a small fraction of the scientific personnel working on this problem who are now devoting their energies to the development of more destructive armaments, we should make great progress in a very short amount of time. What are needed are oral contraceptive pills that will work directly on the ovary or the uterus or the fallopian tubes, that is to say, which will either temporarily stop ovulation or make implantation of the fertilized ovum impossible in the uterus or close off the fallopian tubes from the uterus. Research along these lines is proceeding, but there are only a comparative handful of scientists at work on these problems. Already pills have been tried that inhibit ovulation. These have proved highly effective in the control of conception and are now available at a minimum cost. Such pills undoubtedly are the first genuinely significant contribution to the control of conception throughout the world, and thus constitute the first major step to be realized in the control of population. The first step toward the control of the quality of population is to control its quantity.

## Telegenesis, or Long-Term Artificial Insemination

Men have often thought how wonderful it would be if the sperm of great men could be preserved and then used to inseminate selected women of choice heredity and in this way increase the chances of producing more great men and women like their progenitors. Thus far, this dream has been realized in the pages of science fiction, among the most notable nightmarish realizations of which is contained in Aldous Huxley's *Brave New World*, with its Central Hatcheries and Conditioning Centers and babies who are decanted from glass bottles. Huxley's book was published in 1932. We still have a little way to go before we can decant human babies from bottles, but it can be done with some lower animals, and there can be little doubt that it will be achieved with human beings. Meanwhile, however, the dream of being able to inseminate women with the spermatozoa of men long dead is in the process of realization. Experiments show that the spermatozoa of bulls may now be preserved for several years at the temperature of dry ice, $-79°$ C., in a special glycerol-containing medium and thawed out for use in inseminating cows. At the present time there are thousands of calves who have already been born from such artificial inseminations—whose fathers, in many cases, have been dead for some time before they became fathers!

Interestingly enough, human spermatozoa are much better able to withstand low temperatures than are the spermatozoa of other mammals. It has been possible to preserve human spermatozoa by slow freezing to $-79°$ C. in semen and glycerol, and by subsequent thawing for most of the spermatozoa to recover their motility without the loss of more than one third. Doctors Bunge and Sherman, of Iowa City, have carried this work to its logical conclusion and by means of spermatozoa preserved in this manner succeeded in inseminating three women, who became pregnant as a result and produced perfectly normal children. From the experience with domestic animals there is not the least reason to think that the spermatozoa exposed to these low temperatures for long periods of time have in any way suffered from the experience.

We see, then, that the long-term preservation of human spermatozoa in a thoroughly efficient condition is now a practical possibility, and that long-range posthumous paternity is now probably simply a matter of improving the means of preserving spermatozoa. Nor can there be much doubt that when our scientists have adequately addressed themselves to the task of preserving whole ovaries with their thousands of ova, they will succeed in doing so, and that we can therefore look forward to long-term posthumous maternity as well! The control of human heredity is increasingly being placed within man's power. This is a subject we shall deal with in Chapter 17. Meanwhile, let us proceed to the discussion of heredity and "race."

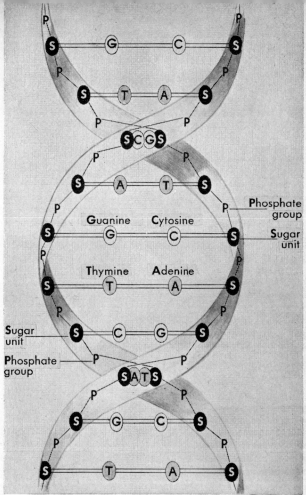

PLATE 1. Model of a Portion of the DNA Molecule, Showing the Two Interlocking Polynucleotide Chains. Only a few of the thousands of turns in the double helix are shown. Each outer helix consists of five carbon-sugar molecules *S* (deoxyribose), alternating with *P* phosphate groups—hence, a pair of sugar-phosphate (deoxyribosephosphate) chains wound in a double helix. The helical polynucleotide chains are united by the closely fitting paired nitrogenous bases AT adenine-thymine and GC guanine-cytosine connected by hydrogen bonds (the double lines). *A*, Adenine; *C*, Cytosine; *G*, Guanine; *P*, Phosphate; *S*, Sugar; *T*, Thymine; = Hydrogen bonds. It has been estimated that if the helical chains (or sugar-phosphate backbones), or tapes, as they are sometimes called, in the human body were placed end to end, they would reach beyond the moon. Each unit phosphate-sugar-base represents a nucleotide; hence, each helix is a polynucleotide chain.

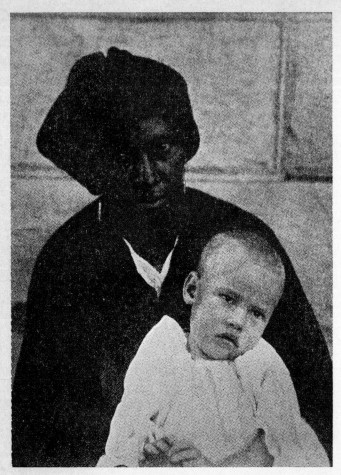

PLATE 2. Segregation in Hybrid Basuto-English Mating. This woman's father was English and her mother Basuto. She married a white man. The two sons both had white skin with brown eyes like the father, and straight brown hair. (After Lotsy.)

PLATE 3. Andalusian Fowl. Crossing of black and splashed-white parents results in blue birds in the first filial generation. The mating of the blue fowl produces ¼ black, ¼ splashed-white, and ½ blue in the second filial generation. In the next generation, the pattern of the mating of the blue fowl is continued; the white and black fowl breed true when mated with birds of the same color. (After Sinnott, Dunn, and Dobzhansky, *Principles of Genetics*.)

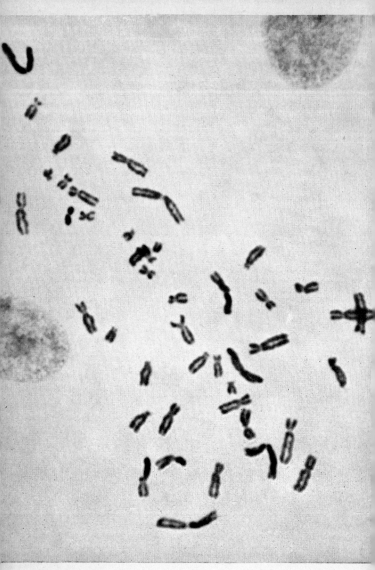

PLATE 4. Normal human chromosomes in diploid number (2n-46) from a newborn male's skin. Enlarged 3,250 times. (Courtesy of Dr. E. H. Y. Chu.)

PLATE 5. Seven Cretins from the Urnatsch Almshouse. Appenzell Canton, Switzerland. The tall man is normal; the height of the woman immediately in front of him is 39 inches. (Courtesy of Dr. J. F. McClendon.)

PLATE 6. Sextuplets. Five boys and a girl born to a West African woman on April 19, 1903, at Accra, Gold Coast. The girl and one boy had a placenta each. The remaining four boys were attached by pairs to two placentae. One boy died two days after birth, four boys died three days after birth, and the girl four days after birth. On her first confinement the mother gave birth to quadruplets, on her second and third to triplets. Thus, in four confinements this woman gave birth to sixteen children. (From H. Cookman.)

PLATE 7. Four sets of English Quadruplets. (Courtesy of *Family Doctor,* London.)

PLATE 8. The Diligenti Quintuplets. (Courtesy of *Family Doctor,* London.)

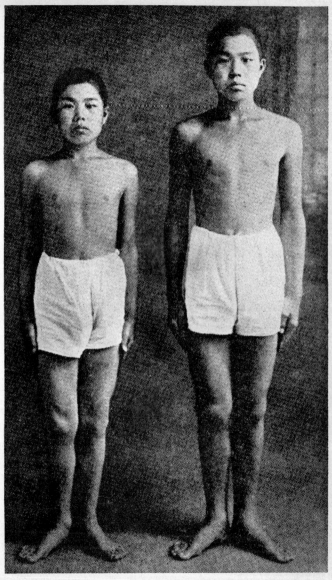

PLATE 9. Difference in Stature of Identical Twins. (From Komai and Fukuoka.)

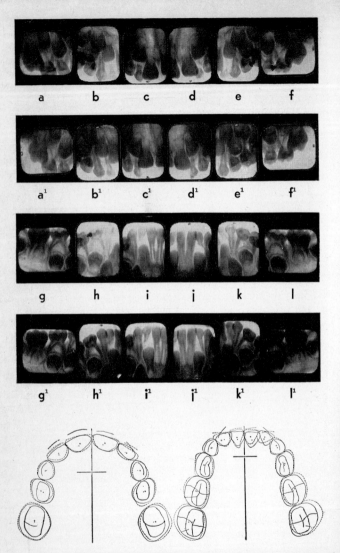

PLATE 10. (1) X rays of the teeth of the upper jaw of identical male twins aged 6 years and 1½ months showing identical pathology of first permanent molar eroding into the second milk molar, f, f¹. (2) X rays of teeth of lower jaws in same identical male twins aged 6 years and 1½ months. (3) A mathematically exact projection of the teeth of same identical male twins aged 7 years and 11 months. Left, upper jaws, right, lower jaws. (From Ashley Montagu.)

PLATE 11. Identical male twins, one born with harelip and cleft palate, and the other perfectly normal. (After von Verschuer.)

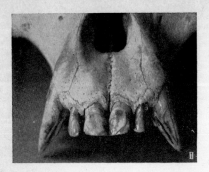

1. Frontal view.

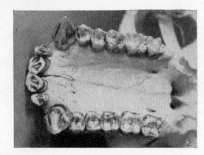

2. Palatine view.

PLATE 12. Premaxillary Diastema in the Gorilla.

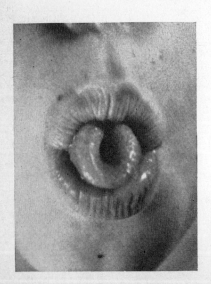

PLATE 13. Ability to roll tongue is inherited as dominant. (Courtesy of A. M. Winchester, from *Genetics*.)

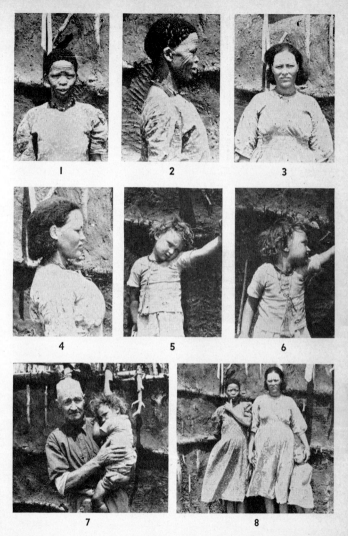

PLATE 14. 1. Bushman mother, from in front, and (2) in profile. 3. Hybrid daughter of white father and Bushman mother, from in front, and (4) in profile. 5. Daughter of white father and hybrid mother shown in 3, granddaughter of woman shown in 1, from in front, and (6) in profile. 7. Husband of hybrid shown in 3 and father of their daughter shown in 5, here with their daughter. 8. Bushman mother, hybrid daughter, and hybrid granddaughter. (Courtesy of Dr. P. V. Tobias.)

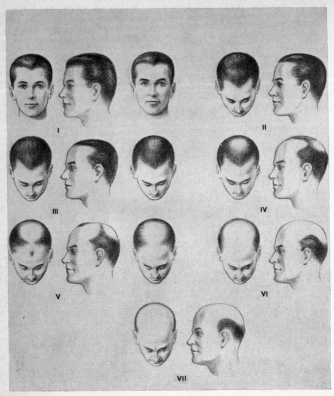

PLATE 15. Patterns of Balding. Types I and II are nonbald. Types III to VII represent a graded progression of common pattern baldness. (Modified after Hamilton.)

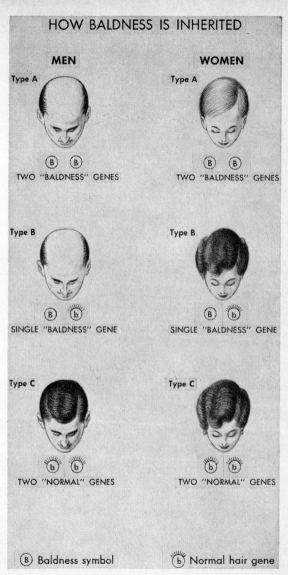

PLATE 16. Baldness is a sex-limited trait due to a dominant gene that is fully expressed in the male, but limited in its expression in the female. (Modified after Scheinfeld, *The New You and Heredity*.)

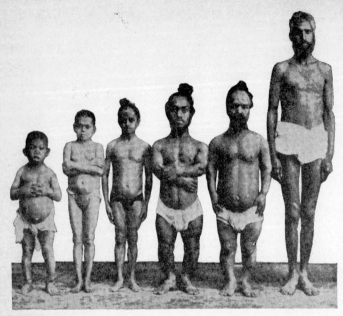

PLATE 17. Three Types of Dwarfs with a Normal Man for Comparison.

|  | Age | Height |
|---|---|---|
| Cretin | 30 | 2 ft. 11½ in. |
| Midget (Ateliotic dwarf) | 20 | 3 " 3 " |
| Midget (Ateliotic dwarf) | 28 | 3 " 4½ " |
| Achondroplastic dwarf | 27 | 3 " 9 " |
| Achondroplastic dwarf | 47 | 4 " 0 " |
| Normal man | — | 5 " 6½ " |

(From Rischbieth and Barrington.)

PLATE 18. Dwarfs and Giants. The first four, three sisters and their brother, are midgets, aged respectively (from left to right) thirty-eight, fifty, thirty-four, and forty-eight. Born in Dresden, Germany, of normal parents, they have normal-sized sisters. The two men in the front row are achondroplastics, aged forty-two and thirty-one respectively. The "giant" is 7 feet 4 inches, and forty-four years of age. Tallness runs in his family. The "fat lady" weighs 540 pounds. Her condition appears to be glandular, for all members of her immediate family are normal. (Courtesy of Amram Scheinfeld, from *The New You and Heredity*.)

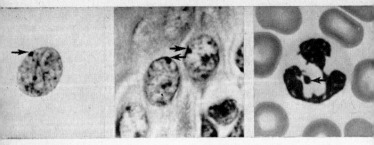

PLATE 19. The Chromatin Body in Female Somatic Cells. 1. Nucleus in an oral smear from a chromosomal female. 2. Cells of the spinous layer in a skin biopsy specimen from a chromosomal female. 3. Neutrophil in a blood film from a chromosomal female. (Courtesy of Dr. Murray L. Barr.)

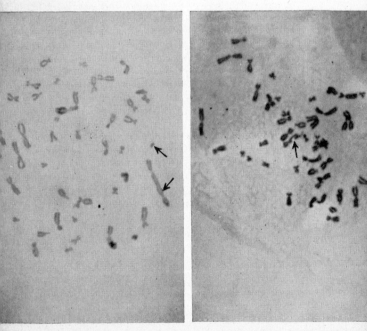

PLATE 20. X-Ray-Induced Chromosome Breakage in Man. Arrows indicate places where breakage of chromatids is evident. Exposure 50r. There were 21.6 percent of cells with such chromosome breaks induced by the X radiation. (Courtesy of Dr. Michael A. Bender, Oak Ridge National Laboratory.)

## 12. HEREDITY AND "RACE"

"RACE" HAS BEEN PUT between quotes at the head of this chapter because it is a questionable term. Most persons who use the term think they know what it means, and what most persons mean by the term is not in agreement with the facts. The idea that "race" means that certain peoples, the so-called races, are characterized by definite inborn physical and mental traits is widespread. It is an understandable enough error. One sees a whole people who look physically unlike any other group of people; they often speak a unique language, or if they don't, they may speak a language in a peculiarly distinctive manner; their customs are different from those of any other people, and they behave in ways that differ from those of other people. Taken together, all these things distinguish this group of people from all other peoples in the world. And this is what is meant by "race." If that is, indeed, what most people mean when they use the word "race," then the word is beyond rescue and it had better be dropped altogether, for its use in the above sense serves only to perpetuate an amalgam of errors that obscures the truth and serves to erect barriers between peoples where such barriers have no scientific basis.

The principal error committed is the assumption that the observed physical differences are innately linked with the observed behavioral differences, and hence the behavior as well as the physical differences are said to be due to "race." The Germans, the French, the English, the Jews, the Muslims, the Negroes, the Chinese, the Japanese, the Italians, and the rest all behave the way they do, it is commonly believed and maintained, because they belong to different "races." Altogether apart from the fact that none of the groups mentioned, even in the strictly biological sense in which the term might be used, constitute "races," it is a common habit to attribute to "race" any form of behavior that is different, and the physical or other characters of the other group provide a convenient peg upon which to hang the differences. Instead of making the effort to understand the meaning of the differences, we too often tend to dismiss them as "inferior," "undesirable," "crude," "repugnant," or at best as "quaint" or "queer." And often enough we settle for what seems to us the inescapable conclusion that we are members of the "superior" group and the others are members, of course, of the "inferior" group. This has always been a comforting decision at which to arrive, for it at once secures one in the belief that one is better than other people and absolves one from the necessity of further inquiry, for are not the facts self-evident? Unfortunately, too often what is evident in facts that are "self-evident" is not the facts but oneself, the self that gets in the way of the facts, distorts them, and prevents us from seeing them as clearly as we otherwise might. Let us, then, consider the facts. A simple and effective technique in dealing with words that contribute to muddled thinking is to drop them altogether. Because the term "race" closes the door on what we are trying to understand, let us use instead the noncommittal term "ethnic group." By the use of such a term we commit ourselves to no particular theory but in effect say that what the differences between ethnic groups may represent is a matter for inquiry rather than for closed judgment, that if we are to obtain a sound view of the facts, we must keep our wits sharpened and our minds open. When we have satisfied ourselves about the facts, we can then decide to use one term or the other. The word "ethnic" is derived from the Greek *ethnos*, meaning a nation, people, tribe, or group.

## Heredity, Ethnic-Group Characters, and the Evolution of Man

We see clearly that some people are black-skinned, that they have broadish noses, smallish ears, black kinky hair, thickish lips, and relatively little body hair. We call such people Negroids. Others have straight black hair, a tendency to projecting upper front teeth, a fold of skin over the inner angle of the eye, and the barest suggestion of a yellowish tinge to the skin. We call such people Mongoloids. Still others have a white skin, a long narrowish nose, all sorts of varieties of hair color and hair form, and a tendency to develop much body hair. Such people we call whites or Caucasoids. And there is a large variety of other groups—no one knows how many—each with its special assemblage of physical traits.

What is the meaning of all these physical differences among human groups? Are they possibly caused by the fact that all the ethnic groups of man are descended from different kinds of monkeys and apes? This question has been seriously considered in the light of the relevant evidence, and it is generally agreed that all the ethnic groups of man must have originated from a single ancestral stock closely resembling, if not identical with, the extinct manlike apes from South Africa known as the Australopithecinae. These were a group of erect-walking apes with the most manlike teeth of any known manlike ape. The Australopithecines were probably the very stock from which man originated. The members of all ethnic groups are far too much alike in their structural and functional characteristics for them to have originated from different apelike forms. And that is precisely the point: the more we study the different ethnic groups of man, the more alike they turn out to be. The likenesses by far outnumber the differences.

If, then, all the ethnic groups of man originated from a single ancestral stock, how did the differences we now observe among them come about? The answer to this question is: probably in much the same way as the differences we observe among different varieties of the same species of any wild animal.

The family of man is a million or more years old. This gives us plenty of time for the development of ethnic-group differences. Human populations were very small in the prehistoric period and can rarely have consisted of more than several hundred individuals. Let us commence with a group of a few hundred individuals. After some time the food supply begins to diminish and a number of families decide to try their lot and *migrate* over the range of mountains. They do so and settle in a region with a good food supply. From this population another group of families decides to splinter off and go seek its fortune beyond the forest. Another group of families decides to cross the water and settle on an island that they can see in good weather. And so it happens that in the course of time, migrating groups of human beings would form settlements, in which they would often remain *isolated* for thousands of years from all contact with other people. Since *mutations* would occur in individuals within each group, and since mutations are always random, some of these fortuitous mutations would gradually establish themselves in the small breeding isolate, and in this way every separated group would in the course of time come to differ physically from every other—that is to say, genotypically in the frequencies of certain genes and phenotypically in the distribution of certain physical traits. Some of the mutations that would establish themselves would undoubtedly be of adaptive value to the members of the group. For example, those possessing a large amount of dark pigment in their skin in a climate in which the sunlight intensity was high would be likely to enjoy a selective advantage over those with lighter skins. The dark-pigmented individuals would tend to leave a larger progeny behind them. In other regions, where the sunlight intensity was low and the skies often cloudy, more lightly pigmented skin would be an advantage in more freely allowing the rays of the sun to penetrate the skin; also, fewer sweat glands would be necessary. This sort of development would be the factor of *natural selection,* or adaptive fitness, at work. Since it is highly improbable that the same kinds of mutations would have occurred in different isolated groups, it will readily be understood how on this basis alone different human groups would come to differ from each other in the course of time. Some of the mutations would be useful, and others would simply become randomly fixed within the group even though they might have no particular adaptive value; such a process is known as *genetic drift*. In the course of time some of the members of these groups might meet, form permanent unions, *hybridize,* and create new settlements. By producing new gene combinations, and allowing for the action of selection, isolation, genetic drift, etc., an altogether new physical type would emerge.

The operation of the six factors mentioned, *migration, isolation, mutation, genetic drift, natural selection,* and *hybridization,* is sufficient to account for the evolution of the genotypic and phenotypic differences that distinguish the ethnic groups of man.

Genetically speaking, an ethnic group is a population that differs in the frequency of some gene or genes and actually exchanges or is capable of exchanging genes across whatever boundaries separate it from other populations of the species. All living mankind is included in the single species *Homo sapiens,* a species that consists of a number of populations that individually maintain genetic and phenotypic differences from one another by means of isolating mechanisms such as geographical and social barriers. In addition to the effects of other influences, these differences will vary as the power of the isolating mechanisms vary. Where these barriers are of low power, neighboring isolates will integrate, or hybridize, with one another. Where these barriers are of high power, such isolates will tend to remain distinct or will replace one another geographically.

## HEREDITY AND "RACE"

Such isolates constitute the *ethnic groups*, which may be defined as arbitrarily recognized groups that by virtue of the possession of a more or less distinctive assemblage of physical traits, derived through a common heredity, are statistically distinguishable from other groups within the species.

It is observed that most of these ethnic groups tend to form certain clusters according to their resemblances in certain characters. For example, black skin yields a whole cluster of Negroid groups; white skin yields another cluster of white, or Caucasoid, ethnic groups, and yellowish skin the cluster of Mongoloids; while chocolate-brown to brownish-white skin associated with abundant, wavy head hair yields the Australoid, or "archaic Caucasoid," cluster. These clusters of characters suggest that some of the ethnic groups exhibiting them may be, in respect of these characters, more closely related than they are to the members of other clusters. Such clusters are termed "major groups."

A *major group* comprises a number of ethnic groups classified together on the basis of their possession of certain common characters that serve to distinguish that major group from others. Four such major groups as stated above are recognized. The usual classification is diagrammed in Fig. 22.

Ethnic and major group differences simply represent more or less temporary expressions of variations in the relative frequencies of genes in different parts of the species range. Such a conception rejects altogether the all-or-none conception of "race" as a static condition of fixed differences. It denies the unwarranted assumption that there exist any hard and fast genetic boundaries between any groups of mankind and asserts mankind's common genetic unity in diversity. Such a conception of the variety of man cuts across national, linguistic, and cultural boundaries and thus asserts the essential independence of genetic factors.

The differences that distinguish the ethnic groups are real and important, important in the sense that they are adaptively valuable. But even between ethnic groups with extreme differences in appearance, it is unlikely that the number of gene differences exceed more than a fraction of 1 percent.

Were an anthropologist to be asked to name the two ethnic groups who physically appear to be most unlike each other, he might mention a Congo Pygmy and a Scandinavian white, or he might mention a white European and a South African Bushman. In fact, these appear to be such extreme

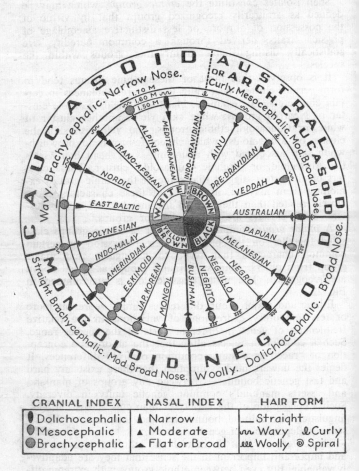

FIG. 22. The Major Ethnic Groups of Man, Showing the General Distribution of Various Traits. (From Ashley Montagu, *An Introduction to Physical Anthropology*.)

HEREDITY AND "RACE" 223

types that a learned German professor, Fritz Lenz, not so long ago declared, "As far as I am aware, neither African Pygmies nor Bushmen interbreed with Negroes or with Europeans; thus, owing to their natural instincts and their habits, they are physiologically isolated." How wrong can a learned German professor be! Neither the "natural instincts" —whatever they may be—nor "their habits" have prevented such interbreeding from occurring. We do not, however, have adequate reports on Pygmy-Negro offspring, but we do have an adequate report on Bushman-European interbreeding. Such a report was published in 1954 by Dr. Phillip Tobias, of the University of Witwatersrand, South Africa. Even though only three generations of a single family are involved, this report throws much light upon the genetic likeness of the ethnic groups of man.

In the first generation of this family, a Bushman woman, shown in Plate 14 (1) and (2), mated with a Dutch white, who was not available for study. The offspring of this union was a daughter, shown in Plate 14 (3) and (4), who presented an appearance intermediate between her Bushman mother and her white father, but with most of her physical characters, which were determined by measurement and inspection, much more closely resembling her white father than her Bushman mother (see photographs). The hybrid daughter was 4⅜ inches taller than her mother, her face was wider and longer, her eyes more closely set, her nose narrower and higher. The hair was long, brown, wavy, and of silken texture, whereas her mother's head hair was of the tightly coiled and tufted variety known as "peppercorn," which is typical of the Bushman. Eyebrows and eyelashes were well supplied with hair, and the body hair was moderately profuse in contrast with the mother's relative hairlessness. Her eyes were brown and her skin light. The ears were without lobes, as in the mother. Eye folds (similar to those seen in many Chinese and Japanese) are present in the mother and daughter, which suggests that both the eye fold and lack of ear lobes behave as dominants. The daughter is in general of much larger and broader build than her mother.

The daughter formed a union with a white Englishman, shown in Plate 14 (7), and by him had two sons and two daughters. The youngest son died at an early age from the bite of a button spider. According to reliable witnesses, the remaining son looks European, one of the girls looks more like her mother than her father, and the only child available for study, the three-year-old girl shown in Plate 14 (5), looks in every way like a European child. This little girl has light brown eyes, very fair skin, large ears with lobes, and a definite eye fold; the nose is narrow and high, but it has a slightly bulbous tip, as does that of her mother and grandmother. Thus, the only features this little girl has in common with her Bushman grandmother are the eye fold and bulbous nose tip. This girl's sister resembled her in being of distinctly European appearance, though her surviving brother, who was not seen and who was European-looking, was said to show rather more evidence of his Bushman grandmother's genes. This is a curious fact, namely, that more European features should appear in the females than in the male. This observation has been recorded for other crossings and is apparently the result of sex influence acting differently upon the genes involved. But our main point here is the demonstration of the small number of gene differences existing between human types that appear extremely different from each other. Were the differences more numerous than they are, we should expect to see more marked differences in the hybrid offspring of such unions, and we should certainly not expect to see virtually all the traits of a white grandfather and father segregating out, as they have done in this little girl whose grandmother was a pure Bushman and whose mother was a hybrid of Bushman-white origin. Nor would we expect to see a first-generation hybrid, such as the mother, so remarkably white and with so few Bushman characteristics if there were many significant gene differences either in number or kind between Bushman and white. The evidence strongly indicates that mankind has obtained the vast majority of its genes from the same common gene pool, and that such mutational differences as have occurred in the course of the evolution of the different ethnic groups have, in comparison with the number of genes that they hold in common, been relatively few.

## Is "Race"-Crossing Harmful?

It has often been asserted that "race"-crossing is harmful: that it succeeds in perpetuating the worst traits of both "races" in the hybrid, that it produces physical disharmonies, that it causes contamination of good stock, that it produces an inferior generation, and so on. All these statements are quite unsound even though they are widely believed. Since half-castes have been more often than not treated as outcasts by society and at best have occupied an anomalous and ambiguous position in society, it is not to be wondered at that they

# HEREDITY AND "RACE" 225

have frequently been regarded by those who forced such a degrading position upon them as "inferiors" or "degenerates." But their "inferiority" has seldom if ever been biological; it is, rather, a social inferiority that has been thrust upon them and justified by the myth of their alleged biological inferiority. The disparities in the social attainments of half-castes or hybrids and whites can be explained by the difference in opportunities as well as abilities. The evidence is full to overflowing on this point in the United States, where Americans of white and Negro origin have earned positions of respect and honor in their communities.

Unfortunately a considerable amount of prejudice has been the rule in this area of conversation, and even scientists have not been guiltless of it. An outstanding example of this is the case of the late Dr. C. B. Davenport, one of the fathers of the science of genetics in America and a distinguished scientist. In a study entitled *Race Crossing in Jamaica,* published by the Carnegie Institution of Washington, D.C., in 1929, in which Davenport interpreted the findings of Dr. Morris Steggerda, the field investigator of "Black," "Brown," and "White" Jamaicans (the "Browns" being the crosses of the "Blacks" and "Whites")", it is with amazement that one perceives Davenport's prejudices at work. Here, for example, is an illuminating illustration of such prejudices at work: "The Blacks," writes Davenport on page 469, "seem to do better in simple mental arithmetic and with numerical series than the Whites. They also follow better complicated directions for doing things. It seems a plausible hypothesis, for which there is considerable support, that the more complicated a brain, the more numerous its 'association fibers,' the less satisfactorily it performs the simple numerical problems which a calculating machine does so quickly and accurately." In other words, when the "Blacks" do better than the "Whites," their superiority must be explained away! And not only explained away but also cited as further evidence of their inferiority!

Some of the hybrids measured by Steggerda exhibited a combination of "long arms and short legs." Upon this, Davenport commented as follows: "We do not know whether the disharmony of long arms and short legs is a disadvantageous one for the individuals under consideration. A long-legged short-armed person has, indeed, to stoop more to pick up a thing on the ground than one with the opposite combination of disharmony in the appendages."

Three out of four brown (hybrid) Jamaicans are cited in

support of this generalization. Here are the figures upon which this generalization is based:

Limb Proportions and Stature in Jamaicans

|  | Black | Brown | White |
|---|---|---|---|
| Arm length in cm. | 57.3 | 57.9 | 56.8 |
| Leg length in cm. | 92.5 | 92.3 | 92.0 |
| Total stature in cm. | 170.6 | 170.2 | 172.7 |

From these figures it will be observed that the arm length of the browns, that is, the hybrids of crosses between the whites and blacks, is, on the average, 0.6 centimeter greater than in the blacks and 1.1 centimeters greater than in whites, and the leg length of the browns is 0.2 centimeter less than in the blacks. It is here that the "disharmony" is perceived by Davenport. It has, however, to be pointed out that the order of the differences in total stature is so small—at most not more than 2.5 centimeters (1 inch) between brown and white —while the average difference between length of arms and legs among the three groups never exceeded more than half an inch, that it could not make the slightest practical difference in the efficiency of stooping.

As the distinguished doyen of genetics in America, William E. Castle, has said, "We like to think of the Negro as an inferior. We like to think of Negro-white crosses as a degradation of the white race. We look for evidence in support of the idea and try to persuade ourselves that we have found it even when the resemblance is very slight. The honestly made records of Davenport and Steggerda tell a very different story about hybrid Jamaicans from that which Davenport and Jennings * tell about them in broad sweeping statements. The former will never reach the ears of eugenics propagandists and congressional committees; the latter will be with us as the bogey men of pure-race enthusiasts for the next hundred years."

The type of evidence that is sometimes wrongly interpreted by investigators as evidence of the ill effects of "race"-crossing is illustrated by a study made by Miss R. M. Fleming in 1939. Studying the offspring of Negro-white unions in the seaports of England and Wales, Miss Fleming found that 10

* Professor Herbert S. Jennings in his book *The Biological Basis of Human Nature* adopted Davenport's interpretations of the ill effects of hybridization, and it is to this that Professor Castle refers.

percent of the hybrids showed a disharmony between the teeth and jaws. The palate was generally well arched, while the lower jaw was V-shaped and the lower teeth slipped up outside the upper lip, seriously interfering with speech, this disharmony "resulting where a well-arched jaw was inherited from the Negro side and a badly arched one from the white side." No other disharmonies were observed.

Miss Fleming states that a "badly arched" jaw was inherited from the white side. In other words, the disharmony was not due to the effects of crossing but to the fact that the condition was transmitted from the white parent to the child. The disharmony was limited to only 10 percent of the cases. It is also possible that while some of these cases merely represent the expression of inherited defects, not necessarily exhibited in the jaws of the parents themselves, still others were due to malnutrition, and that the defect actually bore no relation whatever to the fact that one parent was a Negro and the other a white. Otherwise, it would be expected that more than 10 percent of the hybrids would exhibit disharmonies of the relations of the jaws to each other.

As far as the evidence goes, there is not the slightest reason to suppose that "race"-crossing ever produces disharmonies of any kind. The fact seems to be that the differences between human groups are not large enough to be capable of producing even the slightest disharmonies.

For the same reason, it is difficult to observe any marked evidences of *hybrid vigor* (heterosis) in man as a consequence of mixture of ethnic groups—the gene differences do not seem to be large enough. Hybrid vigor is the phenomenon observed in plants and animals in which, as a result of the crossing of different varieties, offspring tend to be more fertile, numerous, stronger, and larger than the members of the parental stock. This is not to say that hybrid vigor does not occur among human beings, but that if it does occur, it is not as evident as it is among plants and lower animals. Human hybrids tend to show a greater resistance to disease. For example, in Tierra del Fuego the unmixed natives succumbed to the measles, whereas the hybrids were able to resist the disease. The evidence suggests that in man, hybrid vigor expresses itself not in metrical traits but by changes in fertility connected with subtle mechanisms of adaptive value, such, for example, as potential resistance to disease. Differences in fertility within the same populations may also be due to the slight heterotic effects of many individual genes that in combination serve to produce such differences in fertility. It is also probable that levels of intelligence are not only prevented from declining as a result of hybridization but also are "refreshed," as it were, by the introduction of new genes. This greater heterozygote frequency in hybrid populations has also been shown to be associated with a very significantly lower congenital abnormality rate in hybrids as compared with nonhybrid members of the same population (Saldanha).

The principal determining factor in the physical organization of the new human being, the offspring of any union, is the genetic constitution of the parents, and nothing else. Since the evidence leads us to believe that no human ethnic group is biologically either better or worse in any trait, group of traits, or as a whole, than any other, it should be clear that hybridization between human beings cannot lead to anything but a harmonious biological development.

Admixture between hitherto more or less isolated populations will become increasingly more frequent in our shrinking world. This will result in many new gene combinations and greater variety within populations for many generations. And as Penrose has remarked: "This can be regarded as a favorable development because it will increase the number of man's possible inborn reactions, whether physical or psychological, to his rapidly changing civilized environment."

In emphasizing the basic essential likeness of all the ethnic groups of man, let us not forget that fundamental and important as these are, the really valuable qualities that human beings exhibit are not their likenesses but their differences. It is these very differences—the physical differences between the ethnic groups of man—evolved over the long course of human history, that constitute the best testimony to their own value, for it is these physical differences that have enabled the ethnic groups to survive in the environments in which they originated. Hence, there can be no question of "superiority" or "inferiority" in the comparison of these physical ethnic traits; all of them constitute evidence of adaptive fitness, and the genotype of man is such that it is capable of making very rapid phenotypic adjustments to every conceivable environment to which it is called upon to respond. The members of any ethnic group can live in any environment in the world, and by resorting to intelligence and imagination they can adjust to the widest extremes of climate and to virtually any other conditions they are called upon to meet.

## "Race" and Mental Capacities

The notion that "race" represents "something" that is an amalgam of physical and mental traits, so that certain physical traits go together with certain mental traits, is widely held. Indeed, this is the core of what most people understand by "race." In spite of innumerable attempts to prove that such a linkage exists between physical and mental traits no one has succeeded in demonstrating it or even reasonably indicating that such a linkage exists. This is not to deny that some slight differences may exist between some ethnic groups in the frequencies of certain genes underlying mental capacity. It is possible that such slight differences exist, but in spite of all attempts no one has, in fact, ever demonstrated that they do. But one thing seems to be highly probable, and that is if such slight differences exist, then they in no way depend upon physical traits. The evolution of the physical traits of ethnic groups has been in adaptation to kinds of environmental challenges totally different from those that have been involved in the evolution of man's mental capacities.

Whereas the challenges of the physical environments have been very different in different geographic areas of the world, with consequent differences being established in the patterns of genotypic adaptations observable in the physical differences between ethnic groups, the challenges of the many social environments have been fundamentally similar. If we ask ourselves on what qualities in every society, at any particular time, the highest premium was put, and we check this by the facts, we find: the ability to get along with people, the ability to know when to say yes and when to say no, when to keep one's mouth shut whatever one may want to say, the ability to make rapid adjustments to rapidly changing conditions, to know when to stay and fight or when to run away and live to fight another day—in other words, the qualities of maturity of judgment, wisdom, and adaptability. It is this combination of traits, adding up to what we can call the trait of plasticity, that appears to have been at a high premium in all societies—as it still is. This plasticity is not a particular trait but a general one. Instead of leading to fixed responses to the environment, man's evolution has been such as to make him the least behaviorally fixed and the most generally educable or plastic of all living creatures. It is this very plasticity of his mental traits that confers upon man the unique position that he occupies. The acquisition of this capacity freed man from the constraint of the limited range of biologically predetermined responses that characterizes all other animals. In the history of life on this earth, man is the only creature who became capable of substantively controlling his physical environment instead of being controlled by it. He began to live his own life instead of being lived principally by his organic limitations. In all times and in all climes the process of natural selection seems to have favored genotypes that permit greater and greater educability and plasticity of mental traits under the influence of the uniquely social environments to which man has been continuously exposed.

The effect of natural selection in man has probably been to render genotypic differences in personality traits, in mental traits, in genetic potentialities, between individuals and particularly between ethnic groups, relatively unimportant compared with man's phenotypic plasticity. Man's genotype is such that it makes it possible for him to develop the widest range of behavioral adjustments and adaptations. Instead of having his responses genetically fixed, as in other animal species, man is the species that invents its own responses, and it is out of this unique ability to invent, to improvise responses, that the cultures of man are born.

Examining the end effects of the long story of human evolution as they present themselves today—namely, human beings of every ethnic group—the conclusion seems inescapable that natural selection has been operative upon the traits of educability and plasticity in much the same way from the beginnings of man's history, in all human groups, no matter how long they may have been isolated from one another. What, in short, has been at the highest premium in all societies is not some special ability but rather the general ability of behavioral plasticity. Natural selection has favored and continues to favor plasticity. Genetic differences between individuals are by no means washed out, but those differences are retained only if they are of the kind that permit themselves to be eclipsed by phenotypic plasticity. Hence, it is not to be expected that there exist any very significant differences in the distribution of special abilities among the ethnic groups of man.

## "Race" and Intelligence

If the points we have made in the above section are sound, then there should exist few if any really significant differences in the genetic capacities for intelligence of the different ethnic groups. In spite of many attempts to demonstrate that such differences in intelligence exist, there has been complete failure to do so. Certainly differences in intelligence as measured by performance have been found to exist between the ethnic groups of man, but it is one thing to find such differences and quite another to demonstrate that they are due to genetic factors. On the other hand, when it has been possible to demonstrate anything whatever in relation to the intelligence tests on different ethnic groups, it has consisted in the finding that when the environment is equalized, the performance on the intelligence tests becomes equalized.

Intelligence is a function not of genes alone nor of environment alone but of the interaction of the two. Intelligence implies not only innate potentialities but also opportunities for the realization of those potentialities. What applies to different cultures applies to all the individuals as well. Why is it that the Australian aborigines differ from Europeans in cultural attainments as much as they do? Is it because of the differences in the qualities of the intelligence of aborigines and of Europeans? The answer is almost certainly no. The answer is that the differences in the cultural achievements of Australian aborigines and, say, Englishmen are due entirely to the differences in the history of the cultural experiences to which each group has been exposed. It should be recalled that until the eighteenth century the Australian aborigines were an undiscovered people who had been isolated for thousands of years on one of the southernmost islands of the world, and that they had had practically no contacts whatsoever with the outside world. The greater part of the territory occupied by them was (and is) one of the most inhospitable desert regions of the world, and to it they have made a perfect adjustment. What, in comparison, has been the history of the English or, for that matter, any other European nation we wish to consider? It has been a history of continuous cultural contacts with numerous other peoples for several thousand years in an area of the world where the cross-fertilization of ideas, customs, and ways of life has been of the most stimulating kind. Even today, the Australian aborigines have been largely prevented from realizing their magnificent potentialities by discriminatory practices that are not unknown elsewhere in the world. But when Australian aborigines are given the opportunity, they show themselves capable of achievement at least as great as that of the average man anywhere else. Approximately sixty years ago, there was a school in the south-western part of Australia, in the state of Victoria, which was attended entirely by aboriginal children. I quote the Reverend John Mathew: "In schools it has often been observed that aboriginal children learn quite as easily and rapidly as children of European parents. In fact, the aboriginal school at Ramahyuck, in Victoria, stood for three consecutive years the highest of all the state schools of the colony in examination results, obtaining *one hundred per cent. of marks.*" (The italics are the Reverend Mathew's.) The extraordinary artistic ability of the Australian aborigines is only a very recent discovery, for as soon as they were taught to see as Europeans do, both children and many adults soon greatly surpassed in their artistic performances anything that the average European child or adult was able to do.

The story is everywhere the same, as we have already seen on an earlier page; wherever and whenever individuals are given the opportunities to realize their potentialities, we find that human beings everywhere can do what other human beings anywhere have done. Of course there are great variations in the distribution of abilities in every population, but the range of these differences is far greater *within* each ethnic group than *between* different ethnic groups.

# PART TWO

## THE FAMILY ALBUM

*The face in the family album*
*Is the face you have always seen,*
*But it isn't the face of the thousands*
*Of faces you might have been.*

A.M.

In the second part of this book we shall consider the heredity of common physical and functional traits. In some cases the hereditary mechanism is known; in others it is not.

# 13. THE PIGMENTARY SYSTEM OF THE SKIN, EYES, AND HAIR

THE SKIN is the largest organ of the human body and one of the most complex. In addition to its functions of protection, respiration, thermal regulation, sensory and motor response to stimulation, the skin manufactures ergosterol, which is the pro-vitamin of vitamin D, so important in the growth of bones and the development of teeth. To assist it in the execution of its functions the skin is more or less deeply pigmented. The principal pigment of skin, eyes, and hair, called melanin, is a brown-to-black pigment that is produced in cells known as melanocytes. The melanocytes originate in a deep layer of the skin known as the germinative layer, and the melanin they produce passes into the adjacent upper layer, known as the granular layer, or epidermis, of the skin. Melanin is produced in the melanocytes by a reaction between oxygen and the amino acid tyrosine. The enzyme tyrosinase in the melanocytes acts on tyrosine to produce and control the speed of production of melanin. Exposure to the ultraviolet rays of sunlight, for example, activates tyrosinase to convert tyrosine into melanin. There are no differences in the number of melanocytes in the different ethnic groups of man. Difference in pigmentation in populations and in individuals, as well as in different parts of the body, are due to differences in the dispersion and distribution of the melanin particles in the melanocytes.

The amount of pigment in the skin is inherited, but it is not known precisely how many genes are involved, nor is the hereditary mechanism understood. Anything from two to twelve genes have been suggested as involved in skin color, but the truth is that we do not know how many are in fact involved. There must be many. We shall return to this matter again. What we do know with certainty is how, for example, the condition known as *albinism* (Latin *albus*, white), in which there is a complete lack of pigment in the skin, hair, eyes, and nails, is inherited.

Albinism is believed to be due to a biochemical block resulting from the lack of a particular enzyme acting on the tyrosine, which through a series of biochemical steps leads to the formation of melanin, the pigment of our skin, eyes, and hair. Dr. F. Hu and his coworkers have shown that the cells containing melanin pigment, the melanocytes, are present in normal numbers and distribution in the skin of albinos, and that the enzyme that acts on tyrosine, tyrosinase, is also present in these cells, possibly in inhibited form. In addition to the biochemical block, there may be a block in pigment transfer from albino melanocytes to the epithelial cells of the epidermis.

In albinos the skin is typically white or pinkish, the hair straw-colored, and the eyes look pink, owing to the reflection of the blood vessels in them, although in some cases they may appear partially blue. Albinism is known to occur in whites, Negroes, American Indians, and probably in other ethnic groups. It is very common in mice, rats, and rabbits, and has been reported in many different kinds of animals.

Albinism is occasionally associated with pseudohemophilia. There is a prolonged bleeding time with slight reduction in prothrombin consumption. The fact that redheads (see p. 244) are even more likely to be characterized by a prolonged bleeding time suggests that there exists a significant physiological relationship between the factors conditioning the pigment of the skin and those involved in the chemistry of the blood.

In man approximately 1 out of every 20,000 children born is an albino, and in a large proportion of cases the birth of such a child occurs in family lines in which there has never been a previous birth of an albino. In some cases the albino birth may be due to a mutation, but more often than not it is demonstrably due to the coupling of recessive genes for albinism, one derived from the mother, the other from the father, both of whom are carrying them in their heterozygous recessive state. Such a couple, heterozygous for the albino gene, stand a 1-in-4 chance of producing another albino child. The chances of any couple picked at random bearing an albino child are 1 in 20,000. This is a rather surprising figure in view of the fact that 1 out of every 70 persons carries the gene for albinism. Actually the heterozygous carrier is 286 times more frequent than the homozygous albino.

Now, if 1 out of every 70 persons is heterozygous for albinism, why are there so few albinos, only 1 out of every 20,000? The answer is a matter of simple arithmetic. The

THE PIGMENTARY SYSTEM 237

chance of any man carrying the albino gene is 1 in 70; the chance that the woman he marries carries the gene is also 1 in 70. The chance that both carry the gene is 70 times 70, that is, 1 in 4,900. Finally, the chance that the child of such heterozygous parents will be an albino is 1 in 4, and 4 times 4,900 is 19,600, which gives us the approximate estimate of 1 in 20,000 being born an albino.

The frequency of 1 out of every 4 children of a heterozygous couple standing a chance of being an albino derives from the following considerations. Let $C$ stand for the dominant gene for normal pigmentation and $c$ for the recessive

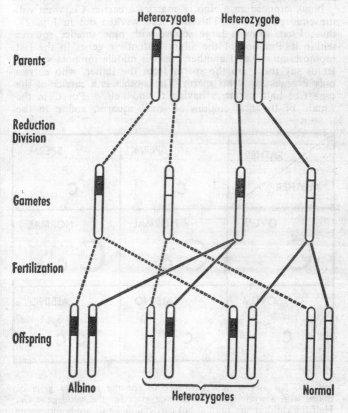

FIG. 23. The Recessive Form of Inheritance of Albinism.

238  HUMAN HEREDITY

gene for albinism. The heterozygous male carrier then has the genotype $Cc$ in his body cells: half his sperm carry $C$ and the other half $c$. The same will be true of the heterozygous female and her eggs: half will carry $C$ and half $c$. If you will take a pencil and paper and try to figure out the possible combinations into which each of these genes from sperm and egg can enter, you will find that there will be one $CC$ child, two $Cc$ children, and one $cc$ child, who will be albino because he will be homozygous for the recessive albino $c$ gene. The mode of transmission of these genotypes is shown in Fig. 23.

Now, suppose an albino $cc$ marries a carrier $Cc$, what will the children be like? This has been worked out in Fig. 24, thus: First, draw a large square with nine smaller squares within it. Put one of the albino parent's $c$ genes in the last topmost square, and another $c$ in the middle topmost square; let us say these are the sperm from the father, who carries only $c$ genes in all his sperm. The mother is a carrier of the type $Cc$: half her eggs having $C$ and half $c$. Put $C$ in the square of the first column of small squares, and $c$ in the

| FATHER / MOTHER | SPERM c | SPERM c |
|---|---|---|
| OVUM C | NORMAL Cc | NORMAL Cc |
| OVUM c | ALBINO cc | ALBINO cc |

FIG. 24. An albino man homozygous for the recessive gene $cc$ mates with a woman who is heterozygous for the same gene $Cc$. Half the children are normal but carriers, and half are homozygous albinos.

middle square under that. Now put $C$ in the middle and last squares of the middle horizontal column of squares and bring down the $c$ from the top squares and place them next to $C$; you should have $Cc$ in each of the two squares. Put small $c$ in the middle and last squares of the lowest level of the diagram and bring down the small $c$ in the topmost squares and put them with the other small $c$'s in the middle and the last of the lowermost row of squares. All this rigmarole simply amounts to saying: mate the first vertical column with the first horizontal column and put the offspring in the four empty squares, as in the diagram above. You will see that from such a mating of an albino with a heterozygote, half the children will be albino and half will be normal, but carriers.

What happens when an albino marries a normal person who does not carry the recessive gene? You should be able to work out the answer to that question for yourself. Such a marriage would involve a $CC$ normal and a $cc$ albino. Since the gene $C$ for normal pigment is present in all the germ cells of the homozygous parent and is a dominant, all the children of such a marriage will be normal, but they will all be carriers of the recessive gene since they will be of the genotype $Cc$. Diagram this for yourself, using a diagram like the one shown in Fig. 24.

You can use this same type of diagram to work out the hereditary mechanism of many other conditions.

While in the majority of recorded cases albinism is transmitted by a recessive gene, in some cases the condition is transmitted by a dominant gene. Thus, we have in albinism an example of the same phenotype being conditioned by two different genotypes. Partial albinism, in which there is congenital absence of pigmentation of some part or parts of the skin and its appendages, is inherited as a simple dominant. It has been reported in Negroes, American Indians, and Asiatic Indians. Ocular albinism, in which the eyes alone are affected, is inherited as a sex-linked trait.

## Skin Color in General

Although the heterozygous parents of albinos appear in every way to be normal, they do, in fact, show one phenotypic evidence of the presence of the defective pigmentary gene, namely, the translucency of the iris to light. If you have a brunet complexion, you know that as a result of more or less prolonged exposure to sunlight your skin can turn many shades darker than it normally is. This change of color is due to an increase in the amount of melanin in your skin, which is activated by sunlight. If you are blond or fair-complexioned, you will know that you tend to burn more readily upon exposure to sunlight than your brunet acquaintances, and that you do not tend to tan as quickly as they do. This is due to the fact that fair-skinned people have much less melanin in their skin than brunets, and hence burn more easily and take longer to increase the melanin content of their skin. From the evolutionary standpoint, this genotypic capacity of the melanin content of one's skin to change has been of great adaptive value in areas of varying sunlight intensity: it allows low-pigmented skins to permit the passage of enough sunlight for the manufacture of vitamin D in the body and, when the sunlight continues, permits the skin to respond with the development of more melanin. Some brunets can turn virtually ebony-black upon prolonged exposure to sunlight; others don't get quite so dark. The same is true of fair-skinned people. It is, I think, fascinating to observe how really dark a fair-complexioned person can get, the blond hair, blue eyes, and very dark skin making a most striking picture. The point, however, of these comments is to draw attention to the fact that all sorts of gradations in skin color can be achieved by the pigmentary system in response to sunlight. This suggests that skin color is the effect of different degrees of pigment intensity controlled by the action of many genes, that there are really no fundamental differences in the kinds of genes that are responsible for the different skin colors of different ethnic groups, but the range of color is merely a matter of the differences in the frequencies of the same kinds of genes. We have already learned that color genes retain their identity and segregate out unaltered in each generation, though in effect they appear to produce blending of pigments. Clearly the interaction of genes is involved, and judging from the whole range of colors of which human skin is capable, it is clear that many genes must be involved, each producing a small effect without dominance. Such genes are called *multiple genes,* or *polygenes.* As I have already said, we do not at present know how many genes may be involved. It is probable that there are at least six pairs of genes at work.

If, for the sake of illustration, we assume that six pairs of genes are involved, and we consider these genes only in Negroes and in whites, then we can assign three pairs of heavy-pigment genes to Negroes, of the genotype $P_1 \ P_1, \ P_2$

## THE PIGMENTARY SYSTEM

$P_2$, $P_3$ $P_3$, and to whites three pairs of light-pigment genes, $p_1$ $p_1$, $p_2$ $p_2$, $p_3$ $p_3$. The more $P$ genes a person has, the darker he will be, for the effects of the $P$ genes are additive; the more $p$ genes a person has, the lighter skinned he will be, for the light-skin genes are negatively additive, as it were. Now, if we say that the first pair of genes in the Negro series have the heaviest pigment, and the second pair a moderate amount of pigment, and the third pair the least heavy amount of pigment, and we say that the first pair of genes in the white series have a small amount of pigment, the second pair a smaller amount of pigment, and the last pair the smallest amount of pigment—remembering that in whites the amount of pigment is much less than in Negroes—it will be seen that on a six-gene basis involving three pairs of genes having different pigment strengths you can get sixty-four, or $2^6$, possible gene combinations, yielding almost as many degrees or kinds of Negro and white skin color. Actually, we know that there are more variations than these, but the choice of number of gene pairs was arbitrary.

The offspring of Negro-white unions will be of the genotype $P_1p_1P_2p_2P_3p_3$; that is, the offspring have three dark-pigment genes and three light-pigment genes, so that they will be intermediate in color between both parents. The sex cells of mulattoes will contain three genes in eight different combinations, that is, in haploid number, as follows: $P_1P_2P_3$, $P_1P_2p_3$, $P_1p_2P_3$, $p_1P_2P_3$, $P_1p_2p_3$, $p_1P_2p_3$, $p_1p_2P_3$, and $p_1p_2p_3$. When mulattoes marry, their offspring, in any population in which there are many such marriages, are likely to show the following distribution of skin colors: about 20, or almost one-third, of the $F_2$ generation out of every 64 individuals will carry three dark-pigment genes and three light-pigment genes, and will therefore resemble their parents in skin color. About 30, almost half, will carry two dark- and four light-pigment genes or four dark-pigment genes and two light-pigment genes, and will therefore have either somewhat lighter or somewhat darker skins than their mulatto parents. About 6 out of the 64 will have five dark-pigment genes and one light-pigment gene, and will be intermediate between Negro and mulatto, and about 6 will have one dark-pigment gene and five light-pigment genes, and will be intermediate between mulatto and white. One out of 64 will have six dark-pigment genes and will be completely Negro in skin color; another one will have six light-pigment genes and will be completely white in skin color. This is diagrammatically illustrated in Fig. 25.

## 242 HUMAN HEREDITY

It should be obvious that a white child cannot be born of the union of two Negroes, nor can a Negro child be born of the union of two whites. If the parents' skin is Negroid, then they carry Negroid genes and they can have children with Negroid skin color only. If the skin of the parents is white, then they carry only light-pigment genes and can therefore have only light-skinned children. If one of the parents carries one dark gene, that is to say, is of the genotype $p_1p_2P_3$, and marries a person who carries no dark genes and is of the genotype $p_1p_2p_3$, then, owing to the fact that the skin-color genes are neither dominant nor recessive, the five light-pigment genes will almost completely "wash out" the effects of the dark-pigment gene in the "blending" of the pigmentary factors. The stories about "black" children suddenly being born to parents who were whites or who had some concealed

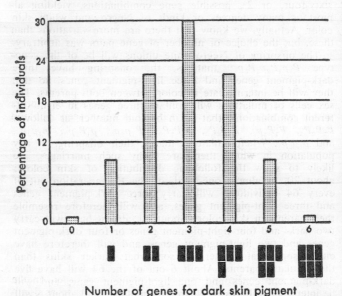

FIG. 25. Inheritance of Skin Color in Negro-White Mating (assuming three pairs of pigment genes without dominance). The heights of the columns show the percentages among the second-generation hybrids of individuals with no color-producing genes, with one, two, three, four, five, and six. (After Dobzhansky, *Evolution, Genetics, and Man*.)

Negro ancestry are complete nonsense. If the parents have white skins their children can have white skins only.

When mulattoes marry Negroes or whites, the skin color of most of their children is intermediate between that of the parents. Three-eighths of the sex cells of mulattoes carry two dark-pigment genes, three-eighths one dark-pigment gene, one-eighth three dark-pigment genes, and one-eighth no dark-pigment genes. Since a Negroid sex cell carries three dark-pigment genes, the offspring of a mulatto-Negro cross would yield three-eighths with five dark-pigment genes, three-eighths with four, one-eighth with six, and one-eighth with three dark-pigment genes. A mulatto-white offspring would yield three-eighths with two dark-pigment genes, three-eighths with one, one-eighth with three, and one-eighth with no dark-pigment genes.

## The Sacral Spot

The sacral spot, sometimes called "the blue spot," or "the Mongoloid spot," is a largish, bluish area found at the base of the spine over the area of the sacrum and adjoining buttocks. It is found in about 90 percent of Negro infants, in many Mongoloid infants, in some American Indians, and in some dark-complexioned whites. It usually disappears within six months to a few years. It is never seen in blond children with blue eyes.

## Finger Smudges

Some infants with some Negroid ancestry show "smudges" of pigment on the backs of the fingers between the joints. Their presence indicates that the child will develop a dark complexion, their absence, that the skin color will develop as white.

## Freckles

These small yellowish-brown pigmented spots seen on the face and other parts of the body apparently behave as dominants in heredity. One or the other parent is usually also freckled. Freckles in some individuals become evident only after they have been exposed to sunlight. In all individuals the intensity of the pigment in them increases following appreciable exposure to sunlight. That is why freckles tend to be reduced in winter and to blossom in summer. Freckles appear to be more frequent in persons with red or a reddish shade of hair. It is interesting to note that many chimpanzees have freckled faces.

## The Skin of Redheads

The skin of redheads is probably lacking in certain pigmentary factors, and this appears to be in some way connected with the inadequacies of their blood-clotting capacities. This is a subject about which we know practically nothing and on which we need much systematic observation. There is some evidence that the bleeding time of redheads is longer than that for brunets or blonds. Some obstetricians take special precautions when a redhead is to be delivered of a baby because of the tendency of some redheads to bleed profusely following delivery. The skin of redheads is very much more susceptible to burning by exposure to sunlight than is that of brunets or blonds. There also appears to be more difficulty in the healing of burns sustained in other ways. Some radiologists routinely give redheads lower doses of radiation than they give their other patients. Red hair behaves as a recessive in inheritance, and so apparently does the gene for the character of the skin that is almost invariably associated with red hair.

That red hair, as distinct from Titian red, has some profound constitutional ramifications is suggested by the fact that tuberculosis in redheads generally runs a much more severe course than in non-redheads. Interestingly enough, the same seems to be true of individuals inclined to freckling. There is also some evidence that redheads have greater difficulties with anesthetics. It would be interesting to know how redheads fare in relation to hemophilia.

## Eye Color

The color of the eyes is determined by the presence in the iris of the very same pigment, melanin, that is so closely related to the color of the skin. The *iris* (Greek for "rainbow") of the eye has received its name from the rainbowlike variety of colors it exhibits in different individuals. The iris is a circular, disclike structure, the central opening of which is the pupil of the eye. The iris acts as a diaphragm controlling the amount of light admitted to the eye. The amount of pigment in the iris also serves a similar function. Hence, it is found that in high-sunlight areas where skins

## THE PIGMENTARY SYSTEM

are deeply pigmented, the iris is too. In low-sunlight areas there is less pigment in the iris. The color of the eye is dependent upon the amount of pigment present in the front and back portions of the iris. The pigment is contained in fine particles, which vary in size and shape.

The iris has a double layer of pigment cells at its back, and a front layer of cells that may or may not be pigmented. The color of the eyes is conditioned by two principal factors: (1) the amount of pigment in the back layers of the iris and (2) the amount of pigment in the front layer of the iris.

If the pigment is present only in the back portion of the iris, the eyes are blue.

If the pigment is also present in the front part of the iris in a moderate amount, the eyes are gray or green.

If the pigment is abundant in both parts, the eyes are brown or black.

The thickness of the iris also modifies eye color.

There are no such things as blue or green pigments in the iris: the different eye colors are due to the different reflective properties of the brown-to-black pigment particles, as determined by their density, distribution, or scattering, and the thickness of the iris. Light of shorter wavelengths is scattered more in passing through tissues than light of longer wavelengths. The veins beneath the skin, the tattooer's black ink, and a piece of black lead beneath the skin appear blue for the same reason. The blueness of the eye comes about in much the same way as the blueness of the sky, which is caused by the reflection and dispersion of light from the particles of dust and other substances in the air. The eyes of all newborn babies of every human variety are blue. The reason for this is that whatever their genetic eye color (which will later become apparent), the pigment particles do not develop in the front part of the iris till some time after birth.

There would appear to be at least one pair of genes, $B, b$, involved in eye color, yielding the genotypes $BB$, $Bb$, and $bb$, $B$ being incompletely dominant so that dark brown would always be $BB$, light brown $Bb$, and blue $bb$. In each of these three genotypes there are many variations not only in shade but also in the distribution of several colors in the same iris, such as green, brown, yellow, and violet in one and the same iris. Also the two irises in the same individual may differ in color. Clearly many more than the pair of major allelic genes are involved to produce these varying eye-color effects; such gene pairs are called *modifiers*. Thus, two blue-eyed parents who are each of the genotype *bb* would give birth to children who were exclusively of the same genotype, but some of these children might in fact have sufficient brown pigment in their irises to cause them to be classified as brown-eyed, and this would normally be due to the action of modifying genes. Metabolic changes during embryonic development conceivably may play a role in some cases.

In albinos, owing to the failure of the general development of pigment throughout the body, the eyes appear pink—the pinkish color being due to the redness of the small blood vessels showing through the iris. It is interesting that the gene for albinism in rats, mice, and rabbits also affects the disposition of these animals. The albino animals are much more gentle, mild, and curious than even their own pigmented litter mates. There is some slight evidence that something of the same sort may also occur in man.

## Color Blindness

This has already been dealt with on pages 187–91.

## Hair Color

A hair is essentially a hollow cylinder, the central canal (medulla) of which is filled with pigment particles, air, and oils; the outer shaft (cortex) contains pigment granules in dark hair, and air in white hair. A single layer of flat scales (the cuticle) overlies the cortex. It is the interaction of all these parts and their contents that, in the presence of light, is responsible for the color of the hair. It is well known that under different varieties of light the color of the hair will appear correspondingly different. It is also known that under different environmental conditions the hair may vary somewhat in color; for example, light brown hair will often turn to blond during the summer as a result of the bleaching effect of the sun. During illness the hair may often lose its glossiness. But there is no evidence whatever, that will withstand a moment's critical examination, that hair is capable of turning gray overnight. None of the claims that have been made have withstood such critical examination.* It is

* For an excellent discussion of this interesting subject, see Bergen Evans, *The Natural History of Nonsense* (New York: Alfred A. Knopf, 1946).

## THE PIGMENTARY SYSTEM 247

extremely unlikely that hair could naturally change in color overnight for the simple reason that the visible hair is a dead structure that contains inert pigment, and it is quite impossible to see how this could be naturally affected. When hair starts turning gray, it is observed that it does so first at the root, while the terminal portion retains its color, so that we have a hair that is gray at the base and pigmented toward the end.

Hair color ranges all the way from the pigmentless straw color of the albino to the deep black to be found among all the peoples of the earth. The number of different hair colors in the human species must be very large indeed. No one has yet made a complete census of them. There is perhaps no other single mammalian species that presents so large a variety of hair colors as man.

Like that of skin color, the genetics of hair color is not at the present time clearly understood. A number of different genes are undoubtedly involved. Dark hair appears to be dominant over blond hair. Dark-haired parents may have children with almost any variety of hair color. Blond-haired parents always have blond-haired children and never have dark-haired children, though they may have red-haired children. These facts indicate that blond hair is recessive to dark hair, so that individuals with dark hair carry either two dark-hair genes or one dark-hair gene and one for another color, whereas blond individuals carry two blond genes.

Red hair seems to be produced by a distinct gene, and this generally behaves like the genes for blond hair, usually being recessive to the darker hair colors, brown and black. Dark-haired parents often produce a redhead. Hence, individuals with red hair carry either one or two genes for redheadedness. The offspring of two redheads will generally, but not always, be redheads. The people who produce the largest number of redheads are the highlanders of Scotland, 11 percent of whom exhibit the trait. In the cases in which only one gene for red hair is present, whether or not the hair of the individual will be red depends upon the "strength" of that gene in relation to its opposite member (its allele). If the gene for red hair is stronger than its opposite member or members, say for dark or blond hair, it will express itself as a function both of its own "strength" and that of the gene or genes for other hair colors. It is for this reason that we often see individuals with different shades of red hair, sometimes with red and brown hair on the same head (to be carefully distinguished from the variety that comes out of a bottle).

The genetics of hair color is complicated by the fact that the hair contains at least two pigments. One of these pigments, a granular melanin, varies in intensity from a light, golden blond through shades of sepia to black. The second of the pigments, carotene, varies qualitatively from yellow to dark red. Hair color is therefore a result of the intensity and the quality of pigment. There are genes that affect intensity, as illustrated by dark hair in relation to light, and genes that affect quality of pigment, as dark hair in relation to red. The fact that there are so many different shades of hair indicates that melanin pigment is dependent upon the action of several genes. Not only this, but these same genes appear to act differently with age. This is seen in many individuals who are blond-haired in childhood and become dark-haired as they grow older. Among the Australian aborigines, who all have dark brown hair as adults, children are frequently born who are fair-haired and who remain so until they approach puberty, when their hair invariably darkens and gradually assumes the dark brown color of the adult. Melanin production in such individuals apparently increases from childhood to maturity. To some extent this also appears to be true of skin color, although the changes in the darkening of the skin appear to proceed much more rapidly than in the case of hair color. The darkening of the hair with the advent of puberty appears to be the result of the action of the increase in the quantity of the sex hormones. The pigmentless hair of albinos is due to the pairing of recessive genes for hair color that lack all pigment-making capacity.

Partial albinism of the hair occurs in some individuals in the form of a white forelock. This is usually limited to the area of the hairline where it meets the forehead. When it is allowed to grow long and is combed backward, it gives the false appearance of a large tract of hair being involved. The condition is inherited as a simple dominant. It is a contemporary fashion among some females to dye the hair in imitation of such a forelock, usually in some blondish color. Partial albinism of the hair is generally associated with some form of partial albinism of the skin, if only a spotting of an extremity or of some other part of the body. A white lock of hair at the back of the head appears to be inherited as a sex-linked trait.

Graying of the hair is also dependent upon the action of a single dominant gene sometimes affected by modifying

## THE PIGMENTARY SYSTEM 249

genes. Graying in identical twins will usually occur at precisely the same time and in the same pattern. The age at which the hair turns gray is inherited as a dominant, as is the tendency to graying of the hair. In some individuals the hair never turns gray, even in extreme old age. I know of no investigations on this subject, but from my own random observation it seems likely that the condition of nongraying is inherited as a dominant. Blonds are said to have more hair on their scalps than dark-haired individuals, and redheads are said to have the least.

### Hairlessness

There are several different kinds of hairlessness (hypotrichosis), ranging from the congenital type to that which is induced by disease or disorder. Conditions of hairlessness, which in the one type are due to recessive and in the other type due to dominant genes, have been reported in man. These conditions are not to be confused with baldness that affects only the hair of the scalp, whereas general hairlessness may affect every part of the body.

### Baldness

Baldness has already been discussed on pages 191–96 as an example of a sex-limited trait.

### Eyebrows

Eyebrows may vary in every conceivable way. They may be thick, thin, scanty, cross over the nose, arched, or straight. The hair usually points away from the nose, but in many instances the hairs lying nearer the nose point in its direction or upward. Such traits are all inherited and, save thin and scanty eyebrows, appear to behave as dominants.

### Eyelashes

The character of the eyelashes, their thickness, length, and curvature are all inherited traits, and again appear to behave as dominants in the case of long lashes and recessive in the case of short ones. If the parents are homozygous for long lashes all the children will have long lashes, but if the parents are heterozygous for these traits some children will have long and some will have short lashes.

An interesting thing about eyelashes is that their color at birth is a good indication of the color the body hair will assume in its final form—excluding gray.

## Face

Mustache and beard hair are under the influence of separate genes, at least with respect to color. In some men the beard may be black or dark brown, but the mustache may be blond or red. The difference in coloration may remain throughout life or until graying sets in. The distribution of the hair on the face, its color, and form are all genetically determined and are traits that appear to behave as dominants. The growth of facial hair in men is largely controlled by the male sex hormones, the absence of these hormones being responsible for the lack of growth and development of facial hair in the female.

## Hypertrichosis of the Ears

The dense growth of hair on the outer ears, for the most part on the helix, or rolled portion, of the ears, known as auricular hypertrichosis, occurs only in men, and until recently was believed to be sex-linked to the Y chromosome and was therefore thought to be transmitted directly only from father to son. The gene responsible may, at least in some cases, be sex-limited, for the author knows of a case of two brothers one of whom exhibits the condition, whereas the other does not. Something of the difficulty of genetic investigation of man is illustrated by the fact that although the author was very friendly with the hypertrichotic brother, when inquiry was made as to whether he would be agreeable to a study of the condition in his family, he promptly and firmly declined with the words: "Why should I be a guinea pig!" Fortunately most people react in a more cooperative manner to such inquiries.

## Whorls and Cowlicks

Some males have a difficult time keeping the hair near the back of the crown from standing up like a clump of untidy hay. This is the hair of the hair whorl. It emerges near the top of the back of the head in a turn usually to the right in a clockwise direction. The clockwise whorl occurs in the majority of individuals, and is inherited as a dominant trait. The counterclockwise whorl found in some individuals is inherited as a recessive. The direction of the whorl determines the side on which the hair will be parted. Individuals with clockwise whorls part their hair on the left, those with counterclockwise whorls part their hair on the right. Parents with counterclockwise whorls will transmit them to all their children. Parents with clockwise whorls will transmit the clockwise whorl to all their children if they are homozygous for the dominant gene, or they will have some children with a counterclockwise whorl if they are both heterozygous for the counterclockwise whorl gene.

Double whorls occur, and are inherited as a recessive trait. They may both go in the same direction or in opposite directions in different individuals. More rarely triple whorls are observed. These, also, are inherited as a recessive trait.

Cowlicks are misplaced whorls. They may occur on the forehead, the extremities, the back, the chest, or on other parts of the body. In most cases they appear to be inherited in much the same way as the crown whorl.

## Body Hair

The growth, development, and distribution of the body hair vary in the two sexes. Body hair is generally much more profusely developed and distributed in the male than in the female and for the same hormonal reasons that influence the growth and development of the facial hair in the sexes. The pattern of growth and distribution of the body hair in both sexes is probably inherited in much the same manner as scalp hair.

## Hands and Fingers

The development of hair on the backs of the hands and fingers is also under genetic control. A good growth of hair on the back of the hand is inherited as a dominant trait; the absence of such a growth is inherited as a recessive. Similarly, the presence of some hair on the middle segment of the middle, ring, and little fingers is inherited as a dominant trait; the absence of such hairs on the middle segments is inherited as a recessive trait. It is extremely rare for hair to be present on the middle segment of the index finger. Why this should be so is a mystery. It is believed that at least

252  HUMAN HEREDITY

five genes are directly involved in influencing the presence or absence of mid-digital hair.

## Hair Form

Hair form is usually classified as straight, wavy, curly, kinky, and coiled (peppercorn tufts, as among the Bushmen of South Africa). When both parents have straight hair the children are all usually straight-haired. Wavy- or curly-haired parents if they are homozygous for waviness or curliness will have only wavy- or curly-haired offspring, but if they are heterozygous for these traits, some of their children will be straight-haired, some curly-haired, and some wavy-haired. Of course, each of these types of hair is variable in its character. Wavy hair may be characterized by low waves, medium waves, and deep waves, and the same applies to curly hair. Straight hair may be coarse, thick, or thin, and

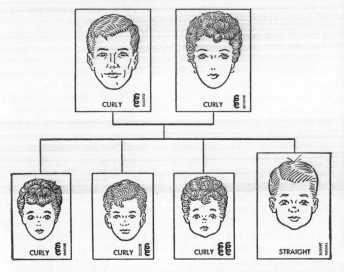

FIG. 26. Inheritance of Hair Form. The symbols stand respectively for curly and straight hair. Curly dominates straight. Here we see the parents heterozygous for curly and straight hair yielding a 3 curly to 1 straight hair ratio in their four offspring.

the number of hairs per square centimeter varies from person to person. The peoples of the Mongoloid major group of mankind have fewer hairs per square centimeter on the body than do members of the other major groups of mankind.

A kinky-haired person from a population that is completely kinky-haired will have offspring with kinky hair regardless of the hair form of the marriage partner. This fact establishes the dominance of kinky hair over all other forms of hair. When a curly-haired individual marries a wavy-haired person, the tendency is for the curly-hair gene to dominate over the wavy-hair gene. The wavy-hair gene tends to dominate over the straight-hair gene. Hence, it seems as if the greater the quality of curliness, the higher the quantity or degree of dominance.

Whereas kinky or woolly hair is a normal characteristic of Negroids, such hair has appeared as a mutation in whites of exclusively white ancestry. Three Norwegian families in which the mutation appeared have been described by Professor O. L. Mohr, of the University of Oslo, and the author of this book has observed the same condition in an American white female of exclusively white ancestry.

## The Adaptive Value of Hair

A feature so widely distributed over the body and so highly differentiated in the various parts of it as the hair in man is almost certainly of considerable adaptive value. The adaptive value of a trait is the extent to which it improves the chance of organisms, in the environments they inhabit, of survival and of leaving progeny, as compared with those of their kind not possessing the trait. What, then, is the adaptive value of the different types of hair on the body and its peculiar distribution? And what is the value of the different forms of hair and their distribution over the various parts of the body in the different major and ethnic groups of man?

These are questions to which the answers can be only conjectural, and our conjectures can be evaluated only in terms of a high or low degree of probability.

In the first place, hair is, as it were, a continuation of the nervous system. It is derived from the same embryonic layer as is the nervous system, namely, the ectoderm.*

---

* The three primary embryonic layers of cells from which the various organ systems of the organism are developed are the endoderm, the mesoderm, and the ectoderm. The innermost layer, the *endoderm*, gives rise to the linings of the digestive and respiratory tracts, of the digestive glands, of the bladder and urethra, and of the thyroid

Among its most important functions is the transmission of tactile sensations from the outside world. The sensory disturbance does not occur in the hair itself, for this has no nervous supply of any kind, but pressure of any kind on the hair is felt at the base of the hair follicle and the adjacent skin, which *is* supplied with sensory nerves, and it is through these that disturbances on the surface of the hairs are transmitted to the brain and spinal cord. Hair, therefore, among its many other functions, constitutes a sentry, as it were, standing guard on the outer ramparts of the body, announcing the arrival of visitors of whom one might not otherwise be aware. Civilized people who habitually wear clothes are not as sensitive to such disturbances as are nonliterate peoples who habitually go unclothed. Individuals living in nonliterate societies, though less hirsute as a rule than those living in civilized societies, are likely to be aware of the presence of an insect on their bodies long before the white would become conscious of it. The adaptive value of hair on the unclothed under primitive conditions as a protection against the unsolicited attentions of various noxious insects should be obvious, for tactile hairs are extremely important in apprising us of the advent upon our bodies of objects long before those objects reach our skins.

Hair in the armpits reduces the friction as the arm moves upon the body, and hair upon the pubic regions of the sexes may serve a similar function when two bodies are involved, as in coitus, as well as serving to intensify sexual sensations.

Head hair has a protective value in many ways: (1) It serves to protect the head from the elements, for example, against (a) excessive sunlight's penetrating rays, (b) heat, (c) cold; (2) it serves to maintain the normal temperature of the brain. The black kinky hair of the Negro is most efficient in these respects under conditions of tropical sunlight. The hair directly absorbs most of the sunlight, acting as a sort of air chamber, while the sunlight that does penetrate beneath the skin activates the scalp sweat glands. The sweat that is then secreted forms a wet blanket. The combination of air chamber and wet blanket serves to protect the under-

---

and thymus glands. The middle layer, the *mesoderm,* gives rise to muscles, bones, cartilage, the dentine of the teeth, ligaments, kidneys, ureters, ovaries, testes, heart, blood, the vessels that carry the blood and lymph, the external covering of the heart, and the linings of the chest and abdomen. The *ectoderm,* the outermost layer, gives rise to skin, hair, nails, oil glands, lining membranes of the nose and mouth, the salivary and mucous glands of the nose and mouth, and the nervous system.

lying tissues from damage from excessive light and heat.

The eyebrows may function to prevent sweat that may be running down the forehead from entering the eyes. The eyebrows may also serve to reduce some of the light entering the eyes by both absorbing some of it and also acting as a shade.

The function of eyelashes is apparently manifold. Eyelashes serve to protect the eyes by preventing particles from entering the eye, the same function as hairs in the nostrils perform, by acting as tactile sensory warning mechanisms, and by absorbing light (persons without eyelashes are forced to squint).

Why men have beards and mustaches and women don't is a mystery. If, as has been suggested by some writers, the beard serves to protect men against freezing temperatures, then why doesn't the biologically more valuable part of the species, woman, have a beard, too? The fact is that in freezing temperatures a beard is a disadvantage, for when the condensed water from one's breath freezes on one's beard, the skin beneath freezes all the more readily. Arctic peoples have very little hair on the face, and what little they may have they either pluck or shave. Mongoloid and Negroid peoples have very little beard or body hair. If the beard owes its origin to sexual selection rather than to natural selection, why are Mongoloid and Negroid men so sparsely endowed with these appendages? So the mystery remains. Perhaps baldness and beards are the purely fortuitous penalties that men must pay for being so constantly perfused with androgens.

# 14. THE FEATURES

## The Nose

THE NOSE is a very complex structure made up of many different parts, so that many genes must be involved in its construction. Yet, as everyone knows, the power of heredity is often strikingly exhibited in this unique projection—for among the Primates, the order of mammals to which man belongs, there is no other member that possesses such a bony cartilaginous prominence. One monkey, the Bornean *Rhinopithecus nasalis*, possesses a remarkable schnozzle, but this lacks the bony eminence of man's nasal bones and is mainly cartilage and soft tissue. Man's nose is due largely to the shortening of the projection of his jaws, which, as it were, has left this peninsula of bone and cartilage projecting into space.

In spite of the fact that different genes are involved in the formation of the different parts of the nose, the form of the nose may be inherited as a whole, which would suggest that the genes responsible for nose form are situated closely together on the chromosome. But this is highly probably not the case. Independent features of the nose are quite as often transmitted as such and not as part of the whole complex of genes derived from a single parent. Because these features are capable of being inherited independently, we shall understand the mechanics of the heredity of nose form much better if we study the individual features of the nose.

THE ROOT. The root of the nose is the part that joins the forehead, and is often miscalled the bridge. The bridge emerges from the root and corresponds to the anterior bony projection of the nasal bones and the cartilage that joins it. It is the bridge of the nose that runs into the tip below. The root of the nose may be flat, as in Mongoloids and to a lesser extent in Negroids, or it may be of medium shallowness, or high, as in the so-called Greek nose, in which the forehead seems to run without any break directly into the nose. The high root appears to be dominant over the medium root; the flat, or concave, root would appear to be recessive.

THE BRIDGE. The angle at which the nasal bones emerge from the root largely determines the degree of projection of the bridge. Hence, genes devoted to the root as well as genes devoted to the projection of the nasal bones participate in giving the bridge its particular form. If one inherits flat, or low, nose-root genes from one parent and high bridge genes from the other, a projecting, or concave, bridge may result. If the root is elevated, the bridge of the nose will tend to be straight. If the root is high, a prominent convex, or Roman nose, may result, and if the root is very high, a Greek straight nose may result. The genes for the prominent convex Roman nose are dominant over the genes for the straight nose. The genes for the high and narrow bridge dominate those for the low and broad bridge.

THE TIP. An amazing variety of nose tips is encountered, from sharp and depending to bulbous and upturned. Some tips exhibit a middle furrow. Round and square tips are encountered in whole families, as are retroussé, or upturned, ones.

THE NOSTRILS. It is not known whether the shape of the nostrils is inherited as an independent trait or whether it is dependent upon the form of the nose. The latter explanation would seem to be the most probable, for there is a definite correlation between the form of the tip of the nose and the wings and nostril shape, small noses with round tips being characterized by small and pear-shaped nostrils, while high bridged noses with square tips have long and slitlike nostrils. Nearly circular nostrils go with bulbous-tipped and concave noses. Such evidence as there is indicates that broad nostrils are dominant over narrow.

THE WINGS, OR ALAE. The wings of the nose are situated on either side of the nostrils, the nasal septum separating one nostril from the other. The wings may be flaring (popularly associated with hot-tempered individuals—an alleged association that has, however, never been scientifically investigated), or they may be more or less closely approximated toward the septum. They may be cut high, so as to expose the septum, or they may be situated at a level lower than the septum.

## The Upper and Lower Jaws

The front teeth of the upper jaw normally bite over the front teeth of the lower jaw, but in some individuals the lower jaw is longer than the upper jaw and the upper front teeth bite behind the lower front teeth. Sometimes, as in the case of the royal house of the Spanish Hapsburgs, in whom the condition has been traced back for more than six hundred years, a protruding lip is associated with this type of undershot jaw. Both conditions are due to independent dominant genes.

THE CHIN. The chin is unique to man; no monkey or ape possesses a protuberant front of the lower portion of the lower jaw, and no man possesses the kind of receding jaw that apes and monkeys do. But some human jaws give the appearance of being more receding than others, while some are definitely more projecting in the chin region than others. There is a common belief that a prominent chin denotes strength of character, but for this belief there is not the slightest scientific evidence.

When the chin protrudes beyond the plane of the face, it is described as prominent. When the chin is in the same plane as the face, it is described as normal, and when it is situated behind the plane of the face, it is described as receding. The receding chin is believed to be due to a recessive gene, the normal chin to a dominant gene, and the prominent chin to a dominant gene. A narrow chin appears to be recessive to a wide chin, and a long chin, of more than two inches from the level of the mouth, appears to be dominant to a short chin.

THE TEETH. Malocclusion of the teeth—that is to say, disordered relations in their biting surfaces—crowding, rotation, reduction in size or number, failure to erupt, impaction, even caries, are usually the consequences of genetic factors. Malocclusion of the teeth can usually be corrected but in some cases the genetic factor is so strong that even when the occlusion has been put into good shape the teeth will return to their former relations. I have known one such case to occur not less than three times—in the case of a dental nurse! Malocclusion appears to be due to a dominant gene that is sometimes incomplete or irregular in its expression. The other conditions named also appear to be due to a dominant gene of a similar nature. There is good evidence that the development of the different parts of each tooth is dependent upon many genes. For example, a tooth may develop perfectly except for complete absence of enamel or dentine or some other structure of the tooth. A search of the individual's pedigree will often indicate similar conditions, thus suggesting that each of these structures of the tooth is genetically conditioned.

It is not possible here to deal with the twenty teeth of the deciduous (shedding or milk) dentition or with all the thirty-two teeth of the permanent dentition. I shall therefore restrict myself to the tooth that is most frequently genetically affected, namely, the upper lateral incisor.

If you will look at the upper jaw of any monkey or ape, you will see that between the canine (eye) tooth and the lateral incisor there is a diastema (space) into which the projecting portion of the lower canine tooth fits. The bone that affords this space is the premaxilla. In the course of evolution man's teeth underwent a reduction in size, especially the canines (though the tips of which still project beyond the occlusal, or biting, level of the other teeth). This reduction of the canines rendered the premaxillary diastema quite unnecessary, and so it eventually disappeared. But in disappearing, the area involved has become somewhat unstable. Because this region of the jaw involves the development, meeting, and subsequent fusion of two separate bones, the maxilla and premaxilla, all within a period from the end of the seventh week to the end of the ninth week of embryonic development, the premaxillary area (in front of the eyetooth on each side) appears to be easily disordered, in some ethnic groups more than in others, with the result that anomalies of development all the way from an undescended lateral incisor to its complete absence or cleft palate and harelip may result.

Among modern whites the upper lateral incisors are missing in about 2.5 percent of the population. They are markedly reduced in size in about the same percentage of cases, are slightly reduced in about 17 percent, rotated in about 4 percent, crowded in about 7.5 percent, and in about 4 out of every 1,000 cases they are duplicated. It is interesting that the Chinese and Japanese exhibit a higher frequency of degenerate lateral incisors (7 and 4.7 percent respectively) but fewer missing lateral incisors (0.15 and 1.1 percent) than whites. Missing lateral incisors are rare among African Negroes, but occur in over 2 percent of mixed American Negroes. It is extremely rare to find missing lateral incisors among the native peoples of the Pacific, such as Polynesians, Melanesians, and Australian aborigines, so that in these peoples this region of the upper jaw and teeth is not an area of instability. Anomalies of the lateral incisor are slightly more frequent in females than in males.

CARIES. Caries, or tooth decay, is known to have a familial incidence. In 1946 Klein published a study of 5,400 individuals, in which he found that the highest incidence of decayed, missing, and filled teeth (DMF experience) occurred in individuals whose parents also had high DMF rates; the individuals with a moderate DMF experience had parents with a similar moderate DMF experience, and those with a low DMF rate had parents with a low DMF experience. When the DMF rate of the mother was low, differences in that of the father were closely related to the rate of the sons but only slightly to that of the daughters. When the DMF rate of the father was low, differences in the DMF rate of the mother were closely related to the rates of both sons and daughters. It was concluded that there exist strong familial bases influencing DMF experience that probably have a genetic basis.

More recently, in 1958, Horowitz, Osborne, and DeGeorge reported on a series of forty-nine like-sexed pairs of white twins drawn from middle-income residents of New York City and ranging in age from eighteen to fifty-five years. Some of these pairs were one-egg twins and others were two-egg twins. It was found that the one-egg twins showed significantly more similarity in their caries experience than did the two-egg twins, thus strengthening the conclusion arrived at in 1940 by Nehls in Germany in a similar study that a hereditary factor exists for susceptibility to caries.

Early decay of the teeth has been recorded in several families in three succeeding generations, the incidence being consistent with a dominant gene as the principal factor. In one family the eleven individuals in three generations who were so affected were all females; the only three males in the pedigree had sound teeth. In this family a dominant sex-linked gene might have been involved.

Early decay limited to a single tooth has also been recorded in several generations. Resistance to decay also appears to be markedly influenced by genetic factors.

## Harelip and Cleft Palate

Harelip and/or cleft palate occurs in approximately 1 out of every 770 births and is about twice as frequent in males as in females. The genetics of the condition in man is quite complicated and has not yet been fully worked out, but there is some reason to believe that the gene for harelip behaves as a recessive and that for cleft palate as a dominant with a variable degree of penetrance. Each condition may be inherited separately or together. There is good evidence that the development of the palate and adjacent structures is extremely sensitive to changes in the prenatal environment, and that severe emotional stress in the mother is capable of producing significant changes in this region (see pages 110–11). Support for this view is furthered by the fact that when one identical twin has harelip or cleft palate or both, in about one-fourth of the cases the other twin shows no signs of the condition. Harelip and cleft palate may be variously inherited as either a recessive, a dominant, or a sex-linked dominant. If the parents carry the genes for harelip and/or cleft palate in the recessive state, the chances are approximately 1 in 10 that they will have a child with a harelip and/or cleft palate. If one parent exhibits the condition, there is about a 2 percent chance of having an affected child.

## The Tongue

Ability to roll the tongue, that is, to bring the sides of the tongue over the top, is inherited as a simple dominant. About 65 percent of people can do it. The inability to roll the tongue is inherited as a recessive.

The ability to taste various substances is also inherited. This fact was discovered when a chemist found the taste of the synthetic chemical substance phenylthiocarbamide (PTC) somewhat bitter, whereas a colleague of his couldn't taste it at all. Various other substances are similarly experienced or not. The ability to taste PTC is due to a dominant gene; the inability to taste it is due to a recessive gene. There are remarkable differences among the ethnic groups of mankind in their ability to taste PTC, as shown in Table XV.

The fact that in some identical twins there is an intrapair discrepancy in the ability to taste PTC, one being a taster and the other a nontaster, suggests that this trait is not as firmly fixed as was at one time believed. Ardashnikov et al. found 3 out of 137 pairs, or 2.2 percent, of identical twins in which one was a taster and the other a nontaster. Verkade et al. investigated 70 pairs of identical twins and found 2 pairs in which one was a taster and the other not, that is, in 2.8 percent. Rife found 3 such pairs in a series of 31 pairs of identicals, that is, in 9.1 percent. Dencker et al. found 2 pairs among 28 pairs of identicals who were discordant for PTC on first investigation, but five months later both were concordant; four months later the same one was pronouncedly discordant again. Discordance in PTC tasting

TABLE XV

ABILITY TO TASTE PHENYLTHIOCARBAMIDE IN HUMAN POPULATIONS

| Population | Place | Investigator | Number Tested | Percent Tasters |
|---|---|---|---|---|
| Welsh | Five different towns | Boyd & Boyd (1937) | 237 | 58.7 |
| Eskimo, Unmixed | Labrador & Baffin Id. | Sewall (1939) | 130 | 59.2 |
| Germans | Copenhagen | Gottschick (1937) | 183 | 62.3 |
| Danish | Copenhagen | Hartmann (1939) | 596 | 62.8 |
| Russians | Zagorsk (n. Moscow) | Boyd & Boyd (1937) | 486 | 63.2 |
| Arabs | Syria (interior) | Hudson & Peter (1934) | 400 | 63.5 |
| Russians | Kharkov | Boyd & Boyd (1937) | 161 | 64.6 |
| American Whites | Montana | Matson (1938) | 291 | 64.6 |
| Yemenites | Yemen, Palestine | Yunovitch (1934) | 59 | 67.7 |
| Armenians | Syria | Berberian (1934) | 294 | 68.0 |
| Ashkenazic Jews | Palestine | Yunovitch (1934) | 245 | 68.5 |
| American Whites | Washington, D.C. | Parr (1934) | 439 | 69.1 |
| Eskimo, Mixed | Labrador & Baffin Id. | Sewall (1939) | 49 | 69.4 |
| American Whites | New York and vicinity | Blakeslee (1932/35) | 400 | 70.0 |
| American Whites | Ohio State University | Snyder (1932) | 3,643 | 70.2 |
| Swiss | Zurich and vicinity | Botsztejn (1942) | 544 | 70.4 |
| Scottish | Glasgow | Riddell & Wybar (1944) | 60 | 71.7 |
| Irish | Dublin | Boyd & Boyd (1937) | 398 | 71.8 |
| Sephardic Jews | Palestine | Yunovitch (1934) | 175 | 72.0 |
| Copts | Cairo, Egypt | Boyd & Boyd (1937) | 110 | 73.6 |
| English | London | Falconer & Fisher (1947) | 629 | 73.7 |
| East Georgians | Tiflis | Boyd & Boyd (1937) | 121 | 74.4 |
| Basques | San Sebastian | Boyd & Boyd (1937) | 98 | 74.5 |
| Egyptians | Cairo | Hickman & Marcos (1934) | 208 | 75.9 |
| American Negroes | Alabama | Howard & Campbell (1934) | 533 | 76.5 |
| West Georgians | Tiflis | Boyd & Boyd (1937) | 218 | 78.0 |
| Mohammedans | Cairo | Boyd & Boyd (1937) | 459 | 78.9 |
| Flathead Indians, Mixed | Montana | Matson (1938) | 442 | 82.6 |
| Mixed Amerindians | Lawrence, Kansas | Levine & Anderson (1932) | 110 | 87.2 |

TABLE XV (CONTINUED)

ABILITY TO TASTE PHENYLTHIOCARBAMIDE IN HUMAN POPULATIONS

| Population | Place | Investigator | Number Tested | Percent Tasters |
|---|---|---|---|---|
| Formosans, Chinese origin | Formosa | Rikimaru (1936) | 5,933 | 89.5 |
| Flathead Indians, Unmixed | Montana | Matson (1938) | 30 | 90.0 |
| American Negroes | Ohio | Lee (1934) | 3,156 | 90.8 |
| African Negroes | Kenya, East Africa | Lee (1934) | 110 | 91.9 |
| Amerindians | Alberta | Matson (1938) | 310 | 92.4 |
| Japanese | Japan | Rikimaru (1936) | 8,824 | 92.9 |
| Unmixed Amerindians | Lawrence, Kansas | Levine & Anderson (1932) | 183 | 93.9 |
| Chinese | Washington & New York | Chen & Chain (1934) | 167 | 94.0 |
| Formosans | Formosa | Rikimaru (1936) | 1,756 | 94.8 |
| African Negroes, Shilluk | Sudan | Lee (1934) | 805 | 95.8 |
| Amerindians | Northern Alberta | Matson (1940) | 559 | 96.9 |
| Navaho Indians | Ramah, New Mexico | Boyd & Boyd (1949) | 269 | **98.2** |

Adapted from Ashley Montagu, *An Introduction to Physical Anthropology.*

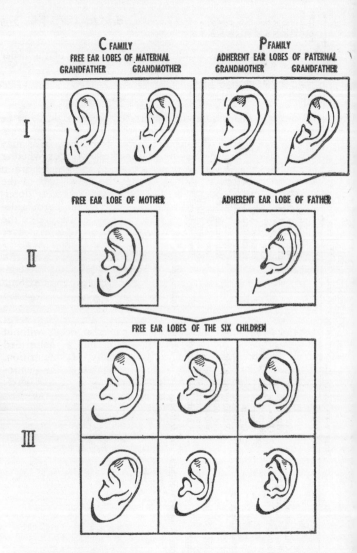

FIG. 27. A Family Pedigree of the Inheritance of Dominant Free and Recessive Adherent or Attached Earlobes. (From photographs after Powell and Whitney.)

among identical twins would appear, then, to be of the order of about 3.5 percent.

## Ears

There are many different features of the ears that can be inherited independently, indicating that a number of genes are involved in determining the form of the ears. Size may vary considerably, whether the rims are rolled or flat, whether a tubercle is present in the upper margin of the rolled portion of the ear, whether a pit is present at the upper part of the front of the ear, and so on. Pits behave as irregular dominants. The lobe of the ear may hang free, and this form is determined by a dominant gene. The lobe that is attached to the head is inherited as a recessive. Large ears are dominant over small ears.

Deformed or misshapen ears are frequently associated with malformations of the organs of the genitourinary system. Frequently a single kidney is seriously affected, but anything from partial to complete failure of development may affect any one or more parts of the system. The condition appears to be inherited as a dominant with approximately 70 percent or more penetrance. Ear deformity can also occur without malformation of the genitourinary tract, but a malformed ear in the presence of enuresis, difficulty of micturition, abdominal colic, or other symptoms referable to the genitourinary tract indicates immediate investigation of the latter.

# 15. THE BODY

## Stature

STATURE IS A COMPLEX THING, and growth in height is dependent upon a large number of genes, for height is made up of growth in a vertical direction involving the segments of the lower extremities, the hips, the various vertebrae of the sacrum and back and their intervertebral disks, the neck, and the height of the skull. Some individuals may be extremely long-legged but have fairly short trunks; others may be short-legged and have long trunks; some have long necks and others short ones; still others have high heads and others low ones. All these traits are controlled by genes, and enter into the determination of stature. Interestingly enough, there is some evidence that genes for tallness and shortness, as such, do in fact exist. Whether or not this be so, another interesting fact is that shortness appears to be rather more dominant than tallness, which, in effect, means that two short parents may have children of all heights, short, intermediate, and tall, because they may be carrying both "short" and "tall" genes, whereas tall parents will tend to have mostly tall children, for they carry paired genes for tallness in the recessive state and are unlikely to be carrying any "short" genes.

It should be fairly clear that if so many different segments of the body enter into the development of stature, this is a condition that must be markedly subject to the influences of the environment. And, indeed, this is precisely what we find —as we have already seen in Chapter 7. Socioeconomic environment, as expressed mainly through the nutritional factor, plays a highly significant role in influencing the development of stature. The average individual raised in a poor socioeconomic environment is likely to be several inches shorter than his offspring who have been raised in a satisfactory socioeconomic environment. When the environment has been fairly equal for parents and their offspring, then it is found that there is a significant correlation between parental stature and the adult stature of their offspring. Short adults are likely to come from short parents, and tall adults are likely to have had tall parents. However, in both tall and short families, the offspring tend to be taller than the shorter parent. You can try this experimentally by asking, in any sizable group of people, all those to raise their hands who are taller than the shorter parent. You will find that the majority will raise their hands. If you then ask those who are taller than the taller parent to raise their hands, you will find the number still appreciable but somewhat reduced.

## Body Form

There is a large variety of body forms, from short and broad to tall and scrawny. Again, the mode of inheritance of body form is quite complex because of the number of parts and genes involved. In general, however, it is clear that body form is inherited in much the same way as stature and is subject to much the same environmental influences. For example, obesity is often due to an excessive intake of food, and leanness to an inadequate intake of food. In addition, obesity and leanness are frequently the result of pathological disorders of a fairly large variety, mostly those affecting the endocrine glands. There is, however, no doubt that obesity and leanness, and most of the conditions in between, as well as body form are traits that may be conditioned by genes, regardless of the diet. There are whole families in which the body form is much the same in all or most of the members. The genetics of the subject is far from understood, but in some cases slenderness appears to be inherited as a recessive and obesity as a dominant, but in numerous other cases the mechanism of inheritance is not as simple as that.

There have been many different schemes by which the body has been classified, but not one of them has thus far proved satisfactory.

Being able to make a biologically sound, fairly reproducible estimate of the body type of an individual would enable us to make studies on the relation of body type to disease, temperament, immunity, and the like. Such studies have been made in recent years, but the results are quite inconclusive in spite of all claims to the contrary. Even the body types, or somatotypes, are quite arbitrarily standardized types. It is impossible to emphasize sufficiently the fact that all such studies are extremely difficult and must be viewed with the greatest caution.

If we are ever to make any progress in the understanding

of the genetics of body form, it is the genetics of the different regions that will have to be studied rather than the body as a whole. The principal fault of most studies that have been attempted in the past is that they have attempted to treat the body as if it were inherited as a whole, when in fact it is inherited as a large variety of distinct units or components in interrelation. The interrelation of these distinct components forms the complex mosaic whole, the body. Because these components are the expression of the modified action of different groups of genes on different autosomes, their varieties and combinations in forming a morphologic whole are virtually unlimited. Hence, any attempt to describe body types on the basis of the gross description of the organism as a whole is foredoomed to failure. Certainly the individual must be studied as a whole, but the description of that whole can satisfy scientific requirements only when the component parts that enter into its formation are analyzed and their interrelations properly understood. This is a task for a whole regiment of investigators rather than for the isolated student here and there.

## Body Type and Temperament

Sheldon and others before and since have attempted to discover whether or not there is any relationship between body type and temperament. It is generally agreed that they have failed to do so.

The riddle of physique and temperament is one that bristles with unsolved problems. It is also complicated by the fact that body type changes with different ages. In adolescence many girls pass through a stage of having excess fat, which they lose in the years prior to middle age and then, often, begin to reacquire. But some never do. The same thing often happens in boys. Bauer has shown that with age the chest tends to become more lateral in type, and there is a tendency toward abdominal fat. The superficial soft parts of the body are well known to be highly subject to environmental influences. Nutrition, occupation, exercise, and numerous other environmental factors will make no difference to a man's blood group or the shape of his nose, but such conditions will to varying degrees affect the size, proportions, and fatty development of the person.

The fact is that the more measurable traits that are included in any attempt to group men together, the more strongly emphasized does the essential individuality of the person become.

In man no genetic relationship has been described between body form or any part of the body and temperament. In rats evidence has been adduced that indicates that the Norway rat, which inherits a black coat color (nonagouti), tends to be tamer and less aggressive than the agouti, or gray-haired, segregates. In these rats the suggestion is that the effect of the genes upon behavior is pleiotropic, that is, it is the result of the action of many genes rather than being due to linkage with specific genes. Nothing comparable to such pleiotropic effects of genes associated with morphological traits is known in man except in certain pathological and abnormal conditions.

## Dwarfism

Extreme reduction in height and other dimensions of the body is usually due to genetic factors. There are two main forms of dwarfism, the strong-man or *achondroplastic* form and the Tom Thumb or *ateliotic* form.

ACHONDROPLASTIC DWARFISM. Achondroplasia (Greek *chondros*, cartilage; *plasis*, a molding) refers to a failure in development of the cartilaginous portions of the bones, resulting in a premature union of the cartilage with the bone. In achondroplastic dwarfism the bones of the upper and lower extremities and the base of the skull are particularly affected. The arm bone (the humerus) and the thigh bone (the femur) are especially affected, so that the trunk and head seem disproportionately large in relation to the extremities. The arms give a curved appearance because they cannot be straightened at the elbow, and the legs are often bowed, so that the individual tends to waddle when he walks. Achondroplastic dwarfism seems always to be inherited as a dominant and is believed to be due to a single dominant gene. Its occurrence in families in which there is no previous history of dwarfism indicates that achondroplastic dwarfism comes into being as a mutation in approximately 1 out of every 25,000 sex cells.

Achondroplastic dwarfs are of normal intelligence, and in medieval times were often employed as court jesters and sometimes attained positions of eminence. They also often served as models to painters. Achondroplastic dwarfs are perfectly fertile and produce an equal number of normal and dwarf children. The women have to be delivered by Cesarian section owing to the smallness of the contracted pelvis.

Among domestic animals the dachshund is an example of an achondroplastic dwarf.

ATELIOTIC DWARFS. Ateliosis (Greek *ateles,* incomplete) is essentially characterized by a failure of the body to grow normally. The skeleton tends to remain in its infantile state. Cartilage in many parts of the skeleton fails to develop into bone, and some of the growing portions of the skeleton, the epiphyses, often fail to appear at all. These are the midgets. Three types have been distinguished: (1) a rare fetuslike midget never exceeding thirty-six inches in height, (2) true midgets, and (3) miniatures, who are essentially miniature adults who have never attained adult proportions but have retained their infantile stature. Charles I of England's famous midget, Jeffery Hudson, was only eighteen inches tall when he was thirty years old. The famous Tom Thumb, a true midget, who was born in 1838, commenced his association with Barnum when he was four years old, at which time he was less than twenty-one inches in height. He grew, however, until in adult life he was somewhere in the vicinity of three feet tall. He married a dwarf, who bore him a child, which, alas, died in infancy.

Ateliotic dwarfism appears to be inherited in most cases as a simple recessive, due to a single recessive gene. Ateliotics seldom attain a height of forty-five inches, are sometimes achondroplastic, and frequently sterile. They are usually of normal intelligence, though sometimes of dull or defective mentality. Ateliotic dwarfism occurs rather less frequently than the achondroplastic variety.

## Pygmies

Pygmies are members of populations in which the height of the individual does not exceed five feet. There are several ethnic groups of Pygmies: the African Pygmies or Negrillos of Equatorial Africa; the Asiatic Pygmies or Negritos, including the Andamanese of the Andaman Islands off the tip of southern India, the Semang of the central region of the Malay Peninsula and East Sumatra, and the Negritos of the Philippine Islands; and the Oceanic Pygmies or Negritos, the New Guinea Pygmies. Although it has been claimed that there is some evidence of achondroplasia in the Congo Pygmies, this has never in fact been demonstrated, nor have there been adequate genetic studies of stature in any group of Pygmies. It is at present assumed that a mutant gene or genes occurred independently in these small, isolated populations and rapidly established itself, resulting in a short-statured population. Precisely what aspects of the skeleton have been affected is yet to be determined.

## Cretinism

We have already discussed cretinism (see pages 111–12) as a disease due to a deficiency in the hormonal secretion of thyroxine from the thyroid gland, and also as arising from environmental deficiencies of iodine in the soil. Thyroid deficiency in the mother during pregnancy is a frequent cause of cretinism in her offspring. However, it seems clear that there is a greater susceptibility to develop cretinism under iodine-deficient conditions in some individuals than exists in others, and this would suggest the possible presence of a recessive gene that renders its possessors more susceptible to cretinism than those who do not possess the gene. Early treatment with thyroxine greatly benefits the cretinous infant, though it seldom succeeds in restoring him to complete normality. The cretin usually remains very short in stature and severely retarded mentally.

## Gigantism

Gigantism is usually due to an oversecretion of the growth hormones from the pituitary gland, and this is a pathological condition. Such a condition is rarely inherited, but cases are on record of extreme tallness affecting most members of a single family. Scheinfeld has described a case of a hereditary, nonpathological giant who was 7 feet 4 inches tall. His maternal grandfather attained 6 feet 7 inches, his father 6 feet 4 inches, his mother 6 feet; two brothers were 6 feet 10 inches and 6 feet 4 inches, and a sister was 6 feet 4 inches. Extreme tallness or gigantism when it is nonpathological would appear to be due to a rare recessive gene, which seems to have been carried by the parents of Jacob Nacken, who is shown in Plate 18.

The human species shows great variability in the distribution of the genes influencing stature. Some ethnic groups show much greater variability than others. For example, Europeans on the whole show greater variability in stature than Pygmies. The Japanese would appear to exhibit a greater frequency of small-stature genes than do Europeans, though there is good evidence that much of the shortness of many

Japanese is due to environmental factors. Greulich, for example, has recently compared the stature and weight of 898 American-born Japanese with a similar number of children in Japan and found the American-born Japanese to be strikingly taller and heavier than the children in Japan. Nevertheless, there can be no doubt of the reality of genes for shortness and tallness.

Nilotic Negroes exhibit a very high frequency of genes for tallness. So do the Hamitic people known as the Watusi of Central Africa, who live in the vicinity of the northeast shore of Lake Tanganyika. Here, both men and women are seldom under six feet tall and many are over seven feet tall. The genes for tallness are well distributed in this group and have enabled them to maintain an aristocratic dominance over the surrounding tribes for centuries. Their contempt for both the white man and other non-Watusi is unconcealed. Among their other accomplishments they happen to be spectacular high jumpers, exceeding their own height at each jump with ease. Among the Pygmies genes for shortness established themselves in an isolated population, and among the Nilotic Negroes and the Watusi the opposite genes for tallness established themselves in these isolates, whereas among the much mixed and mixing, nonisolated Europeans such genes have never had the opportunity to become fixed—there has been too much infusion of new genes all the time for such a thing to happen. Hence, when one plots the distribution of stature for a European population, one finds that it assumes the normal, or bell-shaped, distribution curve. It is found that the majority of the population occupy an intermediate position at the top of the curve, being neither very tall nor very short, with the short and the tall occupying the lower parts of the curve.

## The Head

The size and the shape of the head may be independently inherited. Broad heads (brachycephalic) appear to be dominant to the long heads (dolichocephalic). Mongoloids tend to be brachycephalic, and Negroids tend to be dolichocephalic. (A rough statement of the ethnic distribution of head shapes will be found in Fig. 22, page 222.) Every form of head shape is encountered among whites, although among some peoples, such as the populations inhabiting the Alpine regions of Europe and therefore called the Alpine type, the head form shows a high frequency of brachycephaly. Scandinavians show a large frequency of long-headed forms. Evidently the genes for head form are variously distributed in frequency among the ethnic groups of mankind.

In individual families it is observed that broad-headedness tends to dominate long-headedness and that narrowness of the head tends to be recessive to long-headedness. However, as we have seen in an earlier chapter, head-shape genes are subject to the influence of the environment, the American-born children of long-headed immigrant parents tend to have broader heads than their parents, whereas the American-born children of broad-headed immigrant parents tend to have slightly narrower heads. The reason for this remains obscure.

People often ask whether or not the shape of the head can be permanently altered by artificial means. The answer is that it most certainly can. Children who lie on cradle boards often develop a flat back of the head. In fact, some anthropologists, not being aware of the fact that this occurred among Armenians in their homeland, brought into being a whole new "race," the so-called "Armenoid Race," the principal characteristic of which was supposed to be a flat occiput. Armenians living in America discarded the cradle board and ceased to have flat occiputs! Many American, Middle American, and South American Indian tribes habitually practice artificial cranial deformation upon their infants, whose bones then change form in adaptation to the types of pressures to which they have been subjected. The shape of the head then remains permanently changed after the pressures have been removed.

The next question that is asked is, Are such artificially induced changes in the skull capable of being transmitted to the offspring of such persons? The answer is that they are not. Artificial cranial deformation has been practiced for thousands of years without in any way affecting the genes. It has been widespread in Africa from early times and was practiced among the ancient Egyptians without in any way affecting the heredity of these peoples. *Artificially acquired characters are simply not inherited, no matter what the trait may be that is in question.* For a trait to be inherited, there must be some genetic factor involved. If such a factor is not involved, the trait cannot be due to heredity. It should be added that artificial cranial deformation in no way affects the functioning of the brain.

THE FOREHEAD. The form and characters of the forehead tend to be inherited quite markedly. This involves such features as the height of the hairline, the presence of temporal hair, slants, heights, widths, bulges, bosses (knobs), and su-

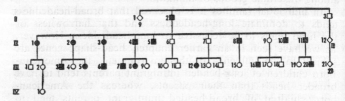

FIG. 28. A Pedigree of Extreme Susceptibility to Nasal Sinus Infection. (After Fraser Roberts, *An Introduction to Medical Genetics.*)

praorbital crests (crests above the eyes). These are quantitative traits, and it appears to be the general case that the larger quantitative expression of these forehead traits is due to dominant genes, the smaller expression to recessive genes.

The genes for frontal bosses, the paired bulges on the forehead, seem to be more frequently distributed among Negroes than among whites.

It remains to be said once more that none of these forehead features is in any way related to intelligence.

OCCIPITAL PROTUBERANCE. At the back of the head one can often feel a projecting area of bone. This is known as the *occipital protuberance*. It appears to be inherited as an incomplete dominant.

SINUSES. So far as I know no one has studied the inheritance of the sinuses of the head. Even now the function of these sinuses is not understood. They are, according to some theories, resonating chambers for the voice; according to others, they exist simply to lighten the weight of the head. Still others claim that they exist to provide ear, nose, and throat doctors with a living. I suspect that when the full story is in, it will be found that large sinuses are inherited as dominants of one kind or another, and that small ones are inherited as recessives.

The heredity of disease of the sinuses has, however, been studied in some striking cases. A case was reported by Gruneberg, cited by Fraser Roberts, of a man who suffered from chronic inflammatory disease of the nasal sinuses. His ten brothers and sisters were completely free of disease, as were their offspring and grandchildren, though among the latter there was one male who suffered from middle-ear disease (acute otitis media) and one who is said to have suffered from catarrh of the frontal sinuses. The brother originally exhibiting the disease, however, transmitted the susceptibility to ten males and six females in two generations. (See Fig. 28.) Thus the susceptibility to acute inflammatory disease of the paranasal sinuses was inherited in one line as a dominant. In the other line of the healthy brother, the descendants had not inherited such a genetic susceptibility, and so were free of the disease in spite of the fact that they lived in similar environmental conditions in the same towns of western Germany.

# 16. HEREDITY AND BLOOD

BLOOD IS A COMPLEX TISSUE. Many constituents enter into its formation. Each of these constituents is genetically determined, and the interaction of these constituents gives rise to characteristics of the blood that are widely and variously distributed among the ethnic groups of mankind. The study of the distribution of these genetically determined characteristics of the blood is proving helpful in studying the relationships of ethnic groups to one another. But more important than that is the practical use to which we are able today to put our growing knowledge—and that knowledge literally increases daily—of the characteristics of the blood. One of the most familiar of the uses to which that knowledge is put is in blood transfusion—a use of knowledge that has already helped to save millions of lives. The characteristics of the blood are also helpful in assisting us to determine whether or not a particular individual can possibly be the parent of a particular child, in forensic medicine, by identifying the characteristics of the blood from stains, and by typing the blood of potential parents prior to treatment, which would otherwise have been overlooked and without which the life of the fetus would have been endangered.

## The A-B-O Blood Groups

The A-B-O substances, which determine the four principal blood groups, were discovered by the Austro-American scientist Karl Landsteiner in 1900, and he distinguished them by

TABLE XVI
THE AGGLUTINOGENS AND AGGLUTININS OF BLOOD GROUPS
A, B, AB, AND O

| Blood Group | Agglutinogen (in red corpuscles) | Agglutinin (in serum) |
|---|---|---|
| AB | A and B | None |
| A | A— | anti-B |
| B | —B | anti-A |
| O | None | anti-A and anti-B |

the letters A, B, AB, and O. The four blood groups are determined by the fact that the red blood corpuscles (erythrocytes) contain two different antigens, denoted by the letters A and B. An *antigen* is a substance that when injected into the blood of an animal results, after some time, in the appearance of antibodies in its blood serum. The antibodies are known as *agglutinins,* and the antigens that produce them are known as *agglutinogens*. The agglutinogens, or agglutinative substances, can be present either singly, as in blood group A or blood group B, or together, as in blood group AB, or be altogether absent, as in blood group O.

The red corpuscles containing the agglutinogens float in the blood serum. In the presence of certain other serums the agglutinogens in blood groups A, B, and AB cause the red blood corpuscles to form clusters or clumps, that is, to agglutinate. The agglutination is produced by the two agglutinating substances, the agglutinins anti-A and anti-B, which are found in the blood serum of some persons. Once the blood corpuscles are agglutinated, the agglutinins then proceed to destroy them (hemolysis).

If a person is of the same blood group as another with whose blood serum some of his own blood is mixed, the blood corpuscles will generally disperse themselves evenly. This is because members of the same blood group do not carry substances that would agglutinate their own blood. This is illustrated in Table XVI.

It should be perfectly understandable why the blood serum could not, and does not, normally carry substances that would cause its own red blood cells to agglutinate, for it would not do to have one's blood cells sticking to one another while circulating through the body. The Landsteiner Rule, therefore, is that if an agglutinogen is absent from the red blood corpuscles of a person, then the corresponding agglutinin is present in the serum of that person. In blood

TABLE XVII
DETERMINATION OF BLOOD GROUPS WITH TWO TEST SERA,
ANTI-A AND ANTI-B

|  | Known Serum Anti-A (Blood Group B) | Known Serum Anti-B (Blood Group A) | Blood Group |
|---|---|---|---|
| Agglutination of the unknown blood corpuscles | − | − | O |
|  | + | − | A |
|  | − | + | B |
|  | + | + | AB |
| + = clumping of red cells |  | − = no clumping |  |

transfusion it is important to avoid introducing blood containing agglutinogens that can react with agglutinins present in the serum of the recipient, for the introduced blood would then be destroyed or agglutinated, blocking the kidneys and even causing the death of the recipient. It will readily be seen that the blood groups can be determined by testing the unknown blood corpuscles with anti-A and anti-B sera, or by allowing the unknown serum to act on known corpuscles of A and B. The manner in which this may be done is shown in tables XVII and XVIII. In Table XIX is shown the usual effect of adding a donor's blood to a receiver's serum.

In populations of European origin the commonest type of blood is O, a condition occurring in about 40 percent of the population. Since blood group O contains no agglutinogens,

TABLE XVIII
DETERMINATION OF BLOOD GROUPS OF SERA WITH KNOWN BLOOD CORPUSCLES A AND B

|  | Known Corpuscles A | Known Corpuscles B | Blood Group |
|---|---|---|---|
| Agglutination by the unknown serum | + | + | O |
|  | − | + | A |
|  | + | − | B |
|  | − | − | AB |

+ = clumping of red cells     − = no clumping

it was formerly given to receivers irrespective of their blood groups; persons of blood group O were therefore called "universal donors." The ab agglutinins of blood group O are generally rendered harmless by dilution or some other mechanism when group O blood is transfused to persons of other blood groups. It happens, however, that some persons of blood group O possess agglutinins of exceptionally high clumping power; the use of their blood is therefore dangerous since, even when diluted, it may destroy the red blood corpuscles of persons of other blood groups. For this reason donors of the same blood group are used whenever possible. Since persons belonging to the least common blood group, AB, approximately 5 percent of the population, possess no agglutinins, they were at one time considered to be capable of receiving the blood of any other group and were therefore called "universal recipients." But the red blood corpuscles of such "universal recipients" were occasionally clumped by the introduced donor's agglutinins. For these reasons the use of

HEREDITY AND BLOOD 279

so-called universal donors and universal recipients is restricted today to special cases. Blood group A occurs in about 40 percent of persons of European stock and blood group B in from 10 to 15 percent.

Three genes are responsible for the four blood groups, which are designated by the italicized letters *A*, *B*, and *O*. There is only one gene on each chromosome for the agglutinable properties of the red corpuscles. At fertilization the double number of chromosomes is produced and two genes for the blood groups now go into the making of each individual. There are now six possible pairs of genes that the individual can inherit, as shown in Table XX: *AA*, *AO*, *BB*, *BO*, *AB*, and *OO*. The blood group to which a person be-

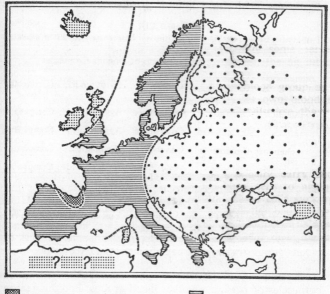

The Basques with High O, very low B and high Rh-negative frequency.

The Celts and other peripheral peoples with high O.

The main Western Europeans with high A.

The Slavs and other Eastern Europeans with high B.

FIG. 29. Blood Groups in Europe. (After Chalmers, Ikin, and Mourant.)

longs depends upon which pair out of these six possible pairs of genes he has inherited from his parents, only one pair of which each parent himself can have possessed. The genetic constitution of human beings with regard to the blood groups is therefore determined in the manner shown in Table XXI.

Genes *A* and *B* are of equal expressive value and therefore the substances that they determine occur together as recognizable agglutinogens. Gene *O* is masked by or is recessive to *A* and *B*, so that *O* is not expressed in the presence of the alleles *A* or *B*. Thus, for a person to belong to group *O* both of the parents must have carried the gene, either in a homozygous condition, where both genes were alike, or in a heterozygous condition, where one gene in each parent was *O* and the other either *A* or *B*. In the former event, all the

TABLE XIX

THE USUAL EFFECT OF ADDING A DONOR'S BLOOD TO A RECEIVER'S SERUM

| Agglutinins in receiver's serum | Agglutinogens of Donor's Corpuscles | | | |
|---|---|---|---|---|
| | (Group A) A | (Group B) B | (Group AB) AB | (Group O) None |
| (Group A) anti-B | Compatible | Agglutinated | Agglutinated | Compatible |
| (Group B) anti-A | Agglutinated | Compatible | Agglutinated | Compatible |
| (Group AB) none | Compatible | Compatible | Compatible | Compatible |
| (Group O) anti-A, anti-B | Agglutinated | Agglutinated | Agglutinated | Compatible |

TABLE XX

THE GENE COMBINATIONS OR GENOTYPES YIELDING THE PHENOTYPES OR BLOOD GROUPS

| Genotype | Phenotype |
|---|---|
| *AA* or *AO* | A |
| *BB* or *BO* | B |
| *AB* | AB |
| *OO* | O |

children would belong to blood group *O*, as, for example, is the case among such South American Indian tribes as the Chulupie of Argentina, the Guarani of Paraguay, and the Ona, Yahgan, and Alacaluf of Tierra del Fuego. Where *O* is carried by both parents in the heterozygous state, the offspring could belong to any one of the four blood groups. The mode of transmission of the genes in the latter case is

## HEREDITY AND BLOOD 281

illustrated in Fig. 30. From these facts it will be seen that the blood-group genes yield six genotypes and four phenotypes, as shown in Table XX.

The account given of the blood groups in the preceding paragraphs is accurate as far as it goes, but in fact blood group A is known to us in two forms, each with its own gene, designated $A_1$ and $A_2$, with $A_1$ apparently dominant over $A_2$, so that there are four rather than three genes involved in the genetics of the blood groups, giving rise to ten genotypes. The recognition of these subgroups is of the greatest

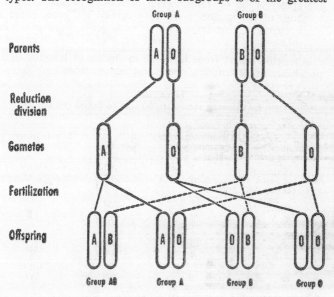

FIG. 30. Chromosome diagram showing the transmission of the genes in the mating of two persons, one of blood group A and the other of blood group B, each being heterozygous for blood group O.

value, for by their means our understanding of the problems they can help to solve is rendered so much more refined. We cannot, however, further develop the significance of the subgroups here.

### The Secreting Factor

The A-B-O blood-group factors may also be determined

from the natural secretions—the saliva, gastric juices, mucous secretions, and urine—of some persons. Such persons are termed "secretors." Persons whose secretions are nearly free of these factors (in water-soluble form) are termed "nonsecretors." The heredity of the secreting factor is extremely simple, two genes being involved, one of which is dominant, $S$, and the other is recessive, $s$, thus giving rise to three genotypes, as follows:

| Genotype | Phenotype |
|----------|-----------|
| SS | Secretor |
| Ss | Secretor |
| ss | Nonsecretor |

Altogether apart from its value in studies of heredity, the possibility of determining the blood factors from such secre-

TABLE XXI
GENETIC CONSTITUTION WITH REGARD TO BLOOD GROUPS

| Sperm containing chromosome carrying gene* | Ovum containing chromosome carrying gene† | Genotype | Blood Group (Phenotype) |
|---|---|---|---|
| A | A | AA | |
| A | O | AO | A |
| O | A | OA | |
| B | B | BB | |
| B | O | BO | B |
| O | B | OB | |
| A | B | AB | AB |
| B | A | BA | |
| O | O | OO | O |

\* These genes could be carried by chromosomes in the ovum.
† These genes could be carried by chromosomes in the sperm.

tions has enabled experts in rare cases to bring more than one criminal to justice from the evidence of his hereditary traits, carelessly left behind him on a discarded cigarette butt or on an envelope!

## Blood Types M, N, and MN

Approximately thirty years after the discovery of the A-B-O blood groups, Dr. Landsteiner and his colleague, Dr. Phillip

Levine, described another system of blood types depending upon two antigens, called M and N, and determined by a single pair of genes. When a chromosome in which gene *M* is located pairs with a chromosome containing gene *M*, the resulting blood type is M. When a chromosome containing gene *M* pairs with a chromosome with the *N* gene, the resulting blood type is *MN*. When pairing is between chromosomes containing *N* genes at each of their corresponding loci, the blood type is N. This is clearly brought out in Table XXII.

The M-N system is quite independent genetically of the A-B-O system, and no one thus far examined has lacked these M-N antigens, which are inherited as characters without dominance. Hence, it is a simple matter to work out the proportion of M-N types in children born to parents whose M-N types are known. Try it before looking at anything but the parents' blood types in Table XXIII.

In man the M and N agglutinogens or antigens seldom have any natural agglutinins or antibodies; hence, it is hardly ever necessary to take them into consideration when making transfusions. The test sera are obtained from rabbits that have been injected with human blood of types M and N and have developed antibodies against these antigens.

An important discovery made in 1947 by Sanger and Race of a new antibody intimately associated genetically with the M-N system has served greatly to enlarge the usefulness of this system for genetic analysis. This new antibody was found to agglutinate 72 percent of type M samples of red corpuscles, 60 percent of MN samples of red corpuscles, and 33 percent of type N samples. The agglutinogen or antigen thus agglutinated has been assigned the letter *S* for the dominant gene and small *s* for the recessive gene. The system, therefore, becomes the M-N-S-s blood-type system controlled, according to Sanger and Race, by two pairs of closely linked genes, with six phenotypes, or blood groups, and ten genotypes, as shown in Table XXIV. Wiener, however, considers that multiple rather than linked genes are involved—the genes $L^s$, *L*, $l^s$, and *l*—with nine phenotypes and ten genotypes, as shown in Table XXV.

## The Rh Blood Types

The significance of the Rh factor has to some extent already been dealt with on pages 97–99. The reader is advised to reread those pages at this point and to return to this page for a more detailed discussion of the genetics of the Rh types.

## TABLE XXII
### THE HEREDITY OF BLOOD TYPES M, N, AND MN

| Sperm containing chromosome carrying gene* | Ovum containing chromosome carrying gene† | Genotype | Blood Type |
|---|---|---|---|
| M | M | MM | M |
| M | N | MN | MN |
| N | M | MN | |
| N | N | NN | N |

\* These genes could be carried by chromosomes in the ovum.
† These genes could be carried by chromosomes in the sperm.

## TABLE XXIII
### THE INHERITANCE OF THE M-N BLOOD-GROUP SYSTEM

| Parents | Children | | |
|---|---|---|---|
| MN x MN | ¼ MM | ½ MN | ¼ NN |
| MN x MM | ½ MM | ½ MN | |
| MN x NN | | ½ MN | ½ NN |
| MM x MM | All MM | | |
| MM x NN | | All MN | |
| NN x NN | | | All NN |

Like the M-N-S-s series, the Rh antigens do not normally have antibodies associated with them in the blood serum, thus differing from the A and B antigens, which do. But the Rh antigens differ from the M-N-S-s series in that when they are introduced into the blood of an individual lacking in them, the Rh antigens will often induce the formation of antibodies. On a first transfusion no untoward results will occur if the donor and the recipient are alike on the A-B-O system. But on a second transfusion the antibodies that may be produced by the first transfusion will tend to agglutinate the red blood corpuscles, which may result in death. This mechanism has already been described in the case of maternal-fetal incompatibility.

That some individuals are born with an Rh factor (Rh positive) and others not (Rh negative) is due entirely to their heredity. It is now known that there are three principal Rh factors in the blood, the original **Rh$_0$**, which is by far the most powerfully and clinically the most important, the **rh′** and

## HEREDITY AND BLOOD

TABLE XXIV
THE M-N-S-s BLOOD-GROUP SYSTEM

| Phenotypes | Genotypes |
|---|---|
| MS | *MSMS* or *MSMs* |
| Ms | *MsMs* |
| MNS | *MSNS*, *MSNs*, or *MsNS* |
| MsNs | *MsNs* |
| NS | *NSNS* or *NSNs* |
| Ns | *NsNs* |

the rh″. Type rh (triple Rh negative) blood lacks all three factors, so that it can therefore be safely used in cases of intragroup incompatibility due to Rh factors.

The three elementary factors or antigens, $Rh_0$, rh′ and rh″, have three theoretically possible contrasting factors, designated $Hr_0$, hr′ and hr″ of which, however, only hr′ and hr″ have been found to exist. With rare exceptions, every blood group contains at least one factor of each of the pairs of factors rh′=rh′, and rh″=rh″. The Hr factors are less antigenic (that is, less capable of stimulating the formation of specific reacting substances) than the Rh factors. The three

TABLE XXV
NOMENCLATURE OF THE M-N-S-s TYPES

| Three M-N Phenotypes | | | Six M-N-S Phenotypes | | Nine M-N-S-s Phenotypes | | |
|---|---|---|---|---|---|---|---|
| Designation | Reaction with serum | | Designation | Reaction with serum | Designation | Reaction with serum | Corresponding Genotypes |
| | Anti-M | Anti-N | | Anti-S | | Anti-s | |
| M | + | − | MS | + | MSS | − | *LsLs* |
| | | | | | MSs | + | *LsL* |
| | | | Ms | − | Mss | + | *LL* |
| N | − | + | NS | + | NSS | − | *lsls* |
| | | | | | NSs | + | *lsls* |
| | | | Ns | − | Nss | + | *ll* |
| MN | + | + | MNS | + | MNSS | − | *Lsls* |
| | | | | | MNSs | + | *Lsl* and *Lls* |
| | | | MNs | − | MNss | + | *Ll* |

Adapted from A. S. Wiener and I. B. Wexler, *Heredity of the Blood Groups*.

elementary Rh factors, it is now known (together with the Hr factors hr′ and hr″), determine eight agglutinogens. Since the chemical constitution of the agglutinogens is unknown, they cannot be determined by chemical analysis, but their presence can be identified by their reactions with specific immune sera. Each of the eight agglutinogens thus recognized is determined by a particular dominant gene. This dominance means that if an agglutinogen is present in an individual, then that particular agglutinogen must be present in at least one of the parents. A single agglutinogen may have from one to three factors and therefore may react with as many sera. One genotype, $R^z r$, will react with all six antisera, which contain the antibodies, being divided into two subclasses, Rh and Hr. The eight allelic genes for the agglutinogens are written $R^0$, $R^1$, $R^2$, $R^z$, $r$, $r'$, $r''$, $r^y$. Since every person has but one pair of $Rh$ genes, a single one derived from the maternal pair and one derived from the paternal pair, there are thirty-six possible ways in which the eight genes can be combined. Using the six most frequently occurring genes ($R^z$ and $r^y$ being very rare), there are twenty-one possible ways in which these six genes can be combined. In other words, twenty-one genotypes are possible, which express themselves in eight Rh blood types, or phenotypes. The Rh type of the individual is therefore the expression of his genotype determined by a single pair of genes, each allele or member of the pair being derived from the opposite parent. The twenty-one possible genotypes produced by this series of six pairs of genes and the eight blood types to which they give rise are shown in Table XXVI. The frequency of the Rh genes and their corresponding agglutinogens as found among whites in the city of New York by Wiener, together with the positive reactions ($+$) and negative reactions ($-$) to the Rh and Hr sera, is shown in Table XXVII. It is a rule that in the case of homozygous individuals where there is a positive reaction to the serum of one set, there is a failure to react with the serum of the other set. On the other hand, in the case of a heterozygous individual, his agglutinogens will react with sera in both sets.

R. A. Fisher originally proposed the theory that the rhesus blood types are controlled by three genes, $D$, $C$, $E$, with their alleles, $d$, $c$, $e$. A single chromosome can carry either a $D$ or a $d$ gene but not both; hence there are three possible genotypes $DD$, $Dd$, and $dd$, and the same is, of course, true for the $C$ and $E$ genes and their alleles. Using this nomenclature an individual who is Rh positive possesses one or two genes that control the presence of the $D$ antigen. An

## TABLE XXVI
### SOME RH GENOTYPES AND PHENOTYPES

| Union of | | Genotype | Rh Phenotype |
|---|---|---|---|
| Sperm containing chromosome carrying gene* | Ovum containing chromosome carrying gene† | | |
| $r$ | $r$ | $rr$ | rh |
| $r$ | $r'$ | $r'r$ | rh' |
| $r'$ | $r'$ | $r'r'$ | |
| $r$ | $r''$ | $r''r$ | rh'' |
| $r''$ | $r''$ | $r''r''$ | |
| $r'$ | $r''$ | $r'r''$ | rh'rh'' (rh$_y$) |
| $r$ | $R^0$ | $R^0r$ | Rh$_0$ |
| $R^0$ | $R^0$ | $R^0R^0$ | |
| $r$ | $R^1$ | $R^1r$ | Rh$_1$ |
| $R^0$ | $r^1$ | $R^0r'$ | |
| $R^0$ | $R^1$ | $R^1R^0$ | |
| $r^1$ | $R^1$ | $R^1r'$ | |
| $R^1$ | $R^1$ | $R^1R^1$ | |
| $r$ | $R^2$ | $R^2r$ | Rh$_2$ |
| $R^0$ | $r''$ | $R^0r''$ | |
| $R^0$ | $R^2$ | $R^2R^0$ | |
| $r''$ | $R^2$ | $R^2r''$ | |
| $R^2$ | $R^2$ | $R^2R^2$ | |
| $r'$ | $R^2$ | $R^2r'$ | Rh$_1$Rh$_2$(Rh$_2$) |
| $r''$ | $R^1$ | $R^1r''$ | |
| $R^1$ | $R^2$ | $R^2R^1$ | |

\* These genes could be carried by the chromosomes in the ovum.
† These genes could be carried by the chromosomes in the sperm.
Adapted from Ashley Montagu, *An Introduction to Physical Anthropology*.

Rh-negative individual does not possess a $D$ gene. The D is by far the most frequently present antigen and producer of its corresponding antibody.

Fisher postulated that the genes controlling the Rh antigens were all closely situated together on the same chromosome—C on the one chromosome and c on the homologous
D                                                              d
E                                                              e
chromosome. Hence, during the reduction division, it is unlikely that these genes will become widely separated from one another on each chromosome and, hence, highly likely that they will reappear in close relation to one another. The exchange of chromosome material during the reduction divi-

sion between homologous chromosomes, called *crossing-over*, has not been observed to occur, a fact that renders the Fisher three-gene hypothesis somewhat dubious. Additional alternative antigens to those already known, such as $C^w$ and $D^u$, are considered to be alleles occurring at the C and D loci. If the Rh blood groups are controlled by three genes segregating together, then the six alleles *C, c; D, d;* and *E, e* (omitting from consideration the rare alleles $C^w$ and $D^u$) result in 8 different combinations or chromosomes, as listed below.

|         |                |
|---------|----------------|
| r       | *cde*          |
| r'      | *Cde*          |
| r"      | *cdE*          |
| $r^y$   | *CdE*          |
| $R^0$   | *cDe*          |
| $R^1$   | *CDe*          |
| $R^2$   | *cDE*          |
| $R^z$   | *CDE*          |

According to this theory, the child inherits the three-gene combination from each parent, so that if the genotype of one parent is *CDE/cde,* then the child receives *either* CDE *or* cde from that parent. Since two chromosomes carrying *R* genes are present in every body cell (the diploid number of chromosomes), this means that the eight different combinations can yield thirty-six different Rh genotypes and the addition of the alleles $C^w$ and $D^u$ make possible ten more combinations, of which only five have been determined, namely $C^wDe$, $C^wde$, $CD^ue$, $CD^uE$ and $cD^ue$, making a total of thirteen and yielding ninety-one genotypes.

The number of ways in which n things can be selected in pairs is derived from the formula $\frac{n}{2}(n+1)$. For example, one-half of each of the six alleles is carried in each of the pair of chromosomes in every body cell, yielding eight combinations of the three genes. We wish to know the number of ways in which these eight combinations of three genes can be arranged in a pair of chromosomes. Applying the formula n=8/2 = 4, and 8 + 1 = 9, and 9×4 = 36, we find there are thirty-six possible genotypes in an eight-combination system of genes segregating together. Similarly, in a thirteen-combination system we write n = 13/2 = 6.5, and 13 + 1 = 14, and 14 × 6.5 = 91.

As a matter of fact, there are at least seventeen allelic genes yielding 153 genotypes. In theory there are many more than these, for we can postulate combinations capable of

TABLE XXVII
THE RH SERIES OF ALLELIC GENES*

| Genes | Gene frequencies among New York City Caucasoids (percent) | Corresponding agglutinogens | Reactions with Rh serums † | | | Reactions with Hr serums † | | |
|---|---|---|---|---|---|---|---|---|
| | | | Anti-$Rh_0$ | Anti-rh' | Anti-rh'' | Anti-$Hr_0$ (hypothetic) | Anti-hr' | Anti-hr'' |
| r    | 38.0  | rh     | − | − | − | + | + | + |
| r'   | 1.4   | rh'    | − | + | − | + | − | + |
| r''  | 0.5   | rh''   | − | − | + | + | + | − |
| $r^y$ | 0.01 | rhy    | − | + | + | + | − | − |
| $R^0$ | 3.2  | $Rh_0$ | + | − | − | − | + | + |
| $R^1$ | 40.4 | $Rh_1$ | + | + | − | − | − | + |
| $R^2$ | 16.4 | $Rh_2$ | + | − | + | − | + | − |
| $R^z$ | 0.1  | Rhz    | + | + | + | − | − | − |

\* After Wiener.

† A single pair of genes is responsible for the individual's phenotype, e. g., genotype $Rh^1Rh^2$ yielding $Rh_1Rh_2$. By analysis of the phenotype in homozygous individuals the effect of a single gene may be inferred.

forming 288 different chromosomes with 41,616 different genotypes!

The distribution of the *Rh* genes among the ethnic groups of mankind is extremely interesting and most helpful in indicating possible origins and relationships. In Table XXVIII are shown the percentage of individuals (rather than the percentage of genes) in the various ethnic groups in whom the various blood types occur. From this table it will be observed that the Basques have an extremely high frequency of Rh negative, 28.8 percent. It is quite possible that the Rh-negative gene originated in Europe from a population of whom the Basques are the present-day representatives. However, there are small isolated populations in Switzerland that are characterized by even higher Rh-negative frequencies than the Basques. Such, for example, are the Western Walsers, who in one village (Tenna) show a frequency of 39.5 percent Rh negative. It will be seen that Mongoloids, Polynesians, Melanesians, and Australoids tend to be entirely Rh positive, and that there are interesting differences between other groups. In Fig. 29 something of the pattern of distribution of the main blood groups in Europe is shown.

## Other Blood Types

It is only possible to mention here some other blood types that have been discovered in recent years and that are under genetic control. These are:

## TABLE XXVIII

### Distribution of the Rh Blood Types

(As Determined by the Use of 3 Sera: Anti-RH₀, Anti-rh', Anti-rh")

| Population | Investigators | Number of Subjects (Rh+) Tested | \% Rh Types rh | Rh₁ | Rh₂ | Rh₁Rh₂ | Rh₀ | rh' | rh" | rh"rh' |
|---|---|---|---|---|---|---|---|---|---|---|
| Papuans | Simmons et al. (1946) | 100 | 100.0 | 0 | 93.0 | 0 | 7.0 | 0 | 0 | 0 | 0 |
| Admiralty Islanders | Simmons & Graydon (1947) | 112 | 100.0 | 0 | 92.9 | 0.9 | 6.2 | 0 | 0 | 0 | 0 |
| Fijians | Simmons & Graydon (1947) | 110 | 100.0 | 0 | 89.1 | 1.8 | 9.1 | 0 | 0 | 0 | 0 |
| Filipinos | Simmons & Graydon (1945) | 100 | 100.0 | 0 | 87.0 | 2.0 | 11.0 | 0 | 0 | 0 | 0 |
| New Caledonians (N&NW) | Simmons & Avias (1949) | 243 | 100.0 | 0 | 77.4 | 2.1 | 20.5 | 0 | 0 | 0 | 0 |
| Loyalty Islanders | Simmons & Avias (Incomplete) | 103 | 100.0 | 0 | 77.7 | 2.9 | 19.4 | 0 | 0 | 0 | 0 |
| Indonesians | Simmons & Graydon (1947) | 200 | 100.0 | 0 | 74.0 | 2.5 | 22.5 | 0.5 | 0 | 0 | 0.5 |
| Australian Aborigines | Simmons et al. (1948) | 234 | 100.0 | 0 | 58.2 | 8.5 | 30.4 | 1.3 | 0 | 0 | 0 |
| American Indians (Mexico) | Wiener et al. (1945) | 95 | 100.0 | 0 | 48.1 | 9.5 | 41.2 | 1.1 | 1.7 | 0 | 0 |
| American Indians (Oklahoma) | Wiener et al. (1946) | 105 | 100.0 | 0 | 40.0 | 17.1 | 39.1 | 2.9 | 0.9 | 0 | 0 |
| Maoris | Simmons et al. (Incomplete) | 32 | 100.0 | 0 | 25.0 | 31.0 | 41.0 | 3.0 | 0 | 0 | 0 |
| Japanese | Miller and Taguchi | 180 | 99.4 | 0.6 | 51.7 | 8.3 | 39.4 | 0 | 0 | 0 | 0 |
| Japanese | Waller and Levine | 150 | 98.7 | 1.3 | 37.4 | 13.3 | 47.3 | 0 | 0 | 0 | 0.7 |
| Chinese | Wiener et al. (1944) | 132 | 98.5 | 1.5 | 60.6 | 3.0 | 34.1 | 0.9 | 0 | 0 | 0 |

TABLE XXVIII

DISTRIBUTION OF THE Rh BLOOD TYPES

(As Determined by the Use of 3 Sera: Anti-RH₀, Anti-rh', Anti-rh")

| Population | Investigators | Number of Subjects Tested | (Rh+) | rh | Rh₁ | Rh₂ | Rh₁Rh₂ | Rh₀ | rh' | rh" | rh"rh' |
|---|---|---|---|---|---|---|---|---|---|---|---|
| Asiatic Indians (Muslims) | Wiener et al. (1945) | 156 | 92.9 | 7.1 | 70.5 | 5.1 | 12.8 | 1.9 | 2.6 | 0 | 0 |
| American Negroes | Wiener et al. (1944) | 223 | 91.9 | 8.1 | 20.2 | 22.4 | 5.4 | 41.2 | 2.7 | 0 | 0 |
| American Negroes | Levine et al. (1945) | 135 | 92.6 | 7.4 | 23.7 | 16.3 | 4.4 | 45.9 | 1.5 | 0.7 | 0 |
| Puerto Ricans | Torregosa et al. (1945) | 179 | 89.9 | 10.1 | 39.1 | 19.6 | 14.0 | 15.1 | 1.7 | 0.5 | 0 |
| White, Americans | Wiener et al. (1946) | 766 | 87.5 | 12.5 | 54.7 | 14.9 | 14.0 | 2.2 | 0.9 | 0.5 | 0 |
| White, Americans | Unger et al. (1946) | 7,317 | 85.3 | 14.7 | 53.5 | 15.0 | 12.9 | 2.2 | 1.1 | 0.6 | 0.01 |
| White, English | Fisher & Race (1946) | 927 | 85.2 | 14.8 | 54.9 | 12.2 | 13.7 | 2.5 | 0.7 | 1.3 | 0 |
| White, English | Murray (1946) | 1,038 | 84.7 | 15.3 | 54.8 | 14.7 | 11.6 | 2.3 | 0.6 | 0.7 | 0 |
| White, Australians | Simmons et al. (1945) | 350 | 85.1 | 14.9 | 54.0 | 12.6 | 16.6 | 0.6 | 0.9 | 0.6 | 0 |
| White, Hollanders | Graydon et al. (1946) | 200 | 84.6 | 15.4 | 51.5 | 12.3 | 17.7 | 1.5 | 1.5 | 0 | 0 |
| White, French | Bessis (1946) | 501 | 83.0 | 17.0 | 51.7 | 13.6 | 13.0 | 3.6 | 0.4 | 0.8 | 0 |
| Basques | Chalmers et al. (1949) | 383 | 69.5 | 28.8 | 55.1 | 7.8 | 6.0 | 0.6 | 1.8 | 0 | 0 |
| Basques | Etcheverry (1947) | 250 | 64.4 | 35.6 | — | — | — | — | — | — | — |
| Swiss (Davos-Sertig) | Moor-Jankowski & Huser (1958) | 34 | 67.6 | 32.4 | — | — | — | — | — | — | — |
| Swiss (Wiesen) | Moor-Jankowski & Huser (1958) | 66 | 65.5 | 34.5 | — | — | — | — | — | — | — |
| Swiss (Tenna) | Moor-Jankowski & Huser (1958) | 38 | 60.5 | 39.5 | — | — | — | — | — | — | — |

Adapted from Ashley Montagu, *In Introduction to Physical Anthropology*.

292 HUMAN HEREDITY

The P antigen due to a dominant gene, *P*, with its recessive allele, *p*.

The *Kell type* due to an incompletely dominant gene, *K*, with the recessive gene designated *k*.

The *Lewis type* due to a recessive gene. The gene for this antigen behaves as a recessive in the adult, but exerts an effect in the infant only in the heterozygous state.

The *Lutheran type* is due to a dominant gene, *Lu*.

The *Duffy type* is due to a dominant gene, $Fy^a$.

The *Kidd type* is probably due to a dominant gene, *Jk*.

The *Henshaw type* is due to a gene, *He*.

The *Hunter type* is due to a gene, *Hu*.

The *Sutter type* is due to a pair of allelic genes, $Js^a$ and $Js^b$.

The *He* and *Hu* types are closely linked to the M-N-S-s system.

In addition to all these "public" types of blood that are widely distributed through populations, there are also some "private" types of blood that occur occasionally in particular families. Nine of the latter have been described, but with further research it may be found that these "private" antigens are not as limited in their distribution as would at the present time appear. However this may be, it is obvious that a great deal more research needs to be done in this fascinating and important field of genetic serology.

One of the great advantages for the study of heredity of the blood groups is that a particular gene is responsible for a particular antigen in the red blood corpuscles, whereas in such traits as skin color, hair form, and the like, several genes are involved. The mechanism of inheritance of a trait in which several genes are involved is much more difficult to unravel than that in which a single gene is at work. In the latter case we are often helped in solving the problems of the workings of heredity because the variables involved are few and comparatively clear cut. A few examples will suffice to show how our knowledge of the blood groups helps us solve not only practical problems but also those of great theoretical interest.

When two things always occur in the presence of each other but are seldom found when the other is not also found with it, we have a right to suspect that they are connected. When this happens in the case of one trait in connection with a gene for some other trait, we may suspect that there is some sort of significant relationship between the genes for these traits. Such a relationship may be one of more or less close linkage or the expression of the multiple effects of a single gene.

Recently doctors Renwick and Lawler in England have brought forward evidence indicating that the abnormality known as the nail-patella syndrome was linked with certain blood-group genes. Individuals affected by this disorder have poorly formed fingernails, particularly of the thumb and index finger, absent or small kneecaps (patellae), and sometimes dislocated elbows and horny growths on the pelvic bones. The condition is due to a single dominant gene. Doctors Renwick and Lawler studied an affected man of blood group $A_1B$ who had conveniently married a perfectly normal woman of blood group O. They had had sixteen children, of whom eleven were living at the time of the investigation. Four of the children had inherited the father's $B$ gene, and it was just these four, two girls and two boys, who showed the nail-patella syndrome. The other seven children had inherited their father's $A_1$ gene and were completely free of the disorder. It would appear, then, that the nail-patella gene is on the same chromosome as the $B$ gene. Further study revealed that not all families showed such a close linkage, for in approximately 10 percent of cases there had been crossing-over, which had separated the two genes. In some cases the nail-patella gene was found to be linked with the $O$ and $A_1$ genes.

Linkages such as these not only enable us to make detailed maps of the distribution of particular genes on the chromosomes and to identify the relation to other genes on the same chromosomes but also enable us to predict the frequency with which such conditions will come into being in any family whose blood groups are known. In this way it is possible to prepare parents for problems, which being anticipated may be dealt with both practically and psychologically in a more satisfactory manner.

## Blood Groups and Disease

In recent years a number of independent investigators have found evidence indicating that persons belonging to different blood groups may differ substantially in their susceptibility to certain diseases of adult life. In 1951 Struthers in England drew attention to a significant relationship between the frequency of blood group A and bronchial pneumonia. In 1953 Aird, Bentall, and Roberts, comparing the blood groups of 3,632 individuals suffering from cancer of the stomach with a control series free of the disease, concluded that "the frequency of blood group A is greater and the frequency of blood group O less in patients suffering

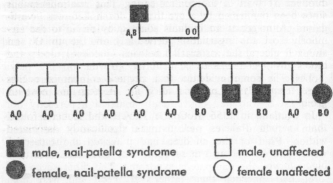

FIG. 31. Pedigree Showing Linkage Between A-B-O Blood Groups and the Nail-Patella Syndrome.

from cancer of the stomach than in the general population of the locality in which they live." For example, in Newcastle they found that blood group A occurred in 43.6 percent of the individuals with cancer, but in only 37.4 percent of the controls who were free of the disease. In other places the figures were as follows: Leeds, 47.9 cancer, 40.3 control; Manchester, 44.5 cancer, 38.4 control; Liverpool, 44.7 cancer, 39.6 control; Birmingham, 57.0 cancer, 44.4 control; London, 46.0 cancer, 42.2 control; Scotland, 36.4 cancer, 32.5 control.

The consistency with which blood group A is in every case significantly more frequent in the cancer victims than in those free of the disease is impressive. However, Speiser in Vienna and Wallace in Glasgow were unable to find any association between cancer and the A-B-O groups in large series of individuals examined by them. Nor have other investigators been able to find such an association. On the other hand, Haddock and McConnell found a significantly high frequency of blood group A in cases in which the cancer arose in the body of the stomach as compared with growths affecting the pyloric antrum of the stomach. Clearly, further careful research is required.

In 1954 Aird and his associates found that in a series of

3,011 cases from three localities in England the frequency of blood group O was significantly higher in patients with peptic ulcers (that is, ulcers affecting either the stomach or duodenum) than in the control series. This relationship has since been confirmed by more than a dozen different investigators. Although there remains some question as to the significance of the association between blood group O and stomach ulcers, the association between duodenal ulcers and blood group O seems to be definitely proved.

There is some evidence that pernicious anemia occurs more frequently in persons of group A than in those of group O.

In Scotland in 1956 McConnell, Pyke, and Roberts found that in men diabetes mellitus was significantly associated with a high frequency of blood group A, but in the case of women sufferers there was no significant correlation with any blood group.

Claims for the association of certain blood groups with other diseases have not thus far been substantiated. Such possible associations are at the present time being investigated by a number of different investigators, and we shall have to await their findings with patience.

Meanwhile, it cannot be too strongly emphasized that the association between blood groups and disease by no means implies a causal relationship. There is no evidence that it is blood of a certain type that produces the susceptibility to disease. It seems rather more likely that if an association exists between blood groups and disease—and this is denied by some authorities, such as Wiener and Wexler, and Manuila—the blood groups may simply represent indicators of some other factor or factors with which they are associated and which are more directly related to the susceptibility to certain types of disease—but what these other factors may be can at present be a matter for conjecture only. A clue to the possible factors involved has been suggested by Dr. A. J. Cain of Oxford. Dr. Cain points out that secretors pour a considerable amount of the antigens from their salivary glands into the commencement of the digestive system, the mouth. This at once suggests that something is being taken into the body, as part of the food, which it is advantageous to neutralize as quickly as possible. In 1948 Renkonen and in 1949 Boyd and Reguera independently discovered that the seeds of many leguminous plants widely used for food, such as the lima bean, contain large quantities of blood-group agglutinins that can be neutralized by secretors. This suggests that these antibodies may have a deleterious effect on some parts of the absorptive epithelial lining of the digestive tract. Cain points out that although such action may be of little importance to civilized man, it might have been of considerable significance under the far more strenuous conditions under which man lived in the prehistoric period. The finding of Clarke and others in 1956 that secretors had a considerably lower incidence of duodenal ulcer as compared with nonsecretors of the A-B-O blood-group antigens strongly suggests the actual nature of the deleterious action. Thus, secretors would have a definite selective advantage in this respect.

# 17. WHAT DO WE DO ABOUT HEREDITY?

## Questions

ACHIEVING SOME UNDERSTANDING of the meaning of heredity, of the manner in which heredity works, is a necessary step in the understanding of some of the fundamental problems that concern man in society. How many children should we have? Should we have any children at all? Should incorrigible criminals or persons affected with certain disorders be sterilized? Is it not true that as a consequence of our immense progress in medicine and sanitation we have enabled untold numbers of unfit individuals to survive who not only would be far better off dead but who also constitute an increasing burden to society and at the same time serve to debase the quality of man's germ plasm? Is not the national intelligence of the country being lowered by the unregulated breeding of the unfit and the mediocre while the fit and the highly intelligent scarcely reproduce themselves? Should not an intelligent society regulate the breeding habits of its members, so that the best are encouraged to reproduce most and the least fit discouraged from breeding altogether? Isn't there some connection between color and culture? It isn't really possible, is it, to make silk purses out of sows' ears? What about birth control? These and a thousand like questions are the questions frequently asked by men and women in our society.

## Ethics and Science

Our understanding of the facts of human heredity enables us to return sound answers to many of those questions, and in those instances in which knowledge is lacking we can frankly recognize our ignorance and take the necessary precautions against uninformed judgments or rash prescriptions for action. In any event, it should be quite clear that what we do about applying our knowledge of human heredity to the solution of human problems, while it may be most strongly supported by a knowledge of the scientific facts, is in the first place a matter of ethics and only secondarily a matter of science.

Ethics is concerned with standards of conduct and moral judgment, with the distinction between right and wrong, with leading, in short, the good life. What is right and wrong for human beings? What is the good life? Countless men have delivered themselves on these matters, and libraries of books have been written in the attempt to answer those questions. Without in any way desiring to slight the answers that have been traditionally given to these questions I will here offer an answer provided by a group of scientists in 1939. During the course of a discussion, the question arose whether it might not be possible to discover a naturally operative principle that governs human conduct. The discussants, professors Edwin Grant Conklin, C. Judson Herrick, Olaf Larsell, and Chauncey Leake, eventually arrived at the following general principle: "The probability of survival of individual, or groups of, living things increases with the degree with which they harmoniously adjust themselves to each other and their environment."

I have independently proposed a principle that is in essential agreement with this. It is that goodness in human conduct consists in behavior that confers survival benefits in a creatively enlarging manner upon others. In reality I think the latter principle complements the first. "Harmonious adjustment" and "in a creatively enlarging manner" are the key ideas in these principles, once we take the desirability of survival for granted. What is meant by "harmonious adjustment" and benefits conferred upon others "in a creatively enlarging manner"?

We mean, quite simply, that human beings stand the best chance of realizing their potentialities when they live cooperatively together, using their natural and social inheritance wisely; when they do not become the spendthrifts of that inheritance, but utilize the gifts both nature and society have brought to them to maximize the happiness of every member of society. Hence, if overcrowding causes misery, disharmony, and renders the harmonic development of the person difficult, then *ipso facto* it must be considered an evil, whether the overcrowding be familial, communal, national, or on a world basis. Therefore even if all individuals were constitutionally healthy, overcrowding or overpopulation must be considered an evil because it renders the harmonious adjustment of the individual to his environment difficult. Consequently we ought to regulate our numbers, on a familial, community, national, and international basis, in such a manner that we would avoid the evils of overcrowding or overpopulation and render possible the optimal development of the individual. As individuals and as a society we ought to think seriously about family limitation for those whose families ought to be limited and of ways of encouraging married couples to have children who ought to have many children. And we ought to think seriously also about the problem of world population, for even if we manage to control the quantity and the quality of our own numbers, unless we take a serious interest in these matters in other countries they eventually may serve to swamp us by the sheer weight of their own numbers. We must remember that thinking seriously about problems is an adaptive trait evolved during man's struggle with the environment in the effort to adjust himself to that environment, and truth and better adjustment are the success of that effort. Principally as a consequence of that effort to think seriously, man today holds the direction of his future evolution in his own hands. The knowledge we have gained of human heredity has placed a great power for good in our hands, but whether or not that knowledge will be used wisely and for good ends will depend largely upon two things: one is the extent to which a knowledge of heredity is understood by the people, and the other is their ability to think critically about that knowledge when it comes to its application to human and social problems. Scientists have, of course, a tremendous amount to learn about human heredity; indeed, compared to the amount that yet remains to be learned, we can truthfully say that only a beginning has yet been made in the understanding of human heredity, but that beginning, as we have seen in the preceding pages, has already been of immense practical value and, little though it is, has placed the control of man's future evolution within his power.

## Science and National Policy

A little knowledge, it has been said, is a dangerous thing—and so it is. The very greatest caution is necessary in dealing with all questions or recommendations that have as their object the influencing of the lives not only of the living but also of those as yet unborn. There are many people, some honestly well meaning and others dubiously so, who maintain that the restriction of immigration to "desirable types" and the "sterilization of the unfit" would greatly benefit our society. Unfortunately the voices of such individuals have sometimes made themselves so effectively heard that they have influenced national and state policies. It is perhaps not as well known as it should be that the United States Immigration Act of 1924 was based on the ill-considered, prejudiced, and unscientific judgments of the late Mr. Harry Laughlin, a worker in The Eugenics Records Office at Cold Spring Harbor, New York, supported behind the scenes by Madison Grant, the author of the notorious *The Passing of the Great Race*, and Charles Benedict Davenport, the director of the Office, to whose remarkable judgment on matters racial in his *Race Crossing in Jamaica* reference has already been made (see page 225). Laughlin was asked to report by the House Immigration Committee in April, 1920, on the relationship of biology to immigration, and their relation to social degeneracy. Laughlin submitted his report, entitled *Analysis of America's Melting Pot,* in November, 1922. The burden of the report was that "the recent immigrants as a whole present a higher percentage of inborn socially inadequate qualities than do the older stocks." On the basis of the listings in order of desirability for immigration submitted by Laughlin, both public and legislative opinion were influenced to produce an immigration act that was unsound and unjust, and that ultimately led to the absurdity of the McCarran-Walter Act of 1952, with its explicit statement that the capacity for Americanization of an individual was determined by his national or "racial" origin.*

In the light of our knowledge of human genetics and of the knowledge we have gained from the social sciences it must be said that the national policy of the United States embodied in its quota system and legalized by its immigration acts is from every point of view biologically and socially unsound. Science, in spite of every attempt to do so, has been able to discover no connection between the biological or genetic traits of any ethnic group or nation and a greater or lesser "percentage of inborn socially inadequate qualities." Nevertheless, and in complete contradiction and disregard of the findings of science, the government of the United States has created laws that discriminate against the members of certain ethnic groups and nations who desire to immigrate to the United States.

This is but one striking example of bad genetics and

* For a good discussion of this subject, see Oscar Handlin's *Race and Nationality in American Life* (Boston: Little, Brown & Co., 1957).

# WHAT DO WE DO ABOUT HEREDITY?

worse social science written into the law of the United States. The racism implicit in the immigration laws stands in strange contrast to the sentiments engraved at the base of the Statue of Liberty:

> Give me your tired, your poor,
> Your huddled masses yearning to breathe free,
> The wretched refuse of your teeming shore,
> Send these, the homeless, tempest-tossed, to me:
> I lift my lamp beside the golden door.

Racism is contrary to all the findings of human genetics and the social sciences.

The "sterilization of the unfit" is yet another of those panaceas frequently urged upon us if we are to save ourselves from the alleged biological degeneration that is destined to overtake our kind unless we resort to such preventive measures. Let us quote from a model Sterilization Law drafted by the same Harry H. Laughlin to whom we have already had occasion to refer in connection with the immigration laws of the United States. In a report entitled *Eugenical Sterilization in the United States,* published in 1922, sterilization of the "socially inadequate" is recommended. A socially inadequate person is defined as "one who by his or her own effort, regardless of etiology or prognosis, fails chronically in comparison with normal persons, to maintain himself or herself as a useful member of the organized social life of the state." The sick of body and mind who are only temporarily so, as well as the aged sick, are exempted from this definition.

We are informed: "The socially inadequate classes, regardless of etiology or prognosis, are the following: (1) Feeble-minded; (2) Insane (including the psychopathic); (3) Criminalistic (including the delinquent and wayward); (4) Epileptic; (5) Inebriate (including drug habitués); (6) Diseased (including the tuberculous, the syphilitic, the leprous, and others with chronic, infectious, and legally segregable diseases); (7) Blind (including those with seriously impaired vision); (8) Deaf (including those with seriously impaired hearing); (9) Deformed (including the crippled); and (10) Dependent (including orphans, ne'er-do-wells, the homeless, tramps, and paupers)."

The reader can judge for himself the kind of individuals who would have been directly affected had sterilization laws been in existence in their day, and also the significant number of individuals who would not have come into being at all had such laws been applied to "socially inadequate persons." One or other of the parents of many great benefactors suffered from one or other of the "socially inadequate" disorders listed by Mr. Laughlin. Had they been sterilized, humanity would have been all the poorer. As for the great men themselves, here are some of the "social inadequates" by the measure of Mr. Laughlin's definition (excluding the feebleminded, of course):

*Insane:* Lucretius, Isaac Newton, Nathaniel Lee, Strindberg, van Gogh, Nietzsche, Pushkin, Emily Dickinson
*Criminalistic:* Villon, Verlaine, Rimbaud, Wilde, O. Henry, Baudelaire
*Epileptic:* Dostoyevsky, Conrad, van Gogh, Caesar
*Inebriate or addicted to drugs:* Tennyson, Coleridge, Lamb, De Quincey, Poe, Modigliani, Dylan Thomas
*Diseased:* Gibbon, Rousseau, Keats, Pope, Pasteur, D. H. Lawrence, Robert Louis Stevenson, Elizabeth Barrett Browning, F. D. Roosevelt, Marcel Proust
*Blind:* Homer, Milton, Helen Keller, Louis Braille
*Deaf:* Beethoven, Helen Keller, Edison
*Deformed:* Pope, Robert Hooke, Byron, Charles Steinmetz, Toulouse-Lautrec
*Ne'er-do-wells:* Socrates, Diogenes, Shelley, St. Augustine, Gauguin, Thoreau
*Homeless:* Jesus, van Gogh
*Tramps:* Vachel Lindsay, W. H. Davies, Walt Whitman, George Orwell
*Paupers:* Jesus, Gandhi, van Gogh, Francis Thompson

Approximately twenty-eight states have adopted a sterilization law, but in most states the law had been only desultorily applied. The states having such laws on their books and the year of their enactment into law are shown in Table XXIX. Those who believe that sterilization would significantly reduce the number of "the unfit" usually operate on the naïve assumption that every individual sterilized would by so much, at least, reduce the frequency of the unfit. But this is not the case at all. Sterilization is not an effective or quick way of eradicating hereditary diseases or abnormalities. For example, in a population such as that of any large city in which random mating is the rule, let us suppose that 1 percent of the members of that population were affected by a simple recessive abnormal condition. Now

## WHAT DO WE DO ABOUT HEREDITY? 303

TABLE XXIX
STATES AND TERRITORIES WITH STERILIZATION LAWS, AND YEAR IN
WHICH A LAW WAS FIRST PASSED

| | |
|---|---|
| Indiana, 1907 | Montana, 1923 |
| California, 1909 | Virginia, 1924 |
| Connecticut, 1909 | Idaho, 1925 |
| Iowa, 1911 | Maine, 1925 |
| Kansas, 1913 | Minnesota, 1925 |
| Michigan, 1913 | Utah, 1925 |
| North Dakota, 1913 | Mississippi, 1928 |
| Wisconsin, 1913 | Arizona, 1929 |
| Nebraska, 1915 | West Virginia, 1929 |
| New Hampshire, 1917 | Oklahoma, 1931 |
| Oregon, 1917 | Vermont, 1931 |
| South Dakota, 1917 | South Carolina, 1935 |
| Alabama, 1919* | Georgia, 1937 |
| North Carolina, 1919 | Puerto Rico, 1937 |
| Delaware, 1923 | |

* Law has been inoperative since 1935, when State Supreme Court rendered an adverse opinion regarding broader sterilization legislation then pending.

let us suppose that all these individuals were sterilized. It would require four generations, one hundred years, to reduce their number to 0.5 percent; seven generations, or from 175 to 200 years, to reduce their number to 0.25 percent. If sterilization were initiated with a disease due to a simple recessive with an incidence of 0.5 percent, it would take six generations, or 150 years, to reduce its frequency to appreciably less than that incidence. The reason for this is that a large number of individuals carrying the gene in its heterozygous recessive state would not exhibit the defect associated with it but would transmit it. In addition, mutant genes would add to the number of individuals, born of apparently normal parents, who were affected.

Sterilizing the healthy relatives of the diseased individuals might serve to accelerate the reduction in the frequency of the abnormality, and of course would also reduce the frequency of the normal offspring that would otherwise have been born to such individuals, whether or not they were carriers of the defective gene.

One percent of the general population is believed to be affected by schizophrenic tendencies. It is considered that about 10 percent of the children of schizophrenics are likely to become schizophrenic. This means that 90 percent of the children of persons who develop schizophrenia will be normal. Does anyone have a right to sacrifice the normal 90 percent in order to prevent the abnormal 10 percent from coming into being?* I think there can be no doubt that no one has such a right nor should anyone ever be granted such a right. To remain with the example of schizophrenia, there is today good reason to believe that this dreadful disease is already becoming amenable to the newer treatments in psychiatry, and there is some promise that cures on a much larger scale than are possible at the present time may be effected in the not-too-distant future. Scientific research, not sterilization, is the proper approach to the problem of schizophrenia, as it is to every other hereditary disorder.

There are relatively few disorders affecting both parents or the genes for which are carried by them in their germ cells that are capable of affecting *all* the children born to them. As we have already seen in such a hereditary disorder as hemophilia, in which nearly every individual suffering from it used to die before reaching adult years, although it is today not possible to cure the disorder, and it may never be possible to cure hemophilia, nevertheless it is possible to control acute bleeding episodes, to relieve the sufferer of much pain and discomfort, and to enable him to carry on his work fairly normally. There is great promise that there will soon be a major breakthrough in the treatment of hemophilia, so that it will be possible to help hemophiliacs precisely as it is today possible to help diabetics. Once the hemophiliac reaches adult years, which, owing to the progress made by modern medicine, he does in many cases, he is able to reproduce. In so doing he increases the frequency of the number of individuals who will suffer from hemophilia. The question then arises, Is medicine doing the right thing by multiplying the possibilities for increasing the number of persons who will suffer from hemophilia?

That is an important question, and so is the answer to it. The answer is that medicine, while increasing the possibility of the number of individuals who will suffer from hemophilia, is by the same means increasing the possibility that hemophilia will become a much less serious disability, for

---

* In point of fact, the total reproduction of schizophrenics is considerably lower than that of any other comparable population. Severely affected schizophrenics are much less likely to produce children than the less severely affected, and as Kallmann has pointed out, in *Heredity in Health and Mental Disorder,* "a schizophrenic who has children is apt to have relatively mild symptoms and therefore a comparatively high degree of resistance to the disease. The factors producing this resistance will be passed on to the children. Where both parents are schizophrenic the children will inherit this capacity for resistance from both parents."

## WHAT DO WE DO ABOUT HEREDITY?

the means that make it possible for hemophiliacs to achieve adult years are the same means that will make it possible for them to live fruitful and happy lives—which large numbers of them are doing today. There are several hemophilic physicians working today in the field of hemophilia who have already advanced our knowledge of the treatment of that condition.

For the sake of those who are dubious about the benefits of sterilization as well as for the sake of those who are not, let me quote the case of a twenty-two-year-old Argentine hemophiliac reported by Dr. A. Pavlovsky, who has attended him since his birth. Dr. Pavlovsky shall present the case in his own words:

> There is no record in his family history of hemophilia until his generation: 3 hemophilic brothers and a sister carrier. The sister married and had 3 children, 2 hemophilic males and a daughter (carrier). He presents one of the severest forms of hemophilia I have ever seen. He has been on various occasions at death's door. For long periods of time this boy was bed-ridden and suffered severe pain during his many hemorrhagic crises. Mercifully today his disease has become less severe, as is generally the case when a hemophiliac reaches adult age. The only sequelae that remain are only slight ankylosis of both ankles and elbows. Always an outstanding pupil, this patient is now studying Law. He is a well-known chess player, and represented the Argentine in the "International Tournament for World Championship" in England. He has many interests in his life, amongst others, one that deserves mention. He uses his free time at week ends to visit lepers to whom he teaches chess. While the Argentine was under the dictator's control this boy gave an example of patriotism and courage. For 5 months he was imprisoned for defending his democratic ideals, and when freed he was immediately gaoled again for protecting the Cathedral of Buenos Aires against the mob sent to sack and burn it. On his release from his second term of imprisonment he at once volunteered to carry bombs and firearms for the revolutionaries. When asked his opinion this young hemophiliac said that in spite of his sufferings; his parent's anxieties; the limitations of his own life; the constant uncertainty of what the morrow might bring forth, he was glad to be alive

and to be able to help others. He added that should he marry, he would have children, as he considers that they should be able to bear what he has borne.*

Either the normal gene *H* had mutated to the abnormal gene *h* in this young man's mother or she was a carrier of the gene inherited in the usual way. Had her carrier state been detected, should she have been sterilized? I think the answer is clearly no. The whole world would have been the poorer had this young Argentine hemophiliac never come into being. Indeed, the world would be a far happier place if it were endowed with more persons of his character, no matter what hereditary disorders they suffered from, rather than millions of physically healthy nondescripts!

## The Mentally Defective

In the scale of humanity every human being has a value, whatever his qualities may be, and unless he is a complete idiot, even if he is feebleminded, as long as he is able to work he can be a useful member of society. As the great geneticist J. B. S. Haldane has said: "I am of the opinion that a man who can look after pigs or do any other steady work has a value to society, and that we have no right whatever to prevent him from reproducing his like." Indeed, many farmers prefer feebleminded men to look after their pigs. Another great authority on the genetics of the mentally deficient, Professor L. S. Penrose, writes in his book *Mental Defect*: "A striking feature of defectives of imbecile and lower . . . grades is their apparent incapacity for being bored with an occupation; and provided some simple manipulation can be taught, the defective is perfectly happy in continuing the same manipulation for days and years without any change. This fact makes possible methods of dealing with patients who might otherwise be difficult to employ. In a regular, even if very monotonous, employment they learn to be useful and worthy people."

Clearly mental defect is not so much a biological as a social problem, for in a society in which occupational status would be measured by ability, places could be found for persons of very limited ability. To a large extent our

* A. Pavlovsky, "The marriage problem and descendants of hemophiliacs," in *Hemophilia and Hemophilioid Diseases*, ed. by Kenneth M. Brinkhous.

society already does this, for the number of persons among us who are something less than bright runs into several millions. It is estimated that at the present time there are well over 2 million mental defectives in the United States. There are also at the present time approximately twenty-eight states scattered throughout the country that have laws permitting the sterilization of mental defectives. More than 60,000 individuals have been sterilized since 1907, when the first of these laws was passed.

Mental defectives do not reproduce themselves efficiently. The idiots, those who belong to the lowest grade of feeble-minded and are unable to do anything for themselves, have a mortality rate that is more than five times greater than that of the general population, and in the younger ages their death rate is nearly ten times greater than that of the general population. The imbeciles, who can learn simple manual tasks, have a death rate that is about twice as high as that of the general population. Among morons, who have a mental capacity reaching to the borderline of low normal intelligence, death rates are also higher at every age than for the normal population. The greater the degree of feeble-mindedness among the mentally defective, the lower is their reproductivity, so that a natural limiting effect is operative upon their multiplication.

The great majority of mental defectives are born to perfectly normal parents; hence, the sterilization of the mentally defective would scarcely result in a significant reduction in the number of mental defectives born.

In the borderline mentally defective, fertility is greater than average, but even this greater fertility may be of advantage to the human species. Such borderline individuals often carry genes in single dose that in the homozygous state produce greater mental ability. The offspring of such abnormal persons are frequently normal and above normal in mental ability, and they help to make up the loss of such persons resulting from the lower fertility of the mentally above average.

There are already promising evidences of new social and biochemical approaches to the problem of mental retardation. For example, Dr. Stephen Zimányi and his coworkers have reported successful initial improvement in the behavior of mentally retarded children treated with powdered RNA. Kirman and others have underscored the new social approaches to the problem of backward children: the mentally retarded should be integrated within the community and not segregated. The future is rich with promise that much will be achieved, both by social and by medical means, to greatly improve the lot of the mentally retarded and to assist them in becoming useful members of society. Preventive measures, genetic counseling and the improvement of the environment, would greatly reduce the incidence of mental deficiences. We can, indeed, already begin to see, in view of the great number of researches that are being carried out in these areas, as well as the interest shown by many authorities, the promise of the future in the process of being realized.

Tests have already been successfully devised that enable us to identify the carriers of the defective genes for a number of hereditary disorders that are unexpressed in the carriers but would be expressed in some of the offspring. For example, in Duchenne-Greisinger disease, or pseudohypertrophic form of progressive muscular dystrophy, which occurs particularly in young boys, and is usually inherited as a sex-linked recessive, it has recently been found by Aebi and his coworkers that the enzyme serum-creatinine-phosphokinase is significantly higher in known carriers of the gene than in noncarriers. This and other findings may soon enable us to identify most carriers and to predict which sisters and other female relatives of affected males would be likely to pass on the disorder to their sons.

It is now possible to determine very early in life which children are likely to develop this form of muscular dystrophy by serum-enzyme estimations. By measuring the amount of serum-aldolase in the blood, even before any clinical evidence of the disorder is manifest, it is possible to predict which children in a sibship containing affected individuals will develop the condition. The children who will develop dystrophy will exhibit a very high level of serum-aldolase.

Similarly, Teller and his collaborators have shown that abnormal qualitative excretion of chondroitin sulfate B in the urine of normal relatives of individuals with gargoylism probably indicates the heterozygous carrier of the disorder.

Dyck and his coworkers have shown that in peroneal muscular atrophy (Charcot-Marie-Tooth disease) carriers of the defective gene, as well as sufferers from its expression, exhibit a low conduction velocity in peripheral nerves. This test provides an extremely simple method of identifying the defective-gene carriers for this disorder.

We have already seen that in phenylketonuria and hemophilia it is now possible to identify carriers of the defective genes. Similar methods of identifying carriers of other hereditary conditions will undoubtedly be rapidly developed in the immediate future. By counseling carriers of such defective genes a very substantial reduction could be achieved in the frequency distribution of such genes and the disorders in which they are expressed.

As our knowledge of genetics develops we can already foresee the therapeutic and biologically creative benefits it will bestow upon man. By its means the abnormal or potentially abnormal individual may be restored to normality. At present this may be done either by compensating for the defect in the genetic material or its metabolic consequences, as in diabetes, which is treated by replacement of the missing material with insulin, or by preventing the overproduction of an undesirable compound, as in megaloblastic anemia, with its accompanying large accumulation of orotic acid, in which case treatment with the end products of orotic acid in the form of yeast nucleic acid has yielded good results.

Once the immunologic difficulties have been overcome, it will be possible by cell or tissue transplantation to introduce normally functioning genetic material into an individual defective in that material. By tissue culture, it will also be possible to replace in an individual's own cells a defective gene by a normal one.

As more is learned about the nature of DNA and the mechanism of gene mutation it will become possible to alter and improve genes within the living cell by a process of controlled mutation. As knowledge grows of the properties of nucleic acid, it may be possible to reconstitute better genes from mixtures of dissociated single-stranded DNA. It may even be possible one day to synthesize DNA enzymatically into new forms and incorporate them into cells.

"Sterilization of the unfit," apart from being a wrongheaded and unsound approach to the problems of the genetically defective, is not a substitute for research.

The truth is that the good will to help is blind and often cruel if it is not guided by true insight based on knowledge. A physician can be of genuine use only when the disease for which he is prescribing has first been carefully investigated. Only then can he be sure what his remedies can effect; otherwise he is a positive danger. Voluntary or compulsory sterilization of "the unfit" is an ill-conceived prescription put forth by physicians with too little knowledge of the disorder for which they are prescribing.

Every so often, some self-appointed pundit, often speaking

with the authority of some institutional affiliation that renders him in his own opinion and in that of the public an expert on the things he is least qualified to judge, delivers himself on the subject of "man, the mechanical misfit." Man's upright posture is blamed for his pendulous abdomen and his back troubles. The obvious fact that man's pendulous abdomen may in large part be due to his poor habits of posture, eating, and exercise, and his back troubles may be due to the beds he sleeps in and the chairs and automobile seats he spends so much of his life in are altogether overlooked. Such authorities also proceed to inform us from time to time that man's social troubles are principally due to his biological deterioration, to the carelessness with which he allows the defective to reproduce and "pollute" the germ plasm of the species.

It is, of course, true that there are many individuals among us who are genetically more or less seriously defective and who are capable of transmitting their defects to a certain proportion of their descendants. In the case of rare recessive defects such as infantile amaurotic idiocy, nothing has to be done since these conditions in large populations eliminate themselves by the early death of the victims. In the extreme cases, such as idiocy, social regulation is necessary, not surgical intervention. Surely, no one will *seriously* suggest that such individuals are responsible for our social ills. Our troubles emanate not from biological idiots but from social ones; and social idiots are produced by society, not by genes. It is therefore social, not biological, therapy that is indicated.

## The Problem of Defectives

Owing to the fact that a certain proportion of normal genes mutate in every generation to the abnormal form, genetically transmissible defective genes will always be coming into being afresh in every human society—but so will the knowledge that will enable us to deal with them. At the present time we are rapidly developing new methods of detecting carriers of defective genes even before those genes have a chance to be transmitted or to express themselves. For example, deficiency in Ac-globulin, the clotting factor, can be picked up by an analysis of the blood of the affected individual. Such a deficiency is due to a recessive autosomal gene. In the homozygous condition the defective gene produces the deficiency disease parahemophilia. Thus, by analysis of the blood of potential parents who are both carrying the gene it would be possible to tell them what the effects are likely to be upon their children if they decide to marry and raise a family. It is similarly possible to do this for many hemophilia-like conditions as well as for several other unrelated conditions, for example, possible erythroblastosis in an infant of Rh-incompatible parents, A-B-O incompatibilities, and the like.

These are considerable achievements. We have every reason to be confident that future advances in genetics will put the control of human heredity even more efficiently within our power. Even the control of mutations is today being discussed by geneticists as a practical possibility. We can therefore look to the immediate future hopefully and for the present realize that there is nothing that such severe measures as sterilization can do that soundly based social control cannot do a great deal better in every way. There is no substitute for sound knowledge humanely applied.

All of us are carriers of some deleterious genes. Such genes are usually recessive. Under a system of random mating such as exists in most civilized countries the accident of marriage sometimes brings such recessive genes together, which will then express themselves to some degree in some of the offspring issuing from such a union. When deleterious genes are carried in the recessive state, there is usually a history in the family of the expression of the gene—in a grandparent or uncle or aunt and in some of their offspring. All of us ought to draw up a pedigree of our families, as far back as we can, and extend it laterally as far as we are able. It is a simple matter to prepare such a pedigree. A sample pedigree is shown in Fig. 32. You can draw up separate pedigrees of the same family for separate traits or you can put them all into one pedigree. Apart from being fun and instructive, such pedigrees ought to be as obligatory as a birth certificate or a marriage license. A couple contemplating marriage could go over their pedigree charts and from them discover whether they stand in need of genetic counsel. In any event, they could, on a purely voluntary basis, take their charts to a genetic counselor,* who could then advise them. While human beings continue to select their mates for reasons that seldom include a mutual inspection of their genes it is foolish to talk of eugenics, or selective breeding from the best stocks.

---

* Genetic counseling services are available in several of our hospitals and at clinics attached to some of our universities. See Appendix D, p. 380–81.

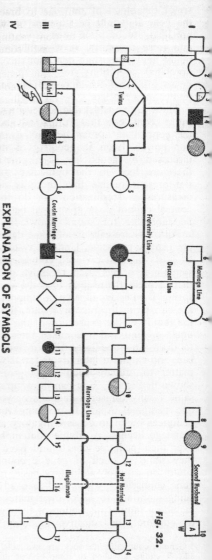

Fig. 32.

## EXPLANATION OF SYMBOLS

**(a) Standard for all pedigree charts:**

☐ = Male; ○ = Female; ◆ = Sex unknown; ● = Still-birth or Miscarriage, × = Children —Number and sex unknown; d.Inf. = Died in infancy;
☐☐ = Parents; ○○ = Twins; ⚡ Points to the propositus or central figure in the pedigree.
○○○○ = Children.

Roman figures to the left indicate generations; arabic figures locate individuals (thus III.7 is the young man in the third generation who married his cousin). Offspring should be listed in the order of birth; for example, in generation III on the chart the four siblings of the propositus are first a brother, then two brothers and a sister in the order of birth in relation to the propositus in which they occurred.

The following letters, placed in or around the individual's pedigree symbol, are standard for certain traits: A-alcoholic; B-blind; D-deaf; E-epileptic; F-feebleminded; I-insane; M-migrainous; N-normal in reference to traits under consideration; Ne-neurotic; P-paralytic; S-syphilitic; T-tuberculous.

**(b)** To fit the particular family and traits (whether physical, mental, or temperamental, good or bad) under consideration, invent special symbols, or select special letters or letters in combination with numbers (in addition to those standardized above) to be placed within or near the particular individual's pedigree symbol, to indicate particular traits and their degree of development.

## WHAT DO WE DO ABOUT HEREDITY? 313

Human beings will continue to breed at random for a long time yet, and it is perhaps just as well that they do. And if under this system of random mating, which has served humanity so well in the past, some intelligent individuals marry some feebleminded ones, it is a real question whether in the long run many of the genes carried by the feebleminded may not be just as valuable to the human race as those carried by the normally intelligent.

We must take the world as we find it and do what we can to leave it a better and a happier place than we found it. Toward this end the most important role the individual can play in any society is that of a parent. The creation of a good human being is the greatest contribution the individual can make to his society. And there can be nothing more gratifying to the individual than the making of such a contribution. This being so, the first consideration of the potential maker of a human being should be not what the biological characteristics of his offspring are going to be, but whether he, the potential parent, is prepared to do his best to make the lives of his offspring as useful and happy as possible, whatever the biological characteristics of his offspring may be. In the final analysis I think this is the long and the short of it.

This means that potential parents who are fully aware of the fact that they will almost certainly bring a certain number of children into the world who will suffer from certain hereditarily determined handicaps should be perfectly free to do so, if they thoroughly understand what they are about and will always be ready to do what needs to be done to help their handicapped children realize their lives as fully as possible. But when the potential parents know themselves well enough to understand that their own personalities are such that they would prove far greater handicaps to a happy family life than any genetic handicaps that their children would have, it might be the path of wisdom for them to desist from having children.

In his excellent little book *Counseling in Medical Genetics,* the director of the Dight Institute for Human Genetics at the University of Minnesota, Dr. Sheldon Reed, writes, "Our serious clients come to us because they are troubled. They show great affection for their abnormal child and give it more than its ordinary share of attention, but the parents are unhappy both for the defective child and for themselves. We have never seen parents who wished to repeat the misfortune." There can be little doubt that in the vast majority of cases this is true, and that most parents will wish to avoid bringing defective children into the world, not because of the genetic hazards to the species but because they are themselves rendered unhappy by the presence of such children. This is, however, not always the case. I know of one husband and wife of more than average intelligence who in spite of the fact that their first-born was a hydrocephalic, the second-born a normal, and the third-born a Down's are still bent on having children, for they have found both the hydrocephalic and the Down's very lovable indeed, and the whole family is a happy one. But this is the exceptional case. I see, however, no good reason why it should not become more frequently the case that parents with defective children are able to accept the condition of their children as a challenge to do the best they can for them, and to be quite happy doing so. The matter, however, is by no means as simple as all that, for as Dr. Bernard Farber has shown in a recent study, the presence of a severely mentally retarded child, especially a boy, can have a seriously disrupting effect upon marital relationships and may adversely affect the personality development of the normal siblings. These are considerations that all parents faced with the possibility of producing a retarded or otherwise defective child must be prepared to give their most careful attention.

Old-fashioned ideas and artificial standards often conspire to make the parent of the defective child feel ashamed, even guilty, as if some offense or solecism had been committed. But in the light of modern genetics and medicine it is today fully understood that such defective children could occur in any family, and that no matter what family they have been born into, they should be no more an occasion for shame or guilt than the development of pneumonia or poliomyelitis in ourselves. Both diseases are avoidable ones, but a great deal of luck affects the avoidability, and so it is with the birth of defective children. What is called for is understanding and the proper treatment. A good environment can make up for many of the defects of the germ plasm.

It is not only a knowledge of human genetics that is necessary but also a knowledge of human nature, especially self-knowledge, when we come to deal with the behavior of ourselves or of other human beings. But a knowledge of genetics, and especially of our own genetic history, is immensely valuable in assisting us to determine what our behavior should be, if not in relation to marriage, that is, to whom we marry, then in relation to whether or not to have children. In this connection one of the most frequent ques-

# WHAT DO WE DO ABOUT HEREDITY? 315

tions that comes up is the matter of *consanguinity*, that is, whether or not to marry a "blood relation." As I have pointed out earlier in this book, blood has nothing whatever to do with determining genetic relationship. What the question "Shall I marry my cousin?" in fact means is, "Shall I marry an individual whose genes are in part derived from the same source as one of my parents?"

Since there are at least twenty-four states that forbid the marriage of first cousins, the question is a legal as well as a genetic one, involving the legal status of offspring, their right to inherit possessions, and their inheritance of possibly deleterious genes.

On this subject there has been a great deal of loose thinking. Yet the facts are quite simple. If the marrying partners come from a good healthy stock, that is to say, one in which no significantly disturbing disorders have been noted, then the chances are high that the offspring will not exhibit any such disorders. We have pedigrees of many distinguished families that bear abundant testimony to this fact. Brother-sister marriage was not uncommon in ancient Egypt and was the rule among the pharaohs. A prince was held to have an unquestionable right to the throne only if he was born of parents who each had equal rights to the throne. By the beginning of the Eighteenth Dynasty (approximately 1580–1350 B.C.), when brother-sister marriage had already been long in practice, there appeared a succession of the most brilliantly gifted rulers Egypt had ever known. A few of the royal personages involved are shown in the following pedigree.

It was Aahmes I and his sister-wife Aahmes-Nefertari who expelled the Hyksos and founded the Eighteenth Dynasty. Aahmes-Nefertari was later deified. Perhaps the most remarkable in the line was Queen Hatshepsut, whose father and mother were half brother and sister, her mother, Aahmes, being the result of two successive marriages between full brothers and sisters. Hatshepsut's husband, Thothmes III, is believed to have been a cousin of the queen.

The Ptolemies, who ruled Egypt from 323 to 30 B.C., usually married their sisters. They are sometimes quoted as a good example of the effects of inbreeding. Upon this score there are two things to be said. One is that the Ptolemies were a mediocre lot to begin with, and the second is that they were by no means as inbred as is sometimes alleged. Finally, the famous Cleopatra (69–30 B.C.), the last of the line, was not the offspring of a brother-sister marriage, as is

316 HUMAN HEREDITY

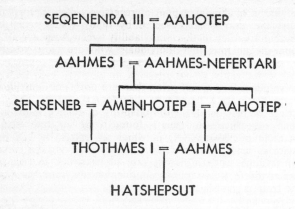

FIG. 33. Portion of the Pedigree of Egyptian Rulers, Illustrating Their Close Inbreeding Practices.

frequently stated, but of Ptolemy XI and an unknown woman.

Although a violent lot—the first Ptolemy was a general in the army of Alexander the Great—the Ptolemies were not particularly known to suffer from any physical or mental abnormalities. Ptolemy I founded the great library at Alexandria and did much to make that city a fountainhead of art and culture.

Before the advent of Moses, the Jews married close relatives by preference. Abraham married his half sister, Sarah (Genesis xx.), and the mother of Moses was his father's paternal aunt (Exodus vi. 20). The Mosaic codification of the laws prohibited all such unions. One cannot help wondering why.

With the beginning of the Christian Era, the pronounced objection to the marriage of near kin appears to have been based on the ground that such marriages tend to produce defects in the offspring. In A.D. 1215 the Church finally decreed that only persons related beyond the degree of third cousins might marry, and this is essentially the law of the Church today.

In general the prohibition placed on the marriage of close relatives is well founded because in the large populations of the civilized world there are distributed many recessive genes for defective traits. It has been estimated that in such populations every individual carries about eight so-called lethal equivalents of this large pool of recessive genes. These lethal equivalents may be represented by many genes, each contributing small reductions in viability when homozygous. It should be obvious that individuals who are close relatives and derive many of their genes from the same sources are much more likely to contribute a matched pair of such recessive genes to their offspring than are unrelated individuals. Hence, fairly rare hereditary disorders would be likely to occur more frequently in the offspring of related individuals than in unrelated—and this is found to be the case. This is illustrated in Table XXX, in which an extremely high frequency of first-cousin marriages is shown among the parents of individuals suffering from rare diseases due to recessive genes. Indeed, we have learned that when a rare hereditary

TABLE XXX

EXTREMELY HIGH RATE OF FIRST-COUSIN MARRIAGES AMONG THE PARENTS OF CHILDREN WITH SOME RARE HEREDITARY DISORDERS

| Disorder | Percent First-Cousin Marriages Among Parents |
|---|---|
| Hepatolenticular degeneration | 38 |
| Pseudohermaphroditism | 37 |
| Deaf-mutism | 33 |
| Congenital afibrinogenemia | 33 |
| Congenital ichthyosis | 30-40 |
| Infantile amaurotic idiocy | 27-53 |
| Microcephaly | 26 |
| Alkaptonuria | 25-45 |
| Xeroderma pigmentosum | 20-24 |
| Albinism | 18-24 |
| Congenital heredoretinopathy | 16 |
| Cystinuria | 15-25 |
| Juvenile amaurotic idiocy | 15 |
| Total color blindness | 11-20 |
| Friedreich's ataxia | 10 |
| Galactosemia | 9-24 |
| Phenylketonuria | 5-14 |
| Gargoylism | 5 |
| Adrenal hyperplasia | 2 |

disorder appears in an individual, it is quite probably going to be found that the parents are close relatives, such as first cousins. The chance that a cousin carries the same recessive gene, say for albinism, in first-cousin marriage is about 1 in 8,* as compared with 1 in 70 unrelated individuals; so it will be seen that first-cousin marriages are quite a hazard. This

* For the method by which this figure is arrived at see Appendix B, p. 375.

should not be surprising in view of the fact that having two grandparents in common, first cousins therefore share more than one-eighth of their genes in common.

The principal effect of first-cousin marriage is to increase the number of individuals who carry a pair of identical recessive genes. A healthy individual carrying a single rare recessive gene will seldom manifest the least evidence of its presence. In double dosage the same gene may produce serious disorders. When individuals mate who carry such a single recessive gene, there is a one-in-four chance that their children, that is, a quarter of their offspring, will inherit both abnormal genes. Since first cousins share the same grandparents, they are much more likely to carry similar genes derived from the same ancestors than are unrelated persons. The children of cousin matings will therefore exhibit conditions due to recessive genes much more frequently than will the offspring of marriages between unrelated individuals.

In the United States and in England about one-half of 1 percent of marriages are contracted between first cousins, but in many localities of the world that are small and in which the populations are isolated either geographically or socially or both, the percentage of marriages between first cousins may be very high. In many nonliterate societies it is obligatory to marry one's first cousin; in still others one marries one's cousin a generation removed. Such communities show the expected results of cousin marriages; many of them are distinguished by a marked absence of hereditary defects, whereas others show a marked presence of such defects. For example, the famous Pitcairn Islanders, the Batz community of Loire-Inférieure, the Hindu community of the Tengger Mountains, Java, and the inbred populations of numerous small islands appear to show no ill effects of such inbreeding. On the other hand, ill effects have occurred in such communities as that of the Nanticoke Indians of Delaware in such physical traits as superior palpebral ptosis (drooping of upper eyelid), the population of Martha's Vineyard in the frequency of deafness in a number of families, the hill folk of New England with the prevalence of feeblemindedness, and the toothless men of the Amil community of Hyderabad, in the province of Sind, Pakistan, where within six generations many families have appeared whose members are the descendants of a single individual with regard to a sex-linked recessive gene responsible for a large number of the males being entirely toothless and without sweat glands.

Clearly it is not inbreeding, no matter how close the inbreeding may be, that is responsible for the appearance of defective traits. Whatever effect inbreeding may have is due entirely to the inheritance received. If that inheritance is good the effect will be good; if it is bad the effect will be bad; if it is indifferent the effect will be indifferent. In this connection we cannot do better than quote the words from East and Jones's classic book *Inbreeding and Outbreeding* (page 244):

> Owing to the existence of serious recessive traits there is objection to indiscriminate, irrational, intensive inbreeding in man; yet inbreeding is the surest means of establishing families which as a whole are of high value to the community. On the other hand, owing to the complex nature of the mental traits of the highest type, the brightest examples of inherent ability have come and will come from chance mating in the general population, the common people so-called, because of the variability there existent. There can be no permanent aristocracy of brains, because families, no matter how inbred, will remain variable while in existence and will persist but a comparatively short time as close-bred strains. But he is a trifler with little thought of his duty to the state or to himself, who, having ability as a personal endowment, does not scan with care the genealogical record of the family into which he enters.

And that will remain at once the best statement of the facts and the best advice on marriage that can be given. Marriage with a relative is a risk, but it should be a calculated risk. For that matter all marriage is a risk, but it is much more of a risk, genetically speaking, when it is contracted between relatives because one can rarely be certain of the absence of deleterious recessive genes carried in the heterozygous state by each partner. When Charles Darwin married his cousin Emma Wedgwood, in 1839, he could have been fairly certain, looking back upon the distinguished backgrounds of both his own and his wife's family, that the children he and his wife would have would be reasonably satisfactory examples of humanity. They were indeed. Darwin's sons became distinguished scientists and leaders of thought in their own right, as have many of his grandchildren, both male and female. (See Fig. 34.) However, Darwin's own family perhaps constitutes an example of some of the dangers inherent in cousin marriage, even when

## 320 HUMAN HEREDITY

such outstanding families as the Wedgwoods and the Darwins are involved. For Charles Darwin and his wife, Emma Wedgwood, had ten children, of whom, his daughter Henrietta Litchfield tells us, many were delicate and difficult to rear, three dying in infancy. The last child, Charles Waring Darwin, born in December, 1856, was without his full share of intelligence and died of scarlet fever in 1858 without ever having learned either to walk or to talk. It requires, however, to be pointed out that Mrs. Darwin was well on toward her forty-ninth year when this child was born, and maternal age may have been the significant factor in this instance.

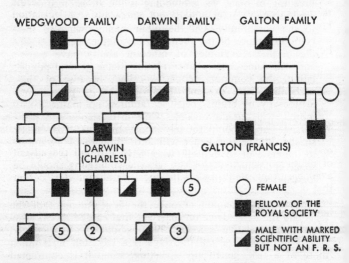

FIG. 34. A Short Pedigree of the Darwin Family.

If there is reason to believe that the inheritance to be passed on to one's offspring in a marriage between relatives is likely to be free of deleterious influences, such judgment being based on a knowledge of the genealogical record on both sides, the genetic risk would be very slight. On the other hand, when the genealogical history on even one side points to the presence of unfavorable recessive genes, it would be wise, if such a union is entered into, to do so with a full understanding of the possible consequences if there are to be any children.

The question whether anyone has the right knowingly to

## WHAT DO WE DO ABOUT HEREDITY? 321

bring children suffering from a hereditary defect into the world is not an easy one to answer. There can be no question that infantile amaurotic idiocy is a disorder that no one has a right to visit upon a small infant. Persons carrying this gene, if they marry, should never have children, and should, if they desire children, adopt them. But what shall we say of such conditions as total color blindness or albinism? Shall persons carrying the genes for these conditions also desist from having children? I think the answer to that question would be best left to the decision of the potential parents themselves, provided they understand as fully as may reasonably be expected the meaning of their decision. It seems to me that what is required is not sterilization or legislation but a more widespread understanding of the genetic facts and the human responsibilities involved, not only by parents but also by the larger society.

A society does not properly acquit itself of its responsibilities to its citizens either by choosing to ignore the existence of defectives and placing the entire burden of their care upon the parents or else by approaching the problem with the crude and ineffectual notion of sterilization. Our society should make much more varied and substantial assistance available than it does at present to the family with one or more defective children. And to forestall the advent of undesired defective children, genetic counseling should be available premaritally and is the most significant clarification that ought to be interestedly sought by the young couple contemplating marriage. The best time and place to prepare individuals in such knowledge-seeking attitudes is in the schools, for an elementary knowledge of the facts of genetics, of human heredity, should become part of the intellectual equipment of every ordinarily educated citizen.

Since in the future it will become increasingly possible, by simple tests, to identify the carriers of recessive genes for various disorders, such carriers can be counseled on the risks to offspring resulting from marriage with another carrier. On the average, one-quarter of the offspring of such marriages will exhibit the disorder, and half will be carriers.

The aspect of eugenics that suggests the negative control of the multiplication of undesirable traits is, in general, misconceived. The objections to negative eugenics are clear, unequivocal, and unanswerable. The recommendations of negative eugenics are (1) theoretically unsound and therefore (2) practically unjustifiable, not to mention the dangers of political misapplication; (3) fertility is generally reduced in those affected by recessively determined abnormalities; (4) even when such individuals breed, their offspring in most cases are normal; (5) even if it were possible, sterilization of the heterozygous carriers would waste an enormous potential of normal births; (6) conditions due to rare dominant genes usually result in infertile individuals, and in (7) other conditions due to unfavorable dominant genes the fertility is low, hence (8) most conditions due to defective genes are self-limiting; (9) carriers of defective genes also carry a large number of normal genes that in homozygous state often give rise to above-average traits; (10) many hereditary disorders are amenable to environmental alleviation; (11) the increase in human mobility and the collapse of innumerable barriers to intermixture between large numbers of the members of populations that were hitherto separated reduce the chances of deleterious recessive genes coming together; (12) sterilizing homozygotes who show a defective trait due to recessive genes still leaves by far the greater number of individuals carrying the gene in the heterozygous state to circulate it freely throughout the population. If, for example, the frequency of a recessive gene in the general population were 1 in 1,000, this would mean that the homozygote in whom the defective trait was expressed would occur in 1 out of 1 million individuals. Supposing the homozygous individual were sterilized, this would still leave 999 heterozygotes to distribute the gene. Clearly a rather inefficient and ineffectual way of dealing with the problem; (13) since the number of genes of any given sort in the human species is usually very large, any artificially induced changes in their number in any local portion of the species is likely to have very little if any effect upon the total frequency. As Penrose has put it, "The wider problem of genetical improvement of the human race must be viewed against the background of gene frequencies in the world population and the relative fitnesses of different phenotypes in different environments"; finally, (14) because negative eugenics overlooks such facts, it would certainly do more damage to the human species than defective genes are capable of doing.

Positive eugenics has somewhat more to be said in its favor insofar as it seeks to underscore the necessity of paying more attention than we do to the choice of mates on the one hand and to the greater encouragement of reproduction of the well endowed on the other. Unfortunately, eugenists in the past have placed the emphasis almost exclusively on intellectual qualities as the principal criterion of the well endowed.

## WHAT DO WE DO ABOUT HEREDITY?

Intellectual qualities are important, but so are other qualities of mind, such as adaptability, integrity, compassion, and balanced temperament. Health may be defined as the ability of the organism to work efficiently and to love—to confer survival benefits upon others in a creatively enlarging manner. I think it can be successfully argued that these traits are ultimately more important than the exclusively intellectual ones for the survival and healthy growth and development of the individual and the species. Intellectual qualities and special abilities are, of course, of the highest value, but they do not constitute all that it takes to make a complete human being or a society. Human beings are not ants, nor should they try to imitate them, in spite of Solomon's injunction, which was simply his way of commending industry. In any event, we do not by any means know enough about man's genetic system to commence modifying it by special types of breeding. Domestic plants and animals having a commercial value may be bred for special traits that improve their desirability, but as every breeder knows, in developing such desirable traits many undesirable ones may turn up and the strains or the specimens exhibiting these undesirable traits have to be discarded.* It is scarcely possible or desirable to follow such a procedure in the case of man. Desirable traits in man, insofar as they are affected by gene differences, appear to be the result of the interaction of many genes, and there is every reason to believe that the infusion of new genes into a stock is, on the whole, to its advantage, so that breeding in the open system of random mating is likely to be of greater benefit to the species than is breeding directed toward the development of some special trait.

We know that musical ability, mathematical ability, and the ability to paint are closely associated with certain genes. These seem to be a few in number. It would be possible to build up whole families of individuals with such special

---

* For example, hornlessness is a valuable trait in goats. In breeding for hornlessness, the breeders have unwittingly increased the frequency of the gene for pseudohermaphroditism, or intersexuality. The dominant genes for hornlessness and the recessive genes associated with intersexuality are somehow linked. The recessive genes under normal conditions remain unexpressed, but under controlled breeding conditions find expression as a result of their unsuspected linkage with the genes associated with the trait considered desirable. Since the intersexual goats, who are reversed genetic females, are sterile, the breeders are introducing a much more undesirable trait into the population of goats than the one they sought to eliminate.

abilities. The Bachs and the Mozart-Weber families are examples of musical dynasties that did this for themselves. The Bernoullis are an example of a mathematical family of great genius; the Pissarros, Camille and Lucien, are an instance of a father and son who were great painters. Anyone interested in creating a line of great musicians, mathematicians, or painters who does not himself possess any of these abilities had better look for a mate who comes from a family that does.

Planned breeding can be left to the individual. I do not think that the State should do anything other than make it possible for the individual to acquire all the knowledge he needs to make the best of his own heredity and to do what he ought and can about the heredity of his descendants—remembering, always, that whatever the genetic inheritance of the individual is, one can do a great deal through the agency of the environment to improve its expression.

# 18. THE BOMB, RADIATION, AND HUMAN HEREDITY

THE ATOM AND HYDROGEN BOMBS brought to the attention of the whole world the fact that Western man now had in his possession a means of bringing death and disaster to vast numbers of human beings in a matter of moments. The progressive increase in the destructive power of the bombs that have since been manufactured, the hydrogen bomb tests and their accompanying fallout, have increasingly served to focus attention upon the dangers to the human species of misused atomic energy.

The constant threat of nuclear war has further served to concentrate attention on the annihilative effects upon human beings and cities, not to mention the sufferings of those who managed to survive. It has been said that in a nuclear war the use of a sufficient number of atomic weapons could physically put an end to the whole of the human species. This is theoretically possible, but practically it would be extremely difficult. Civilized man might certainly succeed in physically wiping himself off the face of the earth, but there are many people in far-removed places of the earth who would not be so eliminated. However, their eventual elimination would come about quite as certainly through the long-term effects of the radiation to which they had been exposed, assuming that this was quantitatively and qualitatively significant.

Let us hope that no individual and no nation will ever be so demented as to use such suicidal devices. Meanwhile, the very fact that such devices are being created in large numbers, that their testing releases dangerous by-products, and that the waste products of atomic-energy plants present serious disposal problems—all constitute conditions that in themselves may seriously affect the immediate or long-term well-being of the human species. The damage done to the hereditary structure of the human species by the energy released from such artificially created sources of radioactivity may in the long run prove far more effective in putting an end to *Homo sapiens* than an outright nuclear war. This is the extreme possibility—the end of the human species, as a consequence of man's being just a little bit too clever for his own good. Anything less than this extreme possibility amounts to such an increase in the number of hereditarily defective individuals of every sort that in terms of the human tragedies involved, it is utterly indefensible for *anyone* ever to permit the use of any device that could produce such avoidable suffering. We are here concerned with matters of life and death, not only as they affect the lives of those now living but also as they will affect the lives of generations of our descendants, of our fellow human beings, who are to follow us from now on into the centuries and the millennia. We of the twentieth century have a grave responsibility placed squarely upon our shoulders, and it is therefore important that we make ourselves as fully acquainted with the facts as we are able, for what we do or do not do about them will largely determine the future of ourselves and of our descendants.

## Radiation and Mutation

Radiation is the process in which energy in the form of rays, waves, particles, beams of particles, or heat moves from one place to another—it is energy on the move. We feel the radiation from the sun in the form of heat and observe its effects on ourselves after more or less prolonged exposure to it on the beach. Getting too hot from the rays of the sun can make us very uncomfortable and even cause us to be sick. Getting burned from the rays of the sun can be pretty serious, and cancer of the skin of the face as a result of overexposure to the sun is a well-known consequence of its disordering effects. These are the kinds of disordering effects of radiation that you can feel and see. By far the most sensitive system in the body to the effects of radiation is the inheritance system, the reproductive cells, the carriers of the hereditary materials, in the gonads, the ovaries of the female and the testes of the male. Any radiation reaching the reproductive cells causes mutations (changes in the material bases of heredity), and such mutations are passed on to succeeding generations. More than 99 percent of mutations are harmful. Thus, any increase in the amount of radiation human beings normally receive from natural sources endangers the adaptive equilibrium attained by the human species in the course of evolution by producing an imbalance as a

# THE BOMB, RADIATION, AND HEREDITY

consequence of the greater number and kinds of mutations the species is suddenly called upon to support.

All matter is made up of atoms. An atom is composed of a nucleus surrounded by a swarm of electrons. The nucleus is composed of protons and neutrons. Electrons are very light, fast-moving particles carrying a negative electric charge. They are sometimes called *beta rays*. Protons are approximately 2,000 times as heavy as electrons and are positively charged. Neutrons are about the same size as protons but are uncharged.

*Alpha particles,* made up of two protons and two neutrons, are the nuclei of helium atoms. They are swiftly moving particles of high energy, carrying a positive electric charge but with little power of penetration.

Every element has a unique and definite number of protons. The atomic weight of an element is determined by the weight of its protons and neutrons. Two atoms having the same number of protons but different numbers of neutrons are known as *isotopes* of the same element. Isotopes are today artificially produced in large numbers and are proving of the greatest value in experimental medicine, although care in their use is now more than ever indicated.

When atoms get overenergetic or "excited," they radiate, that is, emit, the various particles of which they are constituted—those named above as well as others.

An atom gets excited when it is knocked about, as it were. When you throw atomic particles against other particles in the nuclei of atoms, as in an atom-smashing machine, the target nuclei get excited and emit at great speed a considerable variety of particles, including *gamma rays*. Gamma rays are electromagnetic radiations of extremely high energy, having great penetrating power that can traverse the whole body with relatively little absorption. *X rays* are very similar to gamma rays, and are produced by high-energy electrons. X rays are usually produced artificially in special electrical machines, in which a stream of fast-moving electrons is made to strike a metal target. As a result some of the atoms in the target get excited and start emitting some of the energy they have acquired from the electrons that have been hitting them, sending out their own electrons in the form of X rays. The number of electrons sent out will vary in their penetrating power according to the electrical energy used in their production. The biological effects of X rays are brought about by the high-energy electrons that are liberated in the tissues.

## 328 HUMAN HEREDITY

All these types of radiation are known as *ionizing radiations* because in their passage through anything they leave electrically charged atoms and molecules, called *ions*. When an atom is hit by or collides with other atoms or atomic particles, one or more electrons are either gained or lost as a result of the impact.* Such gain or loss occurs during chemical reactions in which electrons are transferred from one atom to another by the action of radiant energy of one kind or another. This process of electron detachment from atoms and reattachment to other atoms is called *ionization*. The physicochemical changes produced in living tissues that result in radiation damage, in the form of the death of cells or altered heredity, are due to the changes initiated by the ionization process. The efficiency of a given dose of radiation in producing biological changes can be related to the number of ions that are produced per unit length of the track through which they pass.

All ionizing radiations, however low the dosage, are capable of producing mutations.

There are some substances that are naturally radioactive, such as radium or uranium. They are not dependent for their radioactivity upon the collision of atoms or their particles, but their atoms are just naturally in a state of excitement. From time to time their nuclei erupt, giving off alpha particles, electrons, or gamma rays.

In addition to naturally radioactive atoms and atoms that are made radioactive by bombardment, there is yet another important natural source of radiation. This is outer space, from which high-energy particles called *cosmic rays* shower upon the earth. It is believed that these cosmic rays have been the principal cause of mutation in all living things.

The intensity or dosage of radiation that the living body receives can be determined by measuring the ionizations produced. The unit of dosage is known as the *roentgen*,† usually abbreviated to the simple letter *r*. In the human body an r corresponds to about 2 ionizations per cubic micron, a micron being 1/25,000 of an inch. An exposure of 1r of radiation over the whole body—which is a small amount of radiation—will result in $10^{17}$ ionizations, that is, 1 followed

---

* The loss of electrons results in a positively charged ion, called *cation;* the gain of electrons in a negatively charged ion, *anion*.

† Named after Wilhelm Konrad Roentgen, the German physicist who discovered the X ray in 1895, for which discovery he was awarded the Nobel Prize in physics in 1901.

by seventeen zeros, or the grand total of 100 million billion ionizations! And that is when only the very slightest proportion of all the atoms in the body are affected. The amount of radiation that each of us receives from our natural surroundings is about 0.1r a year. About a quarter of this amount comes from cosmic rays, somewhat less than this from radioactive elements in our bodies, principally potassium, and about half from the soil and rocks.

Other sources of radioactivity we pick up from our environment are illuminated dials on wristwatches, clocks, automobile dashboards, medical and dental X rays, fluoroscopes, and similar ionizing sources of artificially created radiation. Thus far, the damage done to the hereditary mechanisms of thousands of persons by the misuse of medical and dental X rays has probably been considerably greater than that done by fallout from nuclear explosions.

But ordinary articles that we take for granted are potentially even more risky. It has, for example, recently been shown by Haybrittle in England that the gamma radiation given off by the luminous dials of many ordinary wristwatches is second only to that which diagnostic radiology may contribute to the gonads. Haybrittle found that one modern self-winding wristwatch contained on its dial 2.2 microcuries * of radium. This meant that if the wearer put the watch on his wrist at 8 A.M. and took it off before retiring at midnight, he would receive during those sixteen hours a total of 0.9r per week! This is nearly two-thirds of the present maximum permissible exposure to limited parts of the body, such as hands and forearms. A simple calculation will show that in the course of a lifetime an individual would accumulate a considerable amount of radiation from this source alone! The radiation dose to the gonads would be increased by more than 10 percent above that received from background radiation.

These observations have been confirmed in the United States by Chase and Osol, who calculate that an illuminated wristwatch worn twenty-four hours a day can deliver to the gonads 1.1r units a year if the watch is worn facing the gonads. The potentially harmful magnitude of the radiation from the most active watches, corresponding to 5r in about five years, may be judged in the light of the recommendation

---

* A *Curie* is the amount of material in which $3.7 \times 10^{10}$ atoms disintegrate per second, that is to say, the transformation of 37 billion atoms occurring each second in 1 gram (1/28 ounce) of radium. A *microcurie* is a millionth of a curie.

of the International Commission on Radiation Protection that no one should receive a dose in excess of 5r by the age of thirty. "When," these investigators write, "one further considers that this radiation is several times greater than natural background radiation and exceeds by more than 100 times that presently received from radioactive fallout, the potential hazard to the wearer of a luminous-dial wrist watch raises the question whether the small benefit that may be received from such a watch is worth the hazard."

Dr. Edward W. Webster has ingeniously shown that if one conservatively takes the lowest calculated annual dose to the gonads from the wearing of a luminous wristwatch at 20 mr (mr=1/1,000 of a roentgen), the number of 14-by-17-inch chest X rays one would have to take before exceeding the hazard from such a wristwatch would, in a ten-year period, be 1,200! One could take 1,600 X rays of the head over a similar period! These findings are set out in Table XXXI.

TABLE XXXI
DIAGNOSTIC RADIOLOGY COMPARED WITH RADIUM WRISTWATCHES

| Radiographic Examination | Male Gonad Dose (mr) | Number of Examinations to Equal Dose from Radium Wristwatch in Ten Years* |
|---|---|---|
| Scrotum unprotected, Pelvis | 430 | 0 |
| Scrotum protected, Pelvis | 30 | 7 |
| Lumbar spine | 60 | 3 |
| Intravenous pyelography | 50 | 4 |
| Abdomen | 12 | 17 |
| Barium enema | 70 | 3 |
| Upper gastrointestinal | 8 | 25 |
| Scrotum unprotected, Hip | 2,530 | 0.08 |
| Scrotum protected, Hip | 60 | 3 |
| Scrotum unprotected, Lower femur | 560 | 0.35 |
| Scrotum protected, Lower femur | 60 | 3 |
| Chest 14-by-17-inch film | 0.16 | 1,200 |
| Chest microfilm | 1 | 200 |
| Head or skull | 0.12 | 1,600 |

* 200 mr (milliroentgens) as conservative estimate.
Adapted from E. W. Webster, "Hazards of Diagnostic Radiology."

During the year 1959 a Swiss watch manufacturer marketed wristwatches with luminous dials painted with a compound marketed under the trade name of Lumostabil. After wearing these watches for several weeks, a number of individuals were treated for burns corresponding to the circle of luminous figures. Lumostabil, which was sold without any warning of its radioactivity, merely happened to contain a large amount of strontium 90! Present information is that the sale of Lumostabil has been discontinued. One woman, reported to me by Sir E. Rock Carling, of London, wore such a watch for about eight weeks. It is believed to have been illuminated with about 20 microcuries of strontium 90. It is estimated that this woman received a cumulative dose of 10.15 thousand rad.* The radiation dose from contact with the front of the watch might have been about 10 rad per hour! In December, 1959, several such watches, apparently illuminated with the same compound, were recalled by their vendors in the United States.

In November, 1962, the New York City Board of Health banned the sale of radium-dial pocket watches. Studies by the Health Department's Office of Radiation Control had revealed that the radium-dial pocket watches emitted 75 radiation units a year. This is 150 times more than the permissible, or safe, amount of 0.5 units a year.

It has already been stated that however slight the amount of radiation, it is capable of causing mutations. There is no minimum amount of radiation that must be exceeded before harmful mutations can be induced. About 85 percent of known mutations occurring spontaneously in man are recessive. Of the approximately 75 dominant mutations occurring spontaneously in man, most are detrimental; indeed, about 80 percent are lethal, which means that about two out of ten such mutations will survive to be reproduced in the next generation. Since mutations constitute the raw material of evolution, without which there could be no evolution, how, then, the reader may well ask, does it come about that good effects can result from mutations?

The answer is that it is not mutations of themselves that produce evolution, but rather the process whereby *favorable* mutations, when they do occur, by increasing the adaptive fitness of the organism to meet the requirements of the environment, enable it better to survive and leave a larger progeny behind it than those who do not possess similar favorable mutant genes. This is the process known as *natural selection*.

Most mutations occur spontaneously with a frequency of less than one per million copies. Man's mutation rate throughout his history has been quite adequate to ensure his favorable continuing evolution without the necessity of any

---

* A rad corresponds to an absorbed dose of 100 ergs of energy per gram of air. In 1r the energy absorption amounts to 84 ergs per gram of air.

artificial speeding up of that rate. In fact, the great danger from artificially produced radiation lies in this: by increasing the mutation rate, unfavorable mutations would accumulate in such high frequencies that they would endanger the survival of the species. The compounding hazard being that owing to the extraordinary mastery of the environment man has achieved, in medical care and in everyday living, he tends to prevent the elimination of unfavorable mutations, to interfere with *selection pressure,* to produce a negative selection pressure, and by this means, in the event of any sizable increase in the numbers of harmful mutations, he would contribute further to his own downfall.

It is desirable to be quite clear as to the nature of those mutations. The idea that they would result in numerous monstrosities or freaks is quite erroneous. Undoubtedly some mutations would result in abnormalities of embryonic development, and some developmentally abnormal infants would be born. But these would be the minority. Most mutations are responsible for only slightly detrimental effects.

Paul Dehn, the author of the following parody, is therefore not altogether sound in his contemporary version of an old hymn: *

> Hark, the herald angels sing
> Glory to the new-born thing
> Which, because of radiation,
> Will be cared for by the nation.

However, there are some mutations that in their effects will vary all the way from producing death to causing sterility or some other serious major defect. Such effects will generally be observable within the first generation. The slight detrimental effects would not be so readily observed, for they will remain concealed in the genetic constitution of successive generations. In some individuals the genetic damage would express itself immediately even if the mutant gene were carried in the heterozygous state. Such heterozygous damage, as it may be called, is brought about as a result of the fact that detrimental recessives are not usually completely dominated by the normal gene, for some effect—however slight—is usually produced by the abnormal recessive. The effect is ordinarily much smaller than that which the gene exerts in the homozygous state, and yet as a consequence there may be an appreciable decrease in re-

* *New Statesman,* August 22, 1958, p. 215.

sistance to disease, in length of life, and in fertility in these heterozygous carriers. Every carrier of the mutant gene, and every single descendant who receives it, is in danger of such heterozygous damage. Therefore the number of individuals affected is very great. However weakly expressed, the damage would persist and some of it would be exhibited in each generation in the form of either premature death, the failure to produce the normal number of offspring, or lowered resistance to disease.

The genetic damage done is roughly proportional to the mutation rate. The mutation rate is proportional to the radiation dosage received. If we increase the radiation by 5 percent, then the mutations caused by radiation will also be increased by 5 percent. If, for example, we then add an increase of 7 percent of radiation to the previous 5 percent, we shall have a total dose of 12 percent radiation and a total increase of mutation by another 7 percent to 12 percent. The genetic damage done is cumulative, whether spread out over a short or a long period of time. Large doses of radiation produce more mutations but not more substantially harmful ones.

Clearly, then, if all radiation is capable of producing mutations in the germ cells, there can be no such thing as a safe mutation rate since almost all mutations are genetically damaging. The important consideration is the genetic damage done as a result of the total accumulated radiation dose to the reproductive cells of the individual from his own conception to the time his child is conceived.

In the United States the average age of both parents at the birth of *all* their children is twenty-eight years. Thus, an individual is exposed for approximately his first thirty years (a whole generation) to radiation up to and including the time of the conception and birth of all his children. This is his *total reproductive-life radiation dose*.

The total accumulated thirty-year dose that each individual, on the average, receives has been estimated to be as follows:

| | |
|---|---:|
| *Background Radiation:* Cosmic rays, naturally occurring radioactive substances, etc. | 4.3r* |
| *Medical X rays:* | 3.0r |
| *Fallout:* From weapons tests | 0.1r |
| | 7.4r |

* At high altitudes this would amount to about 5.5r, owing to greater exposure to cosmic rays.

## 334 HUMAN HEREDITY

These estimates do not take into consideration such extra sources of radiation as radioactive dials on watches, dashboards, and other instruments; isotopic tracers used in research; carbon 14 estimation methods; high-power radio and television stations, and the like, not to mention the recently suggested probable mutation effects of increased temperature of the testes in males as a result of wearing trousers!

In 1957 the Swedish workers Ehrenberg, von Ehrenstein, and Hedgran reported the results of an investigation on the scrotal temperature of nude and clothed men. They found the scrotal temperature to be higher by 3.3° C., on the average, in the clothed men as compared with the nude men. They calculate that such an increase in scrotal temperature would increase the mutation rate by about 85 percent! They write, "The fact that our modes of dress have been predominant for several centuries might explain almost half the present load of spontaneous mutations. We thus see how modes of dress based chiefly on sexual taboos might imply genetical hazards 100 to 1,000 times greater than those estimated from different sources of radiation. If the eugenists regard this increase as dangerous, the design of male clothing will have to be reformed, for example, in the direction of the Scottish kilt or of trousers fitted with a codpiece as used in medieval Europe. Garments that tend to press the scrotum against the abdominal skin will be specially hazardous. Further consideration will have to be given to such factors as central-heated rooms and frequent hot baths."

In short, the wearing of trousers, not to mention frequent hot baths and the central overheating to which Americans voluntarily subject themselves for about six months of each year, is probably more dangerous to future generations than the fallout from atomic and hydrogen bombs. The result of all this may be that men will soon be wearing skirts!

Assuming the temperature dependence of spontaneous mutation in men to be the same as observed in the fruit fly, the 85 percent of spontaneous mutations in the male due to 3.3° C. temperature increase would be equivalent to a radiation dose of 40r. As Webster has shown, it would take 250,000 14-by-17-inch X rays and 300,000 head or skull X rays before one would reach a radiation dose equivalent to that resulting from the wearing of trousers! The facts are set out in Table XXXII.

It should be evident from the last two tables that civilized societies have taken a number of practices for granted that may well produce radiation effects more damaging than the

# THE BOMB, RADIATION, AND HEREDITY

TABLE XXXII
DIAGNOSTIC RADIOLOGY COMPARED WITH WEARING TROUSERS

| Examination | Male Gonad Dose r | Number of Examinations for Same Genetic Effect as Wearing Trousers* |
|---|---|---|
| Scrotum unprotected, Pelvis | 0.43 | 90 |
| Scrotum protected, Pelvis | 0.03 | 1,200 |
| Intravenous pyelography | 0.05 | 800 |
| Abdomen | 0.012 | 3,000 |
| Lumbar spine | 0.06 | 700 |
| Barium enema study | 0.070 | 600 |
| Upper gastrointestinal series | 0.008 | 5,000 |
| Scrotum unprotected, Hip | 2.53 | 16 |
| Scrotum protected, Hip | 0.06 | 700 |
| Scrotum unprotected, Left femur | 0.56 | 70 |
| Scrotum protected, Left femur | 0.06 | 700 |
| Chest 14-by-17-inch film | 0.00016 | 250,000 |
| Chest, microfilm | 0.001 | 40,000 |
| Head or Skull | 0.00012 | 300,000 |

* Assumed as 40r equivalent.
Adapted from E. W. Webster, "Hazards of Diagnostic Radiology."

combined effects of fallout and diagnostic radiation put together. We need to reexamine searchingly some of these practices.

If we proceed at the present rate with nuclear tests and medical radiology (and excluding luminous wristwatches), the total accumulated thirty-year dose we are now receiving, on the average, is 7.4r, or 3.1r above the amount of radiation we would normally receive from natural sources. This means that the mutation rate has increased proportionately in countries in which there is considerable exposure to medical X rays, as in the United States. Among uncivilized peoples the exposure is limited to background radiation and fallout, and except near the islands in the Pacific where weapons tests have been conducted, their total additional radiation dosage amounts to no more than 0.1 percent of the total, which is less than 2 percent of the total radiation man is now receiving. It is likely, therefore, that there will be different mutation rates among different populations of man. The reason for this difference underscores the fact that by far the largest dose of radiation, in addition to natural background radiation, is received by civilized man from medical X rays to which he is exposed for diagnostic and therapeutic purposes.

In Vermont, in March, 1963, a set of dishes that had been

in the same family for almost a quarter of a century were found to be radioactive, each dish emitting 3 milliroentgens an hour. At an exposure of three hours a day, the thirty-piece setting would be responsible for an amount of radioactivity exceeding 1,500mr a week. The Atomic Energy Commission recommends 300mr per week as the maximum exposure for those engaged in any form of work involving radioactivity.

How did this set of dishes come to be radioactive? In order to achieve the deep red color that distinguishes these dishes, uranium salts were used in their manufacture.

An amount of less than 2 percent of radiation derived from fallout may seem very little. But in fact when that amount is distributed over a world population of 3 billion, it adds up to a very tangible amount of genetic damage. Dr. James F. Crow, geneticist at the University of Wisconsin, has estimated that an exposure of the world's population to 0.1r would result in at least 8,000 children in the first generation being born with gross physical or mental defects, or a total of 80,000 in the future generations (assuming a world population of 2 billion children). Fetal and childhood deaths would amount to 20,000 in the first generation and 300,000 in the total future generations. There would also be 40,000 embryo and neonatal deaths, a total of 700,000 for all time. Dr. Crow thinks that these figures are probably an underestimate. In addition there would be a larger but unknown number of minor or intangible effects.

What man's natural mutation rate may be is unknown, but Dr. W. L. Russell, of the Oak Ridge Laboratory, has shown in the house mouse that 1r produces about 1 mutation in 4 million genes. Comparing this rate with the spontaneous mutation rate that we know for certain genes, such as the hemophilia gene in man, it is estimated that a dose of between 30r and 60r would double man's natural mutation rates.

It should be emphasized that we are not concerned in this chapter with the effects of radiation upon the body, such as burns, cancer, or leukemia, but with its effects upon heredity.

Having briefly considered the effects of fallout upon heredity, let us now proceed to consider the general effects of possible increased radiation upon the heredity of mankind. This can, perhaps, be best illustrated in the following manner:

It is estimated that between 4 and 5 percent of all live children born in the United States suffer from such defects as mental deficiency, epilepsy, congenital malformations, neuromuscular disorders, disorders of the blood and of the glandular systems, skin and skeletal disorders, or defects of the gastrointestinal or genitourinary systems. About half of these defects, that is, about 2 percent—a total of 2 million —of the total live births, are of genetic origin and appear before sexual maturity.

We know something about mutation rates for certain disorders in various populations (see Table XXXIII), but it is quite another matter for the individual. In every individual,

TABLE XXXIII
ESTIMATES OF SPONTANEOUS MUTATION RATES OF SOME HUMAN GENES

| Trait | Mode of Inheritance | Mutation Frequency of Causal Gene (per million per generation) | Region |
|---|---|---|---|
| Achondroplasia | Dominant | 45 | Denmark |
| Aniridia | " | 5 | " |
| Epiloia | " | 8 | England |
| Microphthalmos | " | 5 | Sweden |
| Partial albinism (with deafness) | " | 4 | Holland |
| Retinoblastoma | " | 4 | Germany |
| " | " | 15 | England |
| " | " | 23 | United States |
| Hemophilia | Sex-linked | 20 | England |
| " | " | 27 | Switzerland & Denmark |
| " | " | 32 | Denmark |
| Muscular dystrophy (Duchenne type) | " | 43 | England |
| " | " | 45 | N. Ireland |
| " | " | 95 | United States |
| Albinism | Recessive | 28 | Japan |
| Amyotonia congenita | " | 20 | Sweden |
| Color blindness (total) | " | 28 | Japan |
| Ichthyosis congenita | " | 11 | " |
| Infantile amaurotic idiocy | " | 11 | " |
| True microcephaly | " | 49 | " |
| Phenylketonuria | " | 25 | — |

Adapted from L. S. Penrose, "The Spontaneous Mutation Rate in Man."

during his lifetime, a certain number of mutations occur due to natural causes. We do not know what the number is. One to four is a fair guess, with the total number of mutant lethal equivalents carried in the recessive state by each individual being eight. But whatever the number is, what would be the radiation dose that would produce an equal number of additional mutations and thus double the total number? It has been estimated that the doubling dose would be somewhere between 30r and 60r. Supposing, now, that the population of the United States were to receive a doubling dose of radiation, what would be the genetic effect upon the population in relation to the genetic defects mentioned above?

The 2 million genetic defectives would eventually be doubled, provided the doubling dose continued in each generation and the population remained stationary. In the first generation there would be an increase of about 200,000 genetic defectives. These would be the individuals with tangible defects responsible for real personal and social distress. But in addition the concealed genetic damage would be considerable.

If the radiation dosage received by the gonads were 10r for the same population, then there would be about 5 million mutants in the 100 million children born in the next generation.

## The Mechanism of Genetic Damage

When a ray of high-energy radiation strikes an atom, it may knock an electron from an inner shell of electrons close to the atomic nucleus or from the outermost shell of electrons orbiting around the atom. If the hole thus produced occurs in the inner shell, the electrons from the outer shell will fall inward and quickly fill the deficiency. In so doing there will be a loss of electrons in the outer shell. This is serious because it is these electrons that determine how one atom will combine with or replace another [as measured by the hydrogen atoms with which they will combine or replace; thus the *valence*, for example, of oxygen is two, i.e., one atom of oxygen combines with two atoms of hydrogen (HOH or $H_2O$)]. Disruption of any of the valence electrons causes a corresponding disruption in the chemical bonds that enter into the formation of a molecule and a disturbance in the balance of forces that maintain the integrity of the molecule. The molecule is fragmented, and those fragments that have absorbed abnormal amounts of energy, greatly exceeding that contained in ordinary chemical bonds, become chemically unbalanced, or unstable. They will therefore attach themselves to virtually any molecule they encounter in order to regain the lost electron and achieve new bonds of stability. The excited atoms involved will continue to behave in this manner until they have calmed down, that is to say, achieved lower energy states.

## THE BOMB, RADIATION, AND HEREDITY 339

High-energy radiation will affect water in the same way, but owing to the simple atomic structure of water, the high energy absorbed will soon be given off as heat and the chemical bonds, two atoms of hydrogen and one of oxygen, are soon reestablished in their normal relationship. But in living cells, which are chemically extremely complex in comparison with water, the effects of radiation produce permanent changes in chemical bonds. Broken bonds become attached to parts of other broken bonds, and all sorts of new and bizarre combinations result. The consequences of this for the organism will vary with the amount of damage of this sort that has been done to the molecules of the cells of which it is constituted. Since the orderly functions of the cells depend upon the orderly arrangement of the bonds within the molecules forming the cells of which the organism is composed, even the disarrangement or rupture of a few bonds may have extensive disordering effects.

Dr. Theodore T. Puck, of the University of Colorado, has recently shown that human cells are far more sensitive to X rays than was formerly supposed, and that a dose as small as 50r is fatal for the reproduction of human cells. Since the maintenance of life is dependent upon cellular reproduction, it will readily be understood why a dose as low as 400r to the whole body results in death in man. But altogether apart from the lethal doses of X rays, any amount of radiation, however low, is capable of damaging the genes responsible for development, as well as the chromosomes in which they are lodged. The evidence is now clear that in man X-ray doses as low as 20r may produce breaks in chromosomes.

In addition to producing changes in the structure of genes, that is, mutations, ionizing radiations produce breaks in chromosomes (see Plate 20). The chromosome breaks, like mutations, are caused by the ionizing radiations in the vicinity of or within the chromosomes. The broken surfaces of the chromosomes remain sticky for some time after the breaks have been produced, and any one of the following four things finally ensues:

1. The broken surfaces unite and healing takes place. At the position where the break originally occurred one or more genes may be damaged or altered and this may give rise to one or more mutations. There may, however, be no damage at the break, in which case no permanent harm has been done.
2. The broken surfaces fail to unite, and the nucleus

of the cell now contains several separated fragments of a chromosome. Such broken fragments can replicate themselves, but as soon as the cell starts to divide and passes through one or two cell divisions the fragmented chromosomes tend to break down, and the cells to which they have been distributed likewise break down and die for want of the necessary genes. Dividing cells are very much more sensitive to radiation than nondividing, or resting, cells.

3. The several broken portions of the *same* chromosome may get shuffled around and their broken ends may stick together in new ways, or one or more of the fragments may be lost. This is known as *chromosome rearrangement*. The consequences of this may be extremely varied, from death to the organism, or every appearance of mutation, or overtly scarcely noticeable.

4. The several broken portions of any two chromosomes, whether paired or not, may accidentally exchange pieces, and the pieces may be of very different lengths. This abnormal process is known as *translocation*. Translocations tend to result in a high proportion of genetic deaths. Since at the maturation division the abnormal translocated chromosomes are unable to pair with the normal chromosomes, many of the gametes will possess a subnormal number of chromosomes and genes. Translocations result either in death or in reduced fertility of those who inherit them, and in either case are inherited like dominant genes.

What happens as a result of chromosome rearrangements and translocation is that the sequences of the purine and pyrimidine bases—adenine, guanine, cytosine, and thymine —between the helices that comprise the DNA, or deoxyribonucleic acid, molecule (see pages 29–48) get disarranged. Since the genetic information that is transmitted by the hereditary mechanism lies coded according to the sequence of these bases, any change in their order or in their structure will be reflected in changes in the information transmitted, and as we have abundantly emphasized, such changes— whether they be in the form of mutations or fragmented chromosome pieces that have attached themselves to pieces from the same or different chromosomes—are usually deleterious in their effects. These changes will occur not only in the reproductive cells but also in the somatic, or body, cells, and of course will cause damage in the body cells that may result in all sorts of more or less serious disorders. In many cases they will have the effect of increasing the frequency of disorders and diseases, and deaths from these.

Chemical defects in the genes, whether in the reproductive or the body cells, result in defective reactions or in the blocking of necessary chemical reactions, so that there tends to be either defective development or a failure to develop altogether as a consequence of the blocking process, which is itself simply a failure in the development of a necessary chemical substance. Such a failure, for example, is believed to be responsible for albinism. In the formation of the pigment melanin, for example, a series of complex chemical steps must occur in an orderly manner. These steps are produced by a series of enzymes. In the case of albinism one of these enzymes fails to develop, and the series of steps that normally leads to melanin formation is blocked and the individual fails to develop any pigment at all. The enzyme thought to be involved is one that acts either on 3,4-dihydroxyphenylalanine or possibly on tyrosine. At its most elementary, the disturbance is essentially one of the intake, consumption, and expenditure of energy, that is, metabolism: the energy relations of the gene are disturbed.

## Some Conclusions

The United Nations Scientific Committee on Effects of Atomic Radiation, in its report published in August, 1958, stated:

> The knowledge that man's actions can impair his genetic inheritance, and the cumulative effect of ionizing radiation in causing such impairment, clearly emphasize the responsibilities of the present generation, particularly in view of the social consequences laid on human populations by unfavorable genes.
>
> Besides increasing the incidence of easily discernible disorders, many of them serious but each comparatively rare, increased mutation may affect certain universal and important "biometrical" characters such as intelligence or life-span.

As these words were being written, two U.S.S.R. scientists, M. N. Livanoy and D. A. Biryukov, reported to the United

## 342 HUMAN HEREDITY

Nations Second International Conference on Peaceful Uses of Atomic Energy, held at Geneva in September, 1958, that profound effects are produced upon the brain and central nervous system by ionizing radiations. Since children are more sensitive to ionizing radiations than adults, the functioning of the nervous system and intelligence would be affected in a large proportion of the young, so that those who had not been hereditarily affected might be brought down to the level of those who had. Of course, a large number of adults at every age would also be detrimentally affected by ionizing radiations affecting the nervous system and intelligence.

So far as the hereditary changes' deleteriously affecting intelligence and the functioning of the nervous system are concerned, these unfortunately are likely to be of the kind that are not detectable in the form of easily seen gene effects. They could, therefore, be picked up only in the form of significant increases in the incidence of such disorders. Exposure to continued small doses of radiation will produce correspondingly small effects in most of the members of the population and large ones in a comparatively small number of individuals. This kind of hidden damage is much more difficult to isolate and deal with. Hence, much more research is necessary, particularly of the kind that will lead us to detect this kind of hidden genetic damage. The danger of rendering ourselves too stupid, as a result of exposure to ionizing radiations, to arrive at such a method is not in the immediate future very great.

Meanwhile, each of us can take certain simple practical steps to reduce the amount of radiation to which we and our children are exposed.

We can begin with all timepieces, clocks, and watches that are radioactive, that is, that have their dials painted with a radium paint—and this includes all radioactive dials on the dashboards of automobiles; they should be jettisoned.

We should see to it that legislation is passed forbidding anyone to use any apparatus producing ionizing radiations who is not properly qualified to use such apparatus.

No X-ray apparatus should be manufactured by companies unlicensed to do so. Licenses to manufacture X-ray apparatus should be issued only to those manufacturers who meet the proper specifications for the manufacture of such installations.

A person shall not be considered qualified to use X-ray

apparatus for diagnostic or therapeutic purposes, no matter what other diplomas he may have, unless he has been specially certified to do so after having undergone the proper training and taken the proper examinations. This would at the present time exclude most medical men and most dentists.

All medical and dental X-ray installations should be subject to a quarterly checking in order to control the dosages, filtration, and the safety in general of those who are exposed to its radiations.

In *The Biological Effects of Atomic Radiation*, a report published by the National Academy of Sciences and the National Research Council, the following recommendations are made:

1. That, in view of the fact that total accumulated dose is the genetically important figure, steps be taken to institute a national system of radiation exposure record-keeping, under which there would be maintained for every individual a complete history of his total record of exposure to X rays, and to all other gamma radiation. This will impose minor burdens on all individuals of our society, but it will, as a compensation, be a real protection to them. We are conscious of the fact that this recommendation will not be simple to put into effect.

2. That the medical authorities of this country initiate a vigorous movement to reduce the radiation exposure from X rays to the lowest limit consistent with medical necessity; and in particular that they take steps to assure that proper safeguards always be taken to minimize the radiation dose to the reproductive cells.

3. That for the present it be accepted as a uniform national standard that X-ray installations (medical and nonmedical), power installations, disposal of radioactive wastes, experimental installations, testing of weapons, and all other humanly controllable sources of radiations be so restricted that members of our general population shall not receive from such sources an average of more than 10 roentgens, in addition to background, of ionizing radiation as a total accumulated dose in the reproductive cells from conception to age 30.

4. The previous recommendation should be reconsidered periodically with the view to keeping the reproductive cell dose at the lowest practicable level. If it is feasible to reduce medical exposures, industrial exposures, or both, then the total should be reduced accordingly.

5. That individual persons not receive more than a total accumulated dose to the reproductive cells of 50 roentgens up to age 30 years (by which age, on the average, over half of the children will have been born), and not more than 50 roentgens additional up to age 40 (by which time about nine tenths of their children will have been born).

6. That every effort be made to assign to tasks involving higher radiation exposures individuals who, for age or other reasons, are unlikely thereafter to have additional offspring. Again it is recognized that such a procedure will introduce complications and difficulties, but this committee is convinced that society should begin to modify its procedures to meet inevitable new conditions.

The committee underscores its conclusion that the public be protected, by whatever controls necessary, from receiving a total reproductive-lifetime dose (conception to the age of thirty) of more than 10r of artificially induced radiation to the reproductive cells.

The public should see to it that it secures such protection.

As Linus Pauling has said, "We are the custodians of the human race. We have the duty of protecting the pool of human germ plasm against willful damage." We must, therefore, do everything necessary to see to it that all atom-bomb testing, which releases so many menacing radioactive substances, will be immediately and forever discontinued.

Among these menacing radioactive substances is carbon 14. Carbon 14 has a life of 8,070 years and is damaging to human beings throughout that time. It has been estimated by Linus Pauling that the carbon 14 thus far released by the bomb tests will in the long run produce about 1 million seriously defective children, about 2 million embryonic and newborn deaths, and minor hereditary defects in many millions of individuals.

These are some of the facts about the effects of radiation upon heredity. The rest of the question is whether you, the reader, are going to remain part of the problem that they present or whether you are going to make yourself a part of the solution.

When adequate precautions are taken, diagnostic and therapeutic radiation can be safely used for the benefit of human beings. Those precautions must be enforced not alone by regulation or law, but even more by the force of the ethical principles involved, for no one has a right to accumulate genetic debts for his descendants to pay. We cannot afford to draw such long-term bills on the future of humanity.

# APPENDIX A
# INHERITED DISORDERS OF MAN

THE CLASSIFIED LIST of inherited disorders presented in the following pages is designed to give the reader some idea of the kinds of disorders that are largely determined by genetic factors, as well as to provide him with a short statement of the mode of inheritance involved. While fuller than any similar list hitherto published, it is to be understood that the present list is by no means a complete census of all known hereditary conditions—not to mention those whose mode of inheritance is unknown. The list is, however, fairly representative.

A word of caution: In consulting this list it should be remembered that many of the conditions listed are sometimes produced in ways that have little or no connection with genetic factors. Intrauterine environmental conditions and extrauterine environmental conditions of many varieties may produce disorders that in every way resemble those principally due to defective genes. Such copies of the genetically induced condition are known as *phenocopies*—and should be carefully distinguished from the genotypically produced disorders. Furthermore, some disorders arise quite sporadically for reasons that are often obscure, and these, too, should be distinguished from the hereditary conditions.

The classification is based fundamentally on the organs or organ systems affected. Thus, if the reader knows which organ or organ system is involved in the condition he wishes to look up, all he has to do is turn to the appropriate heading. If, however, all that he knows is simply the name of the disorder, he may save time by looking at the index, which will tell him on what page the condition is listed. In the index he will find the condition listed separately under both its ordinary and its technical name. However, it has not been possible to classify all the conditions listed under organs or organ systems—for example, allergies or disorders of metabolism —but this should make no great difficulty for the reader.

Wherever possible, the incidence of the condition in the population in general has been given, as well as the sex differences in incidence. When the former is not given, the evidence is not quite clear; when the latter is not given, it may in general be assumed that the condition appears with equal frequency in both sexes; when complete sex-linkage obtains, it is to be understood that the condition appears exclusively in males.

Alternative modes of inheritance can sometimes occur in addition to those given.

NOTE: In this list "dominant" will be abbreviated as D, "recessive" as R, and "familial" as F.

## INHERITED DISORDERS OF MAN 347

## ALLERGIES

*Angioneurotic edema* (*Heredoperiodic edema*) (A highly fatal disorder characterized by acute noninflammatory swelling, primarily involving the skin, subcutaneous tissue, larynx, and gastrointestinal mucosa. Most dangerous when occurring in throat)
D

*Asthma*
Probably D, with about 40 percent penetrance
Risk: With 1 parent asthmatic, 50 percent of their children will be carriers, and of these, 13 percent will exhibit asthma and vasomotor rhinitis and 7 percent hay fever or atopic dermatitis (Besnier's prurigo)

*Urticaria* (Hives)
Possibly irregular R

## BLOOD AND VASCULAR SYSTEM

*ABO hemolytic disease* (ABO incompatibility between mother and child; where, for example, mother is blood group O and the child is A or B, the cells of the fetus tend to be agglutinated by anti-A or anti-B antibodies in the mother's serum. The condition is usually not serious, though it is 3 times more common than rh hemolytic disease)
Incidence: 1 in 71 births, and may arise in 7 percent of AB incompatible mother-child pairs
Risk: Thought to be responsible for the loss of about 25 percent of A children expected from marriage of A × O

*Acanthrocytosis* (Malformed, crenate red blood cells, may be accompanied by atypical retinitis pigmentosa and evidences of involvement of nervous system)
Probably R

*Acatalasia* (Oral gangrene, absence of catalase in blood)
Probably R

*Afibrinogenemia* (*Hypofibrinogenemia*) (Congenital fibrinogen deficiency—blood-plasma globulin—resulting in severe hemorrhagic symptoms)
Probably D with variable penetrance
Preponderance of males
In some individuals carrying the gene in heterozygous state, the gene may express itself in a mild form of the disorder in a relative deficiency of fibrinogen (fibrinopenia) rather than in complete absence of fibrinogen

*Agammaglobulinemia* (Extreme reduction in immunity-conferring gamma globulin)
(i) Infant type
(ii) Adult type
Sex-linked R

*Alder's anomaly* (Coarse granulation of white blood cells)
Probably D

*Anemia* (Defective development of number and structure of red blood cells)
(i) *Familial hemolytic anemia* (Frequently associated with inclusion bodies and defective pigment metabolism)
F

(ii) *Hereditary iron-loading anemia* (Red blood cells unable to utilize normally the readily available iron; fatiguability, upper abdominal pain, enlargement of spleen, and thrombophlebitis; there is a secondary deposit of iron in various organs and tissues)
F and sex-linked. R or incompletely R in female

(iii) *Hyperchromic anemia* (*Fanconi's syndrome*) (Hypoplasia of bone marrow, chronic progressive reduction of red blood cells, often with atrophy of spleen, and accompanying skeletal defects, microcephaly, strabismus, brown pigmentation of skin, undeveloped gonads; usually fatal in childhood.) There is evidence that the condition is due to alterations in one of a pair of chromosomes

There are D and R forms
Preponderance of males

(iv) *Hypochromic, microcytic anemia* (Pallid, small red blood cells, often with enlargement of spleen and liver; frequently terminates fatally at early age)
Probably sex-linked, but may also be autosomal D with sex-limited effect

(v) *Hypoplastic anemia of childhood* (Great reduction in blood cells, terminating in death)
Probably R
Aplastic form: Even more extensive reduction in leukocytes and platelets. Probably R
Incidence: Extremely rare

(vi) *Pernicious anemia (primary anemia, Addison's anemia)* (Chronic progressive reduction of red blood cells)
A genetic tendency evident in many families, but mechanism unclear.
Probably familial D
Incidence: 1-2 in 10,000

(vii) *Sickle-cell anemia* (So called because when drop of blood from affected individual is deprived of oxygen, red blood cells assume sicklelike shapes)
  (i) *Sickle-cell disease* (Severe anemia, usually fatal in childhood.) Homozygous
  D. Incidence: 1 out of 500 American Negroes
  (ii) *Sickle-cell trait* (Mild anemia.) Heterozygous D. Largely limited to peoples of Negroid origin

(viii) *Thalassemia (Cooley's anemia, Mediterranean anemia)*
  (i) *Thalassemia minor* (Mild anemia with increased fragility of red blood cells and increased number of target and oval cells.) Heterozygous, incomplete D
  (ii) *Thalassemia major* (Progressive fatal hemolytic anemia.) Homozygous D

*Arteriosclerosis*
There is a strong tendency in some families for many of its members to develop this disorder, indicating a genetic basis, which is at present not clear

*Bradycardia* (Abnormal slowness of heartbeat)
F

*Atrial septal defect*
(Defect of septum separating atria of heart)
D

*Cardiomegaly, familial*
(Enlargement of heart)
F

*Cardiovascular dysplasia, hereditary* (Enlarged heart, fibrosis of myocardium, hypoplasia of aorta, hypertrophy of interventricular septum, bizarre arrangement of muscle fibers, associated with dyspnea and syncope)
D

*Christmas disease* (Resembling mild to moderately severe hemophilia)
Sex-linked R

*Combined hemophilia and Christmas disease*
Sex-linked

*Combined hemophilia and parahemophilia*
Sex-linked R *plus* autosomal R
Largely limited to males

*Coronary xanthomatosis* (Yellow deposits of nodules in the heart valves and/or the coronary arteries resulting in angina pectoris)
D

*Dextrocardia* (With or without malposition of viscera. Malposition of heart, which occupies right side)
Genetic mechanism unclear
Incidence: About 1 in 29,000

*Endocardial fibrosis* (Breakdown of inner lining of heart; death within first decade)
F, in some cases possibly mutant D

*Eosinophilia, familial* (Increase in circulating polymorphonuclear eosinophils—readily staining leukocytes, or white cells)
Probably D

*Fetal hemoglobin, persistent* (In adult without associated anomalies)
F

*Hageman trait* (Nonhemorrhagic disorder of coagulation of blood with prolonged clotting time of venous blood)
Probably R

*Heart disease, congenital*
(i) *Patent ductus arteriosus* (Failure to close the fetal artery leading from pulmonary to aortic artery.) Probably D
(ii) *Auricular septal defect.* Probably D
(iii) *Lutenbacher's syndrome* (Mitral stenosis and interauricular defect.) Probably R
(iv) *Tetralogy of Fallot* (Pulmonary stenosis or atresia, interventricular defect, and hypertrophy of right ventricle.) Probably R
(v) *Congenital arrhythmia* (Many varieties)
F
Risk: Of any variety of cardiac malformation in children, 1 in 50; likelihood of recurrence of identical malformation, 1 in 100

*Hemangioma of small intestine* (Tumor of multiplied blood vessels of small intestine)
D

*Hemolytic anemia, familial* (Defective development of number and structure of red blood cells, frequently associated with inclusion bodies and defective pigment metabolism)
F

*Hemolytic disease of the newborn* (*Erythroblastosis fetalis*) (Due to incompatibility of the mother's genotype Rh negative and the baby's genotype Rh positive)
Incidence: 1 in 200 pregnancies
Risk: If father carries an Rh-positive gene and an Rh-negative gene, there is an even chance that sibling of affected child will be normal. If father's genes are both Rh-positive, chances of normal future pregnancies are poor.
First child usually escapes

*Hemophilia* (Excessive and prolonged hemorrhage following trivial injury or occurring spontaneously due to deficiency of thromboplastinogen — antihemophilic globulin)
Sex-linked R
Incidence: 1 in 25,000
Mutation rate: 1 in 50,000 in female, 1 in 10,000 in male

*Hemorrhagic disease, autosomal* (Prolonged hemorrhage following injury, with deficiency of antihemophilic globulin)
Unlike hemophilia, this condition is *not* sex-linked, but upon the basis of the few cases recorded seems to be sex-limited occurring only in females.

*Heparinemia* (Oversecretion of heparin from liver into blood, with resulting bleeding)
R

*Hereditary capillary purpura* (*von Willebrand's disease*) (Inadequate contraction of walls of capillaries after injury)
D

*Hereditary labile factor V deficiency* (Prolonged bleeding time, with spontaneous bleeding mostly from mucous membranes)
R

*Homozygous hemoglobin C disease* (Hemolytic anemia, enlarged liver and spleen, etc.)
Probably irregular D

*Hypertension, essential* (High arterial blood pressure of unknown cause.) Incomplete D; in homozygous form resulting in severe hypertension, in heterozygote in moderate elevation of blood pressure. In both cases much under the influence of environmental conditions

*Hypogammaglobulinemia* (Gamma-globulin deficiency; reduction in immunity-conferring gamma globulin)
  (i) *Moderate hypogammaglobulinemia* (Often with enlargement of spleen and decreased number of leukocytes.)
  Probably D
  (ii) *Primary congenital hypogammaglobulinemia* (Lymph nodes almost totally devoid of normal follicles and plasma cells.) Virtually confined to boys. Sex-linked R
  (iii) *Primary adult hypogammaglobulinemia* (Almost complete agammaglobulinemia may be present without enlargement of spleen or reduction in number of leukocytes.)
  Probably R

*Hypoprothrombinemia* (Reduced prothrombin content of blood with resulting mild to serious bleeding)
  (i) *Reduced prothrombin type.* R
  (ii) *Normal prothrombin but reduced free, or active, fraction.* D

*Hypotension* (Low blood pressure)
  Probably D

*Jaundice*
  (i) *Congenital hemolytic* (*Spherocytosis, Gilbert's disease, spherocytic anemia, chronic acholuric jaundice*) (Sphere-shaped red blood cells with marked rate of blood destruction. Controlled by splenectomy.)
  D. Incidence: 1 in 15,000
  (ii) *Familial nonhemolytic* (Jaundice without destruction of red blood cells but with conjugated bilirubin in both serum and bile, thus distinguishing it from (i).
  F

*Leukemia* (Abnormal increase in white blood cells or leukocytes)
  (i) *Chronic lymphatic leukemia* F
  (ii) *Acute leukemia*
  Genetic mechanism unclear
  (iii) *Neonatal familial leukemia* F
  (iv) *Chronic myeloid leukemia*
  (Associated with absence of about a half of long arm of chromosome 21)
  Inheritance unclear

*Neutropenia* (Chronic decrease in number of polymorphonuclear neutrophils, or white cells)
  D

*Ovalocytosis* (*Elliptocytosis*) Oval or elliptical erythrocytes—red blood cells—in blood)
  D
  Incidence: 1 in 3,000

*Parahemophilia* (Deficiency of plasma labile factor V—proaccelerin—necessary for conversion of prothrombin to thrombin; resembles hemophilia clinically)
  R. In the heterozygous state the effects of the single gene can be picked up in the carriers since they have a significant reduction of factor V

*Pelger's nuclear anomaly* (Failure of development of all marrow cell lines, characterized by failure of neutrophils, or polymorphonuclear leukocytes, to develop past the bilobed stage)
  D
  Incidence: Less than 1 in 10,000

*Phlebectasia* (Varicose veins)
  Irregular D

*Plasma thromboplastin antecedent deficiency*) (Bleeding resulting from deficiency of plasma protein required for blood thromboplastin generation)
  R

*Rheumatic heart disease*
  R

*Stuart-Prower clotting defect* (*Factor VIII—stable—deficiency*) (A hemophilialike condition with moderate bleeding tendency and prolonged plasma clotting time)

Incomplete R

*Telangiectasis, hereditary hemorrhagic (Rendu-Osler-Weber disease)* (Dilation of terminal portions of small vessels with repeated bleeding)
    D, with reduced penetrance; in homozygous state may be lethal

*Thromboangiitis obliterans (Buerger's disease)* (Phlebitis with fibrosis involving the accompanying arteries and nerves)
    Probably R

*Thrombocytopathic purpura (Glanzmann's thromboasthenia)* (Normal number but abnormal appearance of platelets, normal coagulation time, poor retraction)
    Probably D

*Thrombocytopenia, familial (Aldrich's syndrome)* (Thrombocytopenic purpura with eczema and undue liability to infection)
    One-third of those affected die before their second year
    Sex-linked R

*Thrombocytopenic purpura, idiopathic (Werlhof's disease)* (Bleeding from and into the skin in small spots and patches, great reduction in blood platelets, poor clotting)
    D

## BONES AND JOINTS

*Achondroplasia (Chondrodystrophy, chondrodystrophia foetalis)* (Stunting of skeletal growth owing to anomalous cartilaginous growth)
    D; some cases possibly R
    Incidence: 1 in 15,000

*Acroosteolysis* (Shrinkage of the bones, usually of the extremities)
    Irregular D
    Onset: Any time from birth

*Amputations, congenital* (Absence of arms and/or legs)
    R, incomplete D, and D genotypes may exist

*Arthritis, chronic rheumatoid* (Inflammatory disease affecting particularly the small joints of hands and feet)
    F

*Arthroonychodysplasia* (Congenital dystrophy of nails, elbows, and absence of kneecap)
    D, of variable expressivity

*Cherubism* (See *Fibrous dysplasia of the jaws, familial*)

*Chondroectodermal dysplasia (Ellis-van Creveld syndrome* (Disturbances in development of hair, teeth, skin, and other ectodermal derivatives; polydactyly; achondroplasia)
    Probably R

*Clubfoot, congenital*
    Irregular D, probably sex-influenced, and exhibiting low penetrance and great variability in expressivity
    Twice as frequent in males as in females
    Incidence: 1 in 1,000
    Risk: For siblings of consanguineous relationship, 25 percent; for sibling with one affected parent, 10 percent; for sibling with unaffected parents, 3 percent

*Deep acetabulum and intrapelvic protrusion*
    Probably D

*Diaphyseal aclasis* (Multiple cartilaginous exostoses, or bony growths)
    D
    Of those affected, 70 percent are males

*Dysplasia epiphysialis punctata (Chondrodystrophia calcificans congenita)*
    Possibly irregular D
    Slightly more frequent in females

*Enchondromatosis (Ollier's disease)* (Multiple deformities of tubular bones)
    Possibly D, with very low penetrance

*Exostoses, multiple hereditary* (Multiple abnormal outgrowths of bone, often accompanied by thyroid, calcium, and phosphorus deficiencies)
    Probably D

*Fibrous dysplasia of the jaws, familial (Cherubism)* (Painless,

progressive disfiguring and enlargement of lower face and jaws; eyes sometimes have upward cast; other bones sometimes involved; remission may occur after adolescence)
Familial, probably R

*Hallux rigidus* (Stiff big toe)
Probably D

*Hallux valgus* (Bunion)
D

*Hammertoe* (The second toe is usually involved)
D

*Hip, congenital dislocation of*
Probably D, with lack of penetrance due to sex and environmental factors; intra- and extra-uterine environmental factors involved; for example, twice as many cases occur in winter as in summer, and maternal age is a significant factor.
85 percent of the affected infants are females.

*Hyperostosis, infantile cortical* (Tender swelling of soft tissues overlying bones of extremities, fever, irritability)
Onset: Before 5 months
Genetic mechanism unclear

*Madelung's deformity* ("Bayonet" hand, shortened forearm, short stature, often with multiple cartilaginous growths)
Probably D

*Marfan's syndrome (Arachnodactyly)* (Generalized disorder of the skeleton characterized by outstandingly long metacarpals, metatarsals, and phalanges; other skeletal and visceral defects are often associated)
D with pleiotropic effect

*Metaphyseal dysostosis, familial* (Maldevelopment of shafts of long bones)
F

*Multiple congenital contractures* (*Arthrogryoposis multiplex congenita, congenital myodystrophy, multiple congenital articular rigidity, dystrophia muscularis congenita, amyoplasia congenita, myodystrophia foetalis deformans, myophagism, congenital arthromyodysplasia*) A syndrome of multiple congenital contractures and joint disturbances)
D and R forms

*Osteitis deformans (Paget's disease)* (Thickening and rarefaction of bones)
Probably irregular D; may be sex-linked in some cases
Slightly greater incidence in males

*Osteoarthropathy of fingers, familial* (Enlargement and loss of mobility of middle and distal joints of fingers; first appears in childhood)
D, with nearly complete penetrance

*Osteochondritis deformans juvenilis (Legg-Calvé-Perthes' disease)* (Disease of the hip joint appearing in children between ages of 8 and 12)
D, sometimes R
Incidence: much higher in boys than in girls

*Osteochondritis dissecans* (Disease affecting several joints)
D

*Osteochondrodystrophia deformans (Morquio-Brailsford disease)* Deformation of skeleton with dwarfing)
Usually R, sometimes sex-linked, and occasionally D

*Osteogenesis imperfecta (Van der Hoeve's syndrome, Lobstein's disease, fragilitas ossium)* Extreme fragility with multiple fractures of bones, blue sclerae of eyes)
(i) *Congenital.* Prognosis very poor
(ii) *Osteogenesis imperfecta tarda.* Divided into two types: I, in which fractures begin at birth, and II, in which fractures begin between 2 and 3 years. In I they continue throughout life; in II they begin to decline in frequency at puberty. Otosclerosis develops in 25 percent of cases.
D, with pleiotropic effect
Incidence: 1 in 50,000

*Osteopetrosis (Albers-Schönberg disease, marble bones)* (Hardening followed by softening and

great fragility of bones)
  (i) *Malignant form.* Probably R
  (ii) *Benign form.* D, with reduced penetrance

*Osteopoikilosis (Spotted bones)* (Islands of compact bone developed in the spongy, or cancellous, portions of bones; symptomless)
  D, possibly R in some cases

*Parietal foramina, enlarged* (Defects in the posterior angles of the parietal skull bones)
  D

*Pes planus* (Convex valgus foot: flatfoot)
  D

*Polyspondyly* (Multiple anomalies of the vertebral column)
  Associated with 45 chromosomes, translocation of greater part of chromosome 23 to long arm of 15

*Radial reduction* (Partial or complete absence of radius and aplasia of radial rays of hands and fingers)
  D
  A very minor defect in one individual may be followed by a much more severe form in later generations.

*Radiohumeral, ulnahumeral synostosis, congenital* (Fusion between the humerus and the ulna or radius)
  D, irregular D, R sex-linked or male sex-limited
  Over 50 percent of cases are bilateral. If not complete, a rudimentary elbow joint may be formed.
  More frequent in males

*Radioulnar defect* (Absence of radius and ulna)
  R or sometimes possibly irregular D
  Incidence: 1 in 75,000

*Rickets, vitamin-D-resistant (Familial hypophosphatemia)*
  Sex-linked D with complete penetrance. Possibly cases occur due to autosomal D or R.
  All of the daughters but none of the sons of affected fathers will be affected; of affected mothers, one-half of the daughters and one-half of the sons are affected. Males are more seriously affected.

*Rickets due to matrix defect, Hypophosphatasia* (Rickets due to defect of calcification of organic bone matrix)
  R, occasionally D

*Scapulae, elevated (Sprengel's deformity)* (High position of the shoulder blades owing to their failure to descend in development)
  D and R genotypes have been described

*Spina bifida* (Cleft spinal column)
  (i) *Spina bifida aperta.* Preponderant in males. Incidence: 1 in 400
  (ii) *Spina bifida occulta.*
  The hereditary mechanism is unknown; it would appear that many genotypes for the condition may exist
  Risk: For siblings, of having spina bifida, anencephaly, or hydrocephaly, 4 percent.

*Split foot*
  Irregular D. Appears to be the effect of the same gene as for cleft hand. Close asassociation exists between syndactyly and cleft foot.

*Spondylitis, ankylosing* (Hardening of the joints of the spinal column and sacroiliac region with rigidity of spine)
  D, with 70 percent penetrance in males and 10 percent penetrance in females. When affected parent is mother, penetrance is almost total in offspring of both sexes

*Tibial defect* (Defective tibia)
  D

*Ulnar defects* (Total or partial absence of ulna with defects of ulnar rays)
  Sporadic

## EARS AND HEARING

*Atresia of auditory meatus* (Developmental failure of earhole to open, usually with rudimentary auditory ossicles)

## 354 HUMAN HEREDITY

**D**

*Auricular appendages* (Congenital tumors on external ear, usually nonmalignant)
  Probably D

*Auricular fistula* (Congenital hole, usually ending in a shallow blind depression where front of rolled part of ear meets skin of face)
  Usually irregular D

*Cat's ear* (Diminished size of ear, projecting sideways)
  D

*Deaf-mutism*
  This condition is dependent upon the way two pairs of genes, *D* and *d*, *E* and *e*, interact in such a way that *D-E* produce normal hearing, whereas any other combination produces children who are born deaf. This is an example of *duplicate recessive epistasis*. When a gene of one pair masks the expression of the genes of another pair, it is said to be *epistatic* to the other pair.
  50 percent of deaf-mutism is due to intrauterine environmental causes.
  Incidence: 1 in 3,000

*Deaf-mutism with complete albinism*
  Sex-linked

*Hypertrichosis auricularum* (Overgrowth of hair on external ear)
  Recent studies confirm the earlier suspicion that the gene for this condition is usually carried on the Y chromosome
  Confined to males

*Labyrinthine deafness* (Inner-ear deafness)
  D

*Malformed ears associated with malformation of genitourinary tract* (Ears show variety of defects and one kidney usually defective, but any part of the genitourinary system may be affected
  D, with about 70 percent penetrance

*Microtia* (Small ears)
  Probably irregular D or possibly sex-linked

*Otitis media* (Middle-ear inflammation)
  Undoubtedly has a genetic basis in many cases, but mode of inheritance not understood.

*Otosclerosis* (Development of spongious bone in labyrinthine capsule, eventually leading to middle-ear deafness)
  D, with reduced penetrance
  More frequent in females
  Incidence: 1 in 500
  Onset: After puberty, usually in adult life

### EYES

*Albinism, ocular.* See *Albinism*, under SKIN

*Amaurosis, congenital* (Total or almost total blindness, involving degeneration of retina and sometimes optic nerve)
  Heredofamilial R

*Anophthalmos* (Congenital absence of eyes)
  Appears to be R; probably occurs in sex-linked R form in association with some form of mental deficiency

*Cataract*
  (i) *Congenital.* Usually D, sometimes sex-linked R
  (ii) *Postnatal.* Generally D
  Whether "senile" cataract is genetically influenced is unknown.

*Choroid defects* (Defects of the middle coat of the eye)
  (i) *Congenital macular dysplasia* (Defective development of choroid.) D
  (ii) *Chorioideremia* (Absence of choroid.) Sex-linked R
  (iii) *Chorioidal sclerosis* (Hardening of choroid.)
    (a) Generalized hardening. Usually D
    (b) Central hardening. Usually R
  (iv) *Sarcoma* (Malignant.) Has been observed as a D

*Color blindness*
  (i) *Partial color blindness, deutan (green) series.* Sex-linked

## INHERITED DISORDERS OF MAN

(ii) *Partial color blindness, protan (red) series.* Sex-linked R

(iii) *Total color blindness.* Sex-linked R

About 8 percent males are color-blind: 75 percent of these defective in green vision and 25 percent in red vision.

Cornea defects
  (i) *Microcornea.* D
  (ii) *Cornea plana* (Flattened cornea) D
  (iii) *Cornea keratoconus* (Conical) D
  (iv) *Corneal vortex veil.* D
  (v) *Corneal opacity, congenital.* R
  (vi) *Macular dystrophy.* R
  (vii) *Granular dystrophy.* D
  (viii) *Latticelike dystrophy.* D
  (ix) *Megalocornea.* Sex-linked, sometimes D

Dyskeratosis (Overgrowth of conjunctiva)
  (i) *Benign intraepithelial* (With overgrowth of oral mucous membrane.) D, with 97 percent penetrance
  (ii) *Congenital* (Zinsser-Cole-Engman syndrome) (Oral-mucous-membrane overgrowth, dystrophy of nails, atresia of lacrimal ducts, anemia, testicular atrophy.) Sex-linked

Epicanthus (Fold of skin over inner angle of eye, as in the Mongoloid major group) D
  This is not a disorder, but a physical trait, sometimes associated with a disorder, as in Down's syndrome

Fundus dystrophy (Degeneration of the back of the interior of the eyeball) D
  Onset: About age 40

Glaucoma (Pressure of fluids in eyeballs, often leading to blindness)
  (i) *Buphthalmia congenital or infantile.* R, occasionally D
  (ii) *Minimal glaucoma.* D

  (iii) *Absolute glaucoma.* D, sometimes R

Hemeralopia (Day blindness) (Inability to see clearly in bright light, usually with color blindness)
  R, sometimes partially sex-linked R

Iris defects
  (i) *Aniridia* (Absence or defect of iris.) D, with variable expression. Incidence: 1 in 100,000
  (ii) *Hypoplasia* (Defective development.) D, sometimes sex-linked R or R
  (iii) *Flocculi iridis* (Woolly iris.) D
  (iv) *Persistent pupillary membrane* ("Pinhole" pupil.) D

Lens defects
  (i) *Congenital dislocation.* D
  (ii) *Delayed dislocation.* D
  (iii) *Congenital dislocation with displaced pupil.* R

Macular dystrophy (Degeneration of the macula lutea, the sensitive yellow spot of the retina)
  D and R forms have been recognized; a sex-linked R form associated with color blindness also is recognized

Microphthalmos (Small eyes)
  (i) *Pure type* (With no associated defects.) Probably R
  (ii) *With cataract.* D and R genotypes occur

Miosis, congenital
  (Complete or partial absence of dilator muscle of iris)
  F

Mirror reading
  Sometimes D

Myopia (Nearsightedness)
  (i) *Extreme.* R
  (ii) *With night blindness.* D and also sex-linked R

Nyctalopia (Night blindness)
  (i) *Congenital stationary night blindness.* D
  (ii) *With myopia.* Sex-linked. Virtually limited to males
  (iii) *With extreme myopia.* R

Nystagmus (Quivering of the eyes)
  Sex-linked R and also irregular D

*Ochronosis, hereditary* (Bluish-black pigment accumulates in various parts of body, uncommon phenylalanine-tyrosine metabolism with excretion of homogentisic acid)
F

*Ocular paralysis* (Paralysis of eye-ball-moving muscles, thus immobilizing eyes)
D and sex-linked genotypes have been described

*Oculo - cerebro - renal syndrome* (*Lowe's syndrome*) (Cataract, hydrophthalmia, rickets, aminoacuduria, renal hypoammonia)
Sex-linked R

*Optic-nerve atrophy* (Degeneration of optic nerve terminating in blindness)
(i) *Congenital*. D
(ii) *Childhood type*. R
(iii) *Adult type* (*Leber's disease*.) Sex-linked R, sometimes D

*Presbyopia* (Farsightedness)
D

*Progressive ophthalmoplegia* (Drooping of an eyelid, then restriction of ocular movement)
D

*Pseudoglaucoma* (A condition resembling glaucoma)
D

*Pseudoglioma* (Blindness, small bulbs, atrophy of iris, secondary cataract, etc., with mental deficiency)
Sex-linked R

*Pterygium* (Patch of thickened, usually fan-shaped conjunctiva extending over part of the cornea)
D, with low penetrance
Onset: After 40 years

*Ptosis* (Drooping of eyelids)
D

*Pupillary membrane, congenital* (Persistence of filaments of pupillary membrane)
F

*Retinitis pigmentosa* (Chronic inflammation of retina with atrophy, etc.)
Usually R, sometimes D, and occasionally sex-linked
Incidence: 1 in 10,000

*Retinoblastoma* (Retinal glioma) (Tumor of retina)
D, with somewhat reduced penetrance
Incidence: 1 in 23,000
Risk: Only 25 percent of affected parents transmit the condition to their offspring because the mutation in many cases has arisen in the somatic cells and not in the reproductive ones. The chance that a child would inherit the trait is about 40 percent, instead of the 50 percent possible if the dominant gene were characterized by complete expressivity.

*Retinoschisis, hereditary* (Splitting of the retina and other disorders, particularly of the vitreous and its vessels)
Sex-linked R
Incidence: Very rare

*Strabismus* (Cross-eyes)
R and D genotypes have been described
Risk: If one child has the condition and the parents are normal, the risk that another child may show the condition is 10 percent; with one parent affected it is 17 percent

*Unpigmented eyes* (The eyes appear to be pink owing to the presence of blood vessels, but the eye is unpigmented. The condition is limited to the eyes alone)
Sex-linked R

*Van den Bosch syndrome* (Choroideremia, acrokeratosis verruciformis, anhidrosis, skeletal deformity, mental deficiency)
Sex-linked R

*Waardenburg-Klein syndrome* (1. wide separation of eyes with lateral displacement of inferior lacrimal punctum, 2. hyperplastic broad nasal root, 3. hyperplasia of medial portions of eyebrows, 4. partial or total heterochromidia iridum, 5. congenital deafness or partial, unilateral deafness, 6. white forelock, rarely with partial albinism)

D
Incidence: 1 in 42,000

## FINGERS AND TOES

*Apical dystrophy* (Absence of terminal portions of second to fifth fingers)
D

*Arachnodactyly* (*Marfan's syndrome*) (Abnormally long fingers)
D, with pleiotropic effect

*Brachydactylia* (Short fingers)
D

*Brachymesophalangy* (Short middle fingers and/or toes)
D

*Brachymorphy with spherophacia* (*Weill-Marchesani syndrome*) Short limbs and digits with small, distorted lens of eye, midaortic necrosis
Incompletely D, possibly R

*Brachytelephalangy of the thumbs* (Short thumbs)
D

*Camptodactyly* (Finger or fingers bent toward palm)
Irregular D

*Clinodactyly* (Bent fingers toward thumb side)
D

*Ectodactylia* (Partial or complete absence of fingers)
D

*Hyperphalangy of the thumb* (Three instead of the normal two phalangeal bones)
D

*Polydactylism* (Reduplication of fingers)
Usually D, with variable penetrance and expressivity

*Split hand* (Lobster hand)
Irregular D
Incidence: 1 in 90,000

*Symphalangism* (Bony or fibrous union of joints of fingers or toes)
D

*Syndactyly* (Fusion of the fingers)
Usually D
Almost twice as frequent in males as females

## GASTROINTESTINAL TRACT

*Gastrointestinal bleeding, familial massive* (Has been described in three siblings during their first year from undetermined site, with spontaneous cessation by end of second year)
Probably R

*Idiopathic megacolon* (*Hirschsprung's disease*) (Enlargement of large, and sometimes portion of small, intestine; nearly always fatal unless surgical correction, by rectosigmoidectomy, is made)
Genetically determined in all cases, but mechanism not understood
Two-thirds of the cases are males
Incidence: 1 in 8,000
Risk: A subsequent brother has a 1-in-5 chance of being affected.

*Meckel's diverticulum* (Persistence of a blind pouch arising from the small intestine a short distance from the cecum—the commonest congenital abnormality of the gastrointestinal tract, often responsible for abdominal symptoms)
Sometimes F

*Peptic ulcer* (Disposition to develop gastric or duodenal ulcer runs in families)

*Polyposis of large intestine*
(i) *Familial intestinal polyposis* (Polyps in colon and rectum, which may become malignant.) D
Incidence: 1 in 5,000
(ii) *Peutz-Jegher's syndrome* (Polyps, mostly in small intestine, but also in stomach, colon, and rectum. There is spotty pigmentation of mucosa of cheek, skin of face, fingers, and toes.) D
Risk: Children of affected individuals have a 1-in-2 chance, in both types of polyposis, of inheriting the condition.
(iii) *Gardner's syndrome* (Polyps of colon associated with bone and soft-tissue tumors.) D
(iv) *Turcot - Déprés - St. Pierre syndrome* (Polyps of

colon associated with brain tumor.)
(v) *Isolated single polyps.* Familial. R
(vi) *Polyposis with polyendocrine adenomatosis* (Endocrine glands affected, peptic ulcer, and intestinal polyps)
Probably R

*Pyloric stenosis, congenital* (Hypertrophic pyloric stenosis)
There is undoubtedly a genetic influence involved but its mechanism is unclear.
Males constitute 80 percent of the cases
Incidence: 3 in 1,000

*Rectal stenosis* (Extreme narrowing of rectum, sometimes with imperforate anus)
Probably R

## GENITOURINARY SYSTEM

*Alport's syndrome* (*Hereditary nephritis associated with deafness*)
Partially sex-linked D
Females are seldom severely affected and have a normal life expectancy. Males usually die before the age of thirty

*Cryptorchism* (Undescended testes)
R
Incidence: 1 in 30 boys under fourteen and 1 in 250 men over twenty-one

*Cystic kidney*
(i) *Congenital*. R. Causes death at or shortly after birth
(ii) *Adult type*. D. Onset: About age forty

*Endometriosis* (Diffuse ectopic growth of lining membrane of uterus)
F

*Eunuchoidism* (Absence of testicles)
Often familial

*Glomerulonephritis, familial* (Disorder of the capillary plexuses of the renal tubules, probably due to a metabolic deficiency, frequently associated with aminoaciduria)
Familial, probably R

*Hermaphroditism, pseudo* (Where only one type of sex-gland tissue is present, either testicular or ovarian but never both)
(i) *Male*. Sex-linked
(ii) *Female*. Probably R
Intrauterine environmental factors undoubtedly important in many cases
7 to 8 times more frequent in male
Incidence: 1 in 1,000

*Hermaphroditism, true familial* (Where both male and female sex-gland tissue is present)
Sex-linked, sex-influenced, and R forms occur, as well as phenocopies of intrauterine origin. Most of the cases are genetic females
Incidence: 1 in 1,000

*Hypospadias* (Opening in underwall of penile urethra)
Irregular R
Incidence: 1 out of every 1,000 males

*Klinefelter's syndrome* (Apparent male who is sterile, often with marked breast development and always with small gonads after puberty)
Usually chromatin positive; characterized by two X chromosomes and a Y chromosome
Incidence: 1 in 1,000

*Lower-urinary-tract obstruction* (Persistent urachus, dilation of bladder, hydroureter, cryptorchism, pulmonary hypoplasia, oligohydramnios, amnion nodosum, talipes deformity, abdominal-muscle dysplasia)
F

*Nephritis, familial* (Inflammation of the kidney; albumin and blood in urine)
Familial, probably R

*Nephronophthisis* (Tubular dysfunction of the kidneys, with defective water reabsorption, accompanied by maldevelopment of bones)
Probably R

*Nephropathy, deafness, and renal foam cells* (Chronic renal disease with associated nerve deafness occurring together or

# INHERITED DISORDERS OF MAN

singly in members of same family. Large numbers of foam cells present in cortex of kidneys)
Sex-influenced D
Males more seriously affected, few surviving beyond thirty. Females have normal life-span

*Nephrosclerosis and essential hypertension* (Hardening of the kidney with increased blood pressure)
D

*Nephrosis, deafness, and urinary-tract anomalies (Braun-Bayer syndrome)* (Kidney disease, deafness, urinary-tract anomalies, digital anomalies of thumbs and great toes, and bifurcation of the uvula)
F
Not all these conditions are found in a single individual, but in the siblings of a single family one or more of them may be found together

*Renal tubular acidosis, familial*
R

*Renal dysplasia* (Abnormal development of kidney)
D

*Renal sclerosis, familial* (Hardening of renal tubules)
Familial, probably R

*Testicular feminization* (Apparent female, with female external genitalia, absent or scanty pubic and axillary hair, no menstruation, and generally incompletely developed vagina; testes are present either in the abdomen, inguinal canals, or the labia majora. An epididymus and vas deferens are commonly present on both sides, and there may be a rudimentary uterus and fallopian tubes)
Chromatin negative, normal number of autosomes, and sex - chromosome constitution XY; testicular feminization is an example of true sex reversal
The condition is familial, and is usually transmitted through the maternal line.
Either sex-linked R or sex-linked D

*Turner's syndrome* (Female with developmental failure of ovaries and complete absence of ova, absence of menstruation, and poor development of secondary sexual characters)
Chromatin negative, but not an instance of sex reversal. Turner's syndrome is due to the presence of only one X chromosome and no other, the total chromosome count being 45.

*Uterus, absence of*
Irregular R

*Uterus, malformation of*
F

## GLANDULAR DISORDERS

*Addison's disease* (Adrenocortical insufficiency, anemia, weakness, low blood pressure, small heart, melanin pigmentation of skin and mucous membrane)
F

*Diabetes, false*
(i) *Diabetes insipidus.* Usually D, sex-linkage also recognized
(ii) *Renal glycosuria.* Usually D
Preponderant in males

*Diabetes mellitus* (Excessive output of sugar in urine, etc.)
Irregular D
The earlier the disorder appears, the greater the familial heredity
Incidence: 1 in 40
Risk: For child of affected parent, 10 to 15 percent; with normal parents, 5 percent; for sibling with one parent affected, 12.5 percent

*Exophthalmic goiter (Graves's disease)* (Overactivity of thyroid gland, tenseness, palpitation, high metabolic rate, protrusion of eyeballs)
R, with partial penetrance and sex-limitation to women
Preponderant in females

*Goiter, simple* (Nontoxic thyroid deficiency)

R, sometimes D, with penetrance of about 1 percent in homozygous males and about 33 percent in homozygous females, and as in Graves's disease, in another one-third of females, with the remainder unaffected. Ovarian hormones may be responsible for high penetrance in female

*Goiter, toxic diffuse*
R

*Gynecomastia* (Development of feminine breasts in males)
D, with incomplete penetrance

*Hashimoto's disease* (Diffuse thyroid enlargement)
D

*Hyperthyroidism, familial* (Excessive activity of the thyroid gland)
F

*Hypogonadism and ichthyosis* (Undescended, undeveloped testes, of pituitary origin associated with barklike skin)
Sex-linked

*Myxedema* (Thyroid deficiency in adult with greatly reduced metabolism, etc.)
Genetics unclear
Preponderant in females

*Pancreatic cystic fibrosis* (Involvement of the pancreas, small intestine, and bronchioles of lungs)
R, in some cases could be due to a single gene
Incidence: 1 in 1,500
Risk: For parents having one or more affected children, 1 in 4

*Parathyroid adenomas, familial multiple* (Tumors of the parathyroid glands, often involving other glands and the skin, with the symptoms of hyperparathyroidism)
F

*Pheochromocytoma, familial* (Tumor due to overgrowth of chromaffin cells in adrenal medulla)
D, with equal degree of high penetrance in both sexes

*Polyendocrine adenomatosis* (Zollinger-Ellison syndrome) (Associated disorder of many of the endocrine glands, usually with peptic ulcer, gastric hyperacidity)
D

## HAIR AND NAILS

*Alopecia, familial congenital; epilepsy; mental retardation; unusual electroencephalogram* (Moynahan's syndrome) (Hairless at birth or scanty, downy hair, associated with small, scanty hair follicles; convulsions may appear within first six months.)
Probably R; in single dose merely inhibiting hair growth during late fetal and early infant life

*Nails, defects of*
(i) *Partial or complete absence.* D or R
(ii) *Thick nails with angular protrusion.* D
(iii) *Thick nails and skin of palms and soles.* D
(iv) *Spotted nails, bluish-white.* D
(v) *Milky - white nails.* Incomplete D
(vi) *Thin nails.* D, with varying penetrance

*Nail-deafness syndrome* (Congenital growth changes in all nails with congenital deafness)
Probably R

*Nail nevi* (Dark-pigmented, longitudinal streaks along the nails of digits and/or toes, sometimes malignant)
F

*Nail-patella syndrome* (Onycho-osteodysplasia) (Poorly formed fingernails with absent or small kneecaps, often linked with one or another of the A-B-O blood groups)
D

*White forelock syndrome* (Waardenburg-Klein syndrome) (Deafness, with localized albinism affecting chiefly the irises, forelock, and inner canthus)
D

*Pili torti* (Twisted hair)
Possibly irregular D
Generally appears in children

who have been previously bald for the first year or two

## HEAD AND NECK

*Acrocephalosyndactylia (Apert's syndrome, acrobrachycephalia)* (Head flattened from back to front with high vault, broad forehead, extreme width between eyes — hypertelorism— and webbing of fingers and toes)
Possibly irregular D or R

*Arrhinencephalia unilaterale* (Disturbed closure of maxillary and nasolacrimal processes. Defects of nasal bone, eyelids, eyeball, orbit, cleft palate, etc.)
F

*Branchial defects* (Defects in the form of fistulas in the neck, often with defects of ears)
D

*Cleidocranial dysostosis* (Maldevelopment of clavicles and incomplete hardening of skull bones)
D, with good penetrance and variable expressivity

*Craniofacial dysostosis (Crouzon's disease)* (Prominence of forehead, beaklike nose, projecting jaws, bulbous eyes— exophthalmos—and squint)
Irregular D, with fair penetrance

*Dacryocystitis* (Inflammation of the tear—lacrimal—sac)
D

*Dacryostenosis, congenital* (Narrowing or stricture of the nasolacrimal duct)
D

*Facial hemiatrophy* (Wasting of one side of face)
In some cases probably R

*Hallermann-Streiff syndrome* (Hypoplasia of jaws, accessory teeth, abnormalities of eyes and ears, dwarfism, thin skin and hair, sterility)
Heredity unknown

*Hereditary white-mouth syndrome* (Epithelial thickening of the mucous membranes of the mouth, principally of the cheeks)
D

*Hydrocephalus* (Congenital progressive enlargement of cerebral portion of head owing to excessive accumulation of fluid in ventricles of brain)
R
Risk: When both parents carry the gene, 1 to 4 per cent of children can be expected to be hydrocephalic
Incidence: 2 in 1,000

*Hydrocephalus with adducted thumbs* (Probably the commonest single-gene form)
Sex-linked R

*Hypertelorism* (Extreme distance between the eyes)
Sometimes D, sometimes R

*Klippel-Feil syndrome* (Shortened neck, etc.)
D, maybe with reduced penetrance.

*Lubarsch-Pick syndrome* (See *Amyloidosis*, under METABOLIC DISORDERS)

*Mandibulofacial dysostosis (Treacher-Collins syndrome)* (Fishlike face, malformed ears, etc.)
Irregular D

*Marshall's syndrome* (Thickened facial bones, depressed nasal root, hypoplasia of jaws, frog eyes, retinal detachment, deafness, dwarfism)
D with high penetrance

*Microcephaly* (Very small head and narrow forehead, large ears, prominent nose)
Probably R

*Micrognathia glossoptosis (Robin's syndrome)* (Hypoplasia of mandible and tongue)
F

*Otomandibular dysostosis* (Agenesis of lower jaw, cleft palate, abnormalities of ear)
Genetic mechanism unknown

*Oxycephaly* (Peaked crown of skull)
Sometimes D, sometimes R
Has been observed in association with hereditary hemolytic jaundice

*Torus palatinus* (Heaping of excessive bone along midline of palate, usually associated with mandibular torus)

## 362 HUMAN HEREDITY

D

### LUNGS AND CHEST

*Bronchiectasis*
 (i) *Congenital* (Cavities of bronchi and bronchioles)
 (ii) *Acquired* (Dilatation of bronchi and bronchioles)
 The familial incidence of this condition has long been known, and it would seem that D and R genotypes occur

*Bronchitis* (Inflammation of mucous membranes of bronchial tubes)
 A tendency to both acute and chronic bronchitis may have a genetic basis

*Emphysema of lungs* (dilatation of alveoli of lungs)
 F incidence has been reported

*Fibrocystic pulmonary dysplasia, familial* (Interstitial inflammation of lungs with later fibrosis and cyst formation)
 F; mode of inheritance uncertain

*Funnel chest* (Conical depression of chest due to sinking of the sternum)
 D

*Pneumothorax, spontaneous* (Generally secondary to erosion of covering — pleura — of lungs, sometimes associated with tuberculosis)
 F, probably D

*Pneumothorax, familial spontaneous* (Entrance of air into pleural cavity due to rupture of pulmonary alveoli leading to air in the interstices of connective tissue of lungs)
 F, probably D

*Kartagener's syndrome* (Situs inversus viscerum, bronchiectasis, and sinusitis)
 F

*Tuberculosis, hereditary disposition to*
 Depends possibly on the presence of a pair of genes; when these are lacking, there is good natural resistance.

### METABOLIC DISORDERS

*Acatalasemia (Takahara's disease)* (Lack of the enzyme catalase, resulting in about half the cases in a peculiar type of gangrene, at present reported only in Japanese people)
 R

*Alkaptonuria* (Congenital disorder of protein metabolism, with excretion of alkapton in urine)
 R in most cases, sometimes D
 Twice as frequent in males as in females

*Amyloidosis, familial primary (Lubarsch-Pick syndrome)* (Disturbance of protein metabolism, marked by deposits of excessive albuminoid throughout organs of the body, often with vitreous opacities in the eyes)
 F, R

*Beta-lipoprotein deficiency* (celiac syndrome, mental and physical retardation, ataxia, atypical retinitis pigmentosa, acanthocytosis, and absence of B-lipoproteins)
 F

*Carboxylase deficiency disease (Maple syrup urine disease)* (Excessive production of amino acids, valine, leucine, and isoleucine, yielding maple syrup odor in urine, associated with gross mental deficiency, probably due to a block at the decarboxylization stage)
 Familial occurrence has been recorded several times
 R

*Congenital virilizing adrenal hyperplasia* (Progressive virilization of offspring due to enzymatic defect in hydroxylation of hydrocortisone precursors)
 R
 Incidence: 1 in 50,000 live births

*Cystinuria (Cystinosis, Cystine storage disease)* (Disorder of cystine elaboration)
 (i) *Childhood type* (Retention of cystine in body and excretion in urine, devel-

opment and growth failure of bone, and degenerative lesions of kidneys, with death from uremia in many cases.) Several genotypes have been described.
(ii) *Adult type* (May pass unnoticed or result in urinary calculi.) Several genotypes have been described.

*Diabetes insipidus, hereditary hypophysial* (A lifelong disorder marked by excessive thirst and polyuria, due to deficiency of posterior pituitary secretion
D, sometimes sex-linked R

*Familial cholesterol ester storage disease* (Associated with excessive amounts of cholesterol and adequate amount of high-density lipoprotein, enlarged liver and spleen. Probably due to enzymatic defect in cholesterol esterase system)

*Familial Mediterranean fever* (Disorder characterized by short attacks of fever, peritonitis, pleuritis, and arthritis)
R
Persons with amyloidosis (q. v.) are usually heterozygous for the *FMF* gene

*Favism* (Acute hemolytic anemia after eating broad beans)
Incompletely sex-linked

*Fructose intolerance* (Due to deficiency of liver aldolase. Intake of fructose produces sharp rise in blood fructose and precipitous fall in blood glucose; may be accompanied by severe gastrointestinal, neurological, and mental symptoms)
R

*Fructose and galactose intolerance, familial* (Intake of fructose and/or galactose produces symptoms of varying severity similar to those seen in fructose intolerance. Unless recognized early, severe cerebral damage may be done in children as a consequence of frequent hypoglycemic episodes)
F

*Fructosuria, essential (Levulosuria)* (Defective conversion of fructose into glycogen, benign)
Probably R
Incidence: 1 in 130,000

*Galactosemia (Galactosuria)* (Defective conversion of galactose —a component of milk sugar —into glycogen; disorder of infants. Severe symptoms develop unless galactose eliminated from diet)
F, R

*Glucose-6-phosphate dehydrogenase deficiency* (Deficiency of enzyme concerned in red-cell metabolism resulting in sensitivity to sulpha and antimalarial drugs such as primaquine phosphate. Present as hemolytic anemia)
Probably incompletely sex-linked

*Glucoglycinuria* (Abnormal excretion of glucose and aminoacetic acid, without increased excretion of other amino acids in urine. Often associated with cystic fibrosis of pancreas)
D

*Glycorusia, renal* (Sugar excreted in urine without presence in blood to any abnormal extent, resistant to insulin; benign, no subjective symptoms)
D

*Gout (Gouty arthritis)* (Disturbance in purine metabolism)
Incomplete sex-limited D
Males constitute 90-95 percent of sufferers from gout
Incidence: 88 in 100,000

*Hartnup disease* (Pellagra-like rash, temporary cerebellar ataxia, nystagmus, incontinence, mental confusion, and constant excretion of amino acids from kidney. Manifests itself in childhood; Condition improves with increasing age
Probably rare R

*Hemochromatosis (Pigmentary cirrhosis, Bronze diabetes)* (Enlargement of spleen and liver with cirrhosis, slate-blue-gray color of skin, diabetes, and cardiac disease, with enormous accumulation in tissue of iron containing pigment)
Probably D with variable penetrance and variable expressivity
20 males to every 1 female

*Hepatolenticular degeneration* (*Wilson's disease*) (Tremor, cirrhosis of liver, eye changes, emotional lability, greenish pigment on undersurface of cornea, moderate degree of intellectual impairment, excessive secretion of amino acids in urine suggesting metabolic defect of kidney)
R
Onset: In second decade

*Histidinosis, familial* (Excessive amounts of histidine in blood)
F

*Hypercalcemia* (Hypersensitivity to Vitamin D, with generalized vascular calcification)
Genetic mechanism not established

*Hyperlipemia, familial* (Defective clearing of fat from the blood, associated with early onset of atherosclerosis, a senile type of arteriosclerosis)
D, with incomplete penetrance
Onset: From late second decade
Incidence: 2.5 in 100

*Hyperoxaluria* (Continuous excretion in urine of large amounts of oxalates; death usually occurs in childhood or early adult life from renal failure)
Probably R

*Hyperprolinemia, familial* (Defective metabolism of amino acid L-proline. May be accompanied by cerebral dysfunction, hematuria, polynephritis, and nerve deafness)
F

*Hyperuricemic nephropathy, familial* (Gout with severe renal disease)
F

*Hypoglycemia, spontaneous* (Low blood sugar with tremors, spasms, sweating, sometimes convulsions, cross-eyes, and mental retardation)
R

*Hypokalemia, congenital* (Hypokalemic alkalosis, failure to thrive and inability of kidney to conserve potassium, with polyuria, polydipsia, and inability to concentrate and acidify urine, with constipation)
Probably R

*Hypolipemia, with retarded development and steatorrhea* (Defective lipid production, red-scaling eruptions of skin, white-patched nails, and tapetoretinal degeneration)
F

*Hypophosphatasia* (Rickets due to defect of calcification of organic bone matrix)
R

*Hypophosphatemia, familial* (See Rickets, Vitamin-D resistant, p. 353)

*Jaundice, congenital nonhemolytic* (*Crigler-Najjar syndrome*) (Nonobstructive jaundice without destruction of red blood cells due to absence of glucoronosyl transferase of liver)
F

*Jaundice, hemolytic neonatal* (Hemolysis of red cells and jaundice in neonate, often associated with deficiency of enzyme 6-PDG)
F

*Lipoidosis* (*Lipid proteinosis*) (Nodular hard lesions distributed over the head and neck and extremities; the mucous membranes of the lips, mouth, pharynx, and larynx show yellow-white patches consisting of fatty deposits. These may so narrow the larynx that the child is unable to breathe, and dies unless a tracheotomy is performed)
R

*Lipomatosis, multiple* (Overgrowth of fatty tissue; may occur in any part of body)
Probably D

*McArdle's syndrome* (Absence of muscle phosphorylase, resulting in muscle weakness)
R

*Methemoglobinemia* (Presence of the transformation product of oxyhemoglobin — methemoglobin—in the blood; methemoglobin is incapable of carrying oxygen; sometimes associated with mental defect)

Sometimes D, and in some families R
(i) *Methemoglobin reductase deficiency* resulting in excess methemoglobin in total hemoglobin. R
(ii) *Hemoglobin M disease* (Due to presence of abnormal hemoglobin Hgb. M.) D
(iii) *Erythroglucose deficiency disease* (Inability of red cells to utilize glucose for reduction of methemoglobin) D

*Oculo - cerebral - renal syndrome* (Syndrome of Lowe, Terrey and MacLachlan) (Early infantile development of cataracts, later development of mental retardation, slow growth, abnormal globulins, granular casts)
Sex-linked

*Oxalosis, familial* (Deposition of calcium oxalate crystals in renal tubules and interstitial tissue, with extensive destruction and renal failure terminating in death)
D
Onset: May occur in infancy, but usually in second decade. Many more males affected than females.

*Oxaluria* (Abnormal excretion of oxalates in urine)
Sex-limited D
Manifests itself exclusively in males

*Pentosuria* (Excretion of pentose type of sugar, benign, often mistaken for diabetes)
D and R genotypes have been described
Twice as many males affected as females
Incidence: 1 in 50,000

*Phenylketonuria* (*Phenylpyruvic oligophrenia*) (Failure of transformation of phenylalanine into tyrosine, with elimination of phenylpyruvic acid in urine—never found in normal individuals—with mental retardation and pigmentary and constitutional peculiarities)
R
Incidence: 1 in 5,000
60 percent are idiots, 30 percent imbeciles, remainder of somewhat higher mentality; epileptiform seizures frequent

*Polyseroseritis, familial recurring* (Periodic episodes of severe abdominal, chest, or joint pain, corresponding to serosal membrane involved, with fever and leukocytosis)
F

*Porphyria* (Disturbance of pigment metabolism, with excretion of porphyrins and reddish or brown coloration of teeth; extreme and dangerous sensitivity to sunlight)
(i) *Congenital.* Probably R
(ii) *Acute intermittent.* Irregular D
(iii) *Porphyria cutanea tarda.* F
(iv) *Hereditary coprophyria.* F

*Periodic paralysis, familial* (of masticatory and facial muscles at intervals of 1 to 3 months)
(i) *Hypokalemic familial periodic paralysis* (Low serum potassium levels during attacks) D
(ii) *Hyperkalemic familial periodic paralysis* (*Adynamia episodica hereditaria*) (Excessive serum potassium levels during attacks)
D
(iii) *Normokalemic familial periodic paralysis* (Normal serum potassium levels during attacks)
D

*Pseudocholinesterase deficiency* (Deficiency of pseudocholine enzyme, with resulting hypersensitivity to various substances)
Probably R

*Pseudo - pseudohypoparathyroidism* (Differing from pseudohypoparathyroidism only in that serum calcium and phosphorus are normal)
F

*T-substance anomaly* (Excessive excretion of abnormal product of purine metabolism in urine, with retarded physical and often mental growth)

R
*Skeletal muscle phosphorylase deficiency* (Muscular weakness and wasting due to absence of phosphorylase activity. In some members of family, phosphorylase activity may be reduced without accompanying weakness or wasting)
F
Onset: In fifth decade
*Tangier disease* (Disorder of cholesterol metabolism, with low cholesterol and very low alpha-lipoprotein)
F

## NERVOUS SYSTEM
### Neurological Disorders

*Acoustic neuroma, bilateral* (Tumor of acoustic nerve)
D
Regarded as a form of neurofibromatosis, or von Recklinghausen's disease.
*Altzheimer's syndrome* (Cortical degeneration of the brain)
D
Onset: In adult life
*Amyotrophic lateral sclerosis* (*Charcot's disease*) (Progressive muscular atrophy, increased reflexes, spastic muscular irritability)
  (i) *Infantile*. Rare. Probably R
  (ii) *Adult*. D, with incomplete penetrance, and twice as common in males as in females
*Analgesia, congenital generalized* (Congenital indifference to pain, superficial and deep)
Has been found associated with trisomy of groups 13 to 15
*Anencephaly* (Absence of the brain)
Possibly R
Occurs about one-third more frequently in females
Incidence: 1 in 1,000
Risk: There is a 2-7 percent risk that later pregnancy will result in child with defects of nervous system. There is a 20 percent chance of aborting
*Ataxia-telangiectasia* (Cortical cerebellar degeneration, dilatation of external blood vessels of eyes and butterfly area of face, frequent pulmonary infections, and peculiar eye movements) D
*Auditory imperception, congenital* (Inability to understand meaning of sounds without affection of hearing)
D
*Brain degeneration, growth retardation, with white stubby hair*
Sex-linked
*Cerebral sclerosis, diffuse* (*Schilder's disease, encephalitis periaxialis diffusa, subcortical encephalopathy*) (Hardening of the brain)
  (i) *Merzbacher - Pelizaeus form* (Rotary movements of head and eyes, spasticity of extremities, dementia, and Parkinsonism.) R, D, and sex-linked forms.
  Onset: No later than third month; may live into old age. Considerably more frequent in males. Rare.
  (ii) *Scholtz form* (Deafness, blindness, weakness and spasticity of legs, dementia.) Sex-linked R
*Cerebromacular degeneration* (*Juvenile Tay-Sachs disease, Spielmeyer-Vogt disease, Sjögren's disease*) (Degeneration of brain and retina leading to complete blindness and death, usually associated with generalized imidazole aminoaciduria. There is a deficiency in the metabolism of amino acids of the histidine group)
Aminoaciduria, D
Cerebromacular degeneration, R
These two traits probably different manifestations of the same gene
Onset: About 6 years of age
*Disseminated sclerosis* (Patches of hardening in brain, causing more or less paralysis, tremor, nystagmus, disturbances of speech, etc.)
D, with low penetrance; other forms of inheritance possible.

Incidence: 1 in 2,500
In the majority of cases if one of a pair of identical twins is affected, the other is not

*Dysautonomia, familial* (Multiple nervous symptoms, defective lacrimation, disturbed swallowing reflex, skin blotching with excitement or eating, poor motor coordination, relative indifference to pain, excessive sweating, increased salivation, emotional lability, reduction in intellectual capacity, periodic spells of vomiting, etc.)
F, probably R

*Epilepsy* (Chronic nervous disorder characterized by periodic convulsive attacks)
Genetic factors undoubtedly present, but at present not understood. R and irregular D cases have been described, but there are many other genotypic patterns involved.
Incidence: 1 in 250

*Filum terminale brevis*
(Shortening of the terminal strand at the base of the spinal cord, putting severe tension upon nerves to lower extremities, with resulting pain and functional deformities)
Possibly D

*Hemangioma of cerebellum and retina* (*Lindau's disease*) (Blood-vessel tumors and cysts of "little brain"—cerebellum—and inner cell layer of eyeball)
Irregular D

*Huntington's chorea* (Progressive muscular spasm with increasing mental deterioration)
D
Onset: About 35 years on the average, earlier onset in females than in males

*Migraine* (Periodic severe, usually unilateral headache, with nausea, vomiting, and visual aura)
(i) *Simple migraine* (Headache preceded by photophobic and visual aura, followed by nausea and vomiting.) D

(ii) *Short hemiplegic migraine* (As in simple migraine, except visual aura replaced by neurologic deficits such as hemiplegia, hemisensory loss, and aphasia.) F

(iii) *Prolonged hemiplegic migraine* (Headache may precede visual aura, and neurologic signs may last for several hours or days.) F

*Neurofibromatosis* (*von Recklinghausen's disease*) (Multiple tumors of nerve sheaths situated over the surface of the skin, a benign but disfiguring disorder)
D, with high penetrance but variable expressivity

*Pain insensitivity with anhidrosis* (Insensitivity to noxious stimuli, with exception of corneal stimulation and inhalation of ammonia. Sweat glands of normal structure but functionally inactive, probably related to inability to tolerate high environmental temperatures) Genetic mechanism unclear

*Pick's disease* (*Pick's lobar atrophy*) (Progressive degeneration of brain)
D
Onset: From 39-55 years

*Specific dyslexia* (*Congenital word blindness*)
D, with high penetrance
Males much more frequently affected than females
Incidence: 5 in 100

*Sydenham's chorea* (*St. Vitus' dance*) (Involuntary movements of limbs and facial muscles)
R
Onset: Childhood
Females affected three times more frequently than males

### Nerve, Muscle, and Organ Changes

*Arthralgia, periodic* (Joint pains at periodic intervals)
D, with high penetrance and minimal variability of expression

*Bell's palsy* (Unilateral paralysis of facial muscles supplied by seventh nerve)

F; sometimes D, with low penetrance

*Bonnevie-Ullrich syndrome (Pterygium syndrome)* (Abducens and facial paralysis combined with anomalies of ear, ocular adnexa, muscles of limbs—especially hands — maldeveloped mandible, thoracic malformations, maldevelopment of gonads)
  Irregular D. There are 44 autosomes but only a single sex chromosome

*Cerebral diplegia, congenital* (Spastic weakness of legs at birth, sometimes with ataxia and mental defect from birth, with tendency to improvement)
  Probably one or more R genes

*Dystonia musculorum deformans, idiopathic (Paroxysmal muscular deformation, torsion spasm, dystonia lenticularis, dystonia lordotica progressiva)* (Disease of basal ganglia. Paroxysmal twisting and writhing movements of body and muscles)
  D, with almost complete penetrance and variable expressivity. Recessive forms may exist

*Hallervorden and Spatz syndrome* (Increasing rigidity of limbs with difficulty of speech and progressive dementia)

*Heredopathia atactica polyneuritiformis (Refsum's disease)* (Night blindness, atypical retinitis pigmentosa, later polyneuritis, muscular weakness, and is terminated by sudden death)
  R
  Onset: In the second to fourth decades

*Muscular dystrophy, progressive*
  (i) *Facioscapulohumeral type*
    D. Onset: Usually between 7 and 20 years
  (ii) *Childhood progressive muscular dystrophy (Duchenne, or pseudo-hypertrophic type)*. Sex-linked R
    Onset: Usually at about third year; invalids by 9-12; death often before 25 from respiratory infections. Mutation rate: About 1 in 10,000 male births
  (iii) *Childhood autosomal recessive type* (Occurs also in females.) Onset: 5 to 13 years, with slower progression than sex-linked type

*Muscular weakness, atrophy, and mental deficiency*
  Sex-linked R

*Myotonia congenita (Thomsen's disease)* (Muscle spasm or rigidity)
  D, with reduced penetrance in some families

*Myotonia dystrophica* (Extreme weakness, usually of muscles of face and neck, which may spread to other parts of body)
  Irregular D
  Cataract common in otherwise normal parents and relatives, as well as tendency to baldness

*Paralysis agitans (Shaking palsy)*
  D, with penetrance of about 60 percent

*Paramyotonia* (Muscular spasm or rigidity in reaction to cold)
  D

*Parkinsonism, familial*
  (Shaking palsy)
  Familial D

*Peroneal muscular atrophy (Charcot-Marie-Tooth disease)* (Degeneration of muscles of legs, feet, and hands. Low conduction velocity in peripheral nerves)
  D, R, and sex-linked genotypes have been described
  Onset: Neither parent affected, at 11 years; one parent affected, at 19 years

*Photosensitivity, familial* (Seizures or their equivalents induced by light stimuli)
  F

*Progressive bulbar paralysis* (Paralysis of muscles of tongue, lips, larynx, and pharynx, in later life)
  D, R, and sex-linked
  R forms have been described

*Progressive hypertrophic polyneuritis* (Shooting pains and numbness in extremities, muscular weakness and atrophy in hands, forearms, and legs)

D

*Spastic diplegia with idiocy or imbecility, congenital* (Diplegia-bilateral paralysis)
    Probably R
    Incidence: 1 in 1,000-2,000 live births
    Onset: At end of first year

*Spastic paraplegia* (Spasmodic paralysis of the lower extremities)
    D, R, and sex-linked genotypes have been described; multiple alleles or more than one gene pair may be involved
    More frequent in males

*Spinocerebellar ataxia* (Loss of muscular coordination)
    (i) *Friedrich's ataxia* (*Spastic ataxia*) Usually R, sometimes sex-linked R
    (ii) *Marie's ataxia*. Usually D
    Several alleles or more than one pair of genes may be involved in these and other forms of hereditary ataxia.

*Torsion dystonia* (Turning and twisting movements of trunk and proximal parts of limbs, with loss of muscle tone)
    R

*Tuberous sclerosis* (Development of brain tumors, mental deterioration, epilepsy, tumors of skin and viscera)
    D, with modifying genes

## Mental Deficiencies

*Amaurotic idiocy, familial* (Abnormal storage of fats, mental impairment leading to idiocy, progressive blindness, paralysis, wasting, and death)
    (i) *Infantile type* (*Tay-Sachs*). R. Incidence: 1 in 250,000. Onset: At 1-6 months; death within 2 years.
    (ii) *Juvenile type*. R. Incidence: 1 in 40,000. Onset: About 6 years; death at 15-20.
    (iii) *Adult type*. R.

*Cataract with idiocy or imbecility, congenital*
    Probably R
    Incidence: 1 in 50,000 live births
    Risk for siblings: 16 percent

*Cretinism* (Congenital arrest of mental and bodily growth at infantile level)
    The type that does not respond to thyroid treatment is probably genetically determined, but in a manner unknown.

*Cretinism, familial goitrous* (Congenital cretinism due to the lack of an oxidative enzyme that aids in the conversion of iodide to iodine)
    R

*Down's syndrome* (*Mongolism, Mongolian idiocy, Trisomy 21*) (Associated with presence of extra chromosome No. 21. Mental retardation associated with various physical anomalies)
    Incidence: 1 in 700 births
    Risk: Increases with increasing age of mother. From the age of 40, the chance of having a Down's child is from 1 to 6 percent. Risk of recurrence at any age is almost fourfold (3.7)

*Depression* (Several different forms due to different biochemical abnormalities, at present unknown, but the existence of which is clearly indicated)
    F

*Epiloia* (*Bourneville's disease*) Tumors of brain, skin, and viscera associated with mental defect in 70 percent of cases, with epilepsy in 80 percent)
    Irregular D
    Incidence: 1 in 50,000

*Feeblemindedness* (IQ range, 50-69)
    Many different genotypes involved

*Idiocy* (IQ range, 0-19)
    Apparently many different genotypes involved

*Imbecility* (IQ range, 20-49)
    Many different genotypes involved

*Laurence-Moon-Bardet-Biedl syndrome* (Underdevelopment of external genitalia, obesity, retinitis pigmentosa, polydactyly, and mental deficiency)

R, with variable penetrance and expressivity
Almost twice as frequent in males as in females

*Mental deficiency, cerebellar ataxia, bilateral primary optic atrophy, speech disorder, and esophagal achalasia, or inability to relax*
D

*Mental deficiency, sex-linked familial*
Appears in males only
Sex-linked R

*Microcephaly, familial* (Familial "pinheadedness")
Probably R
Incidence: 1 in 25,000-50,000 births
To be distinguished from forms of microcephaly that are caused by irradiation *in utero*, toxoplasmosis, and similar environmental conditions

*Microphthalmos and mental deficiency* (Small eyes with mental deficiency)
Sex-linked R
Incidence: 2.5 in 100,000
Risk: With both parents unaffected where a child is affected, for sibs, 9 percent; in those cases in which the microphthalmic child is mentally normal, risk for future siblings very low

*Mongolism.* See *Down's syndrome*

### Mental Disorders

*Autonomic dysfunction (Riley's syndrome)* (Crying without tears, excessive sweating, skin blotches, emotional instability, motor incoordination, etc.)
Probably R

*Manic-depressive psychosis* (Mental disorder characterized by marked emotional oscillation between manic and depressed states, usually cyclical)
D, with incomplete penetrance
Incidence: Slightly less than 1 percent

*Neurotic temperament* (Functional nervous disorder)
In many cases largely environmentally caused, in still others probably a condition involving a number of genes that will express themselves only in the appropriate environment

*Schizophrenia* (Disturbance in reality relationships, fragmented personality)
Some R, but with variable expressivity; others D, with variable expressivity; some cases not genetic at all
Incidence: 1 in 100 to 1 in 200

### NOSE

*Anosmia* (Total absence of sense of smell)
D

*Choanal atresia* (Occlusion of the openings of the nasal fossae into the pharynx)
Probably R

*Epistaxis* (Nosebleeding, often due to varicose enlargement of nasal veins)
Probably D

*"Potato nose"* (Upper part of nose distended like a balloon)
Probably irregular D

*Rhinitis* (Inflammation of nasal mucous membrane)
(i) *Acute catarrhal rhinitis* (In childhood.) D
(ii) *Chronic hypertrophic rhinitis* (Chronic catarrh) D
(iii) *Atrophic rhinitis* (Chronic catarrh with thinning of mucous membrane.) D
(iv) *Polyposis* (Polyps of nose and sinuses.) Genetic mechanism unclear

*Rhinophyma (Rosacea of the nose)* (Reddish-blue, swollen, large dilated pores of lower part of nose)
Has been observed in over three generations; genetic mechanism not clear

### SKIN

*Acanthosis nigricans* (Hypertrophy with pigmentation of papillae of outer layer of skin)

## INHERITED DISORDERS OF MAN

(i) *Benign.* D and irregular D
(ii) *Malignant.* Possibly R
(iii) *Pseudoacanthosis with obesity.* Probably environmentally conditioned

*Acrodermatitis enteropathica* (Disorder of infants commencing as skin eruption on an extremity or around an orifice, followed by hair loss and diarrhea and other gastrointestinal disturbances. Often ends fatally. Probably due to enzyme deficiency)
F

*Albinism* (Congenital absence of pigment in skin and its appendages)
(i) *Generalized* (Albinism of the whole body.) R. Also occurs as D. Incidence: 1 in 15,000-20,000
(ii) *Partial* (Albinism of forehead, neck, linea alba, parts of the body, or white forelock.) D
(iii) *Ocular* (Albinism limited to eye.) Sex-linked R
(iv) *Partial with deaf-mutism* (with probable sublethal effect). Sex-linked R

*Angiokeratoma corporis diffusum* (*Fabry's disease*) Numerous small blood tumors covered by wartlike growths over entire body. Sometimes associated with disease of kidneys
Sex-linked R

*Eczema* (Inflammation of the skin with multiple lesions)
Irregular D

*Epidermodysplasia verruciformis* (*Lewandowsky-Lutz disease*) (Flat polygonal papules first appearing on limbs and then spreading to face, hands, and feet, and occasionally to whole body)
Probably R

*Epidermolysis bullosa* (*Goldscheider's disease*) (Blistering of the skin with generalized pathology)
(i) *Simplex type.* D, with reduced penetrance. 59 percent of those affected are males. Onset: In first year
(ii) *Dystrophic hyperplastic type.* D, with high penetrance. Onset: Anytime between birth and puberty
(iii) *Dystrophic hypoplastic type.* R, occasionally incompletely sex-linked R. Onset: Usually soon after birth

*Erythema nodosum, familial* (Hypersensitive skin reactions due to allergens associated with many diseases, with accompanying constitutional upset)
Probably D, with variable penetrance

*Erythemato-keratotic-phacomatosis* (Reddening and darkening of cheeks, extensor aspects of arms, glove areas, lower abdomen, "trouser" distribution and feet, with hyperkeratosis of palms, soles, and dorsa of hands)

*Hyperelastosis cutis* (*Ehlers-Danlos syndrome*) (Extreme laxity of skin and joints and fragility of skin and blood vessels)
D, with variable penetrance and expressivity
Somewhat more frequent in males

*Ichthyosis* (Horny plating and fissuring of skin)
(i) *Ichthyosis foetalis.* R. Death at birth or shortly afterward
(ii) *Ichthyosis congenita.* R
(iii) *Ichthyosis simplex.* D, with variable expression and manifestations; sometimes sex-linked R

*Keloids, multiple* (Fibrous tumors of the skin)
D
Onset: Any age, usually at end of second decade. Occurs about nine times more frequently in Negroes than in whites

*Keratosis*
(i) *Keratosis follicularis* (*Darier's disease*) (Numerous follicular papules over body.) D
(ii) *Keratosis follicularis spinulosa* (A mild form of keratosis follicularis.) Sex-linked R

*Lupus vulgaris* (Tuberculosis of

the skin, generally of the face or mucous membranes of nose, mouth, pharynx, and throat)
F

*Nevi* (Birthmarks)
D when hereditary

*Pachyonychia congenita* (Heaped-up, clawlike nails, thickening of skin, excessive sweating, etc.)
D
Males affected four times more frequently than females

*Piebaldness* (Forehead blaze with nonpigmented spotting on limbs and trunk)
Usually D

*Pityriasis rubra pilaris* (Papular disease of skin with scaling)
D, with high penetrance

*Porokeratosis* (Keratinization, or horniness, of skin)
D
Males affected twice as frequently as females

*Pseudoxanthoma elasticum (Gröndblad-Strandberg syndrome)* (Papular eruption, usually on neck and in armpits, but may appear anywhere on body; associated with blood streaks in retina)
Probably R, with partial sex-limitation to female

*Psoriasis* (Chronic red-brown scaly papules of skin)
Irregular D, sometimes R
Risk: For siblings of affected child, parents normal, 2-7 percent; one affected parent, 9-20 percent. For children of affected parent, 15 percent

*Recurrent bullous eruption of feet* (Blistering of soles of feet, particularly in hot weather)
D
Preponderance of males

*Systemic lupus erythematosus* (Tuberculosis and reddening of skin of whole body)
F

*Tylosis* (Thickening of skin of palms, soles, and flexor surfaces of digits)
D, with high penetrance
Onset: May start at birth or be delayed for five to six years

*Urticaria pigmentosa* (Red-brown pigmented macules, or blemishes, of skin)
Possibly R

*Xanthelasma* (Fatty skin growths, usually on eyelids)
D

*Xanthomatosis*
(i) *Tuberous* (Yellow nodules in skin.) D
(ii) *Coronary* (Yellow nodules in heart.) D
(iii) *Familial hypercholesteremia* (Yellow nodules in skin associated with high cholesterol concentrations in blood. Nodules occur also on periosteum and tendons.) D

*Xeroderma pigmentosum with multiple basal-cell cancer* (Freckles, warty growths, superficial ulcers, and malignant changes; earliest signs: reddening and excessive freckling during first summer of infant's life; death generally before 30)
Incompletely sex-linked R

## TEETH AND MOUTH

*Ankyglossia* (Tongue-tie)
D

*Caries* (Tooth decay)
Probably D

*Central incisors, absence of all*
Incomplete D or sex-linked R

*Central incisors, very large*
D

*Chin, receding (Angle's Class II deformity)*
Irregular D

*Cleft palate (without harelip)*
D, with variable penetrance
Many more females than males affected
Incidence: 1 in 2,500
Risk: Normal parents with no affected relatives, 1 in 80; with both parents normal and an affected relative, about 1 in 8; with one parent affected, about 1 in 6; children of an affected parent, married to a normal, about 1 in 15

*Dentin dysplasia* (Rootless teeth)
D

*Dentinogenesis imperfecta* (Imperfect formation of dentin)

# INHERITED DISORDERS OF MAN

D
Incidence: About 1 in 8,000
Enamel, defective development of (Amelogenesis imperfecta)
  (i) *Agenesis* (Only thin covering of enamel present; molars may fail to develop.) Sex-linked D
  (ii) *Hypocalcification* (Loss of enamel shortly after eruption of teeth.) D
  (iii) *Hypomaturation* (Soft enamel, cloudy agar jelly appearance.) Sex-linked R
  (iv) *Pigmented hypomaturation* (Soft-pigmented enamel.) Probably R
  (v) *Local hypoplasia* (Brown, pitted, or lined undeveloped areas, mostly on incisors and pre-molars. Milk teeth alone may be affected.) D, with incomplete penetrance
  Incidence of defective enamel development: About 1 in 14,000
Frenulum, hypertrophied
  (With pseudocleft of upper lip, tongue, palate, and jaws, with mental deficiency, anomalies of hand, and trembling)
  D, apparently limited to females, possibly lethal for male
Gingival hyperplasia (Overgrowth of gums)
  D
Harelip (with or without cleft palate)
  R, sex-linked D genotypes, and incomplete D genotypes have been described
  Twice as many males affected as females
  Incidence: 1 in 1,000
  Risk: When both parents normal, 1 in 20; for siblings when one parent affected, 1 in 10; children of an affected parent, married to a normal, 1 in 50
Jaw, fibrous swelling of
  Possibly D
Jaw, protrusion of lower (Angle's Class III deformity)
  D
Lateral incisors, absence of
  D
Molars, absence of

D
Premolars, absence of
  D
Teeth, absence of all
  D
Teeth, supernumerary
  D

## MISCELLANEOUS

Cancer of breast
  F, transmitted through maternal side
Carotid-body tumors, bilateral (Glandlike structures attached to branch of occipital artery arising at bifurcation of common carotid artery in neck, associated with reflex changes in respiration)
  F
  Familial incidence: 26 per 100
  Non-familial incidence: 3-5 per 100
Ectodermal dysplasia
  (i) *Anhidriotic type* (Absence of sweat glands, peculiarities of skin, and developmental failure of teeth to complete absence of teeth.) Usually sex-linked R
  (ii) *Idrotic type* (Sweat glands normal, other symptoms present.) Usually D
Epithelioma adenoides cysticum (Multiple benign cystic adenomas)
  D, with partial sex-limitation to females
Gargoylism (Hurler-Pfaundler syndrome, dysostosis multiplex)
  (Dwarfism, enlarged liver and spleen, corneal opacities, and mental deficiency; death from heart failure usually before 20)
  Usually R. The less severe form, without clouding of the cornea, indicates sex-linked R inheritance
Gaucher's disease (Disorder of the reticuloendothelial system —spleen, liver, lymph nodes, bone marrow)
  (i) *Infantile type.* Probably R. Onset: Acute, at about sixth month; ends in early death
  (ii) *Adult type* (Benign)

Probably D. Onset: In advanced age

**Goiter-deafness syndrome** (Inability to form organic iodine compounds at normal rate, accompanied by perceptive deafness)
R

**Hepatic fibrosis, congenital** (Hardening of the liver, usually with cystic kidneys, and progressive dilatation of bile ducts)
F, probably D

**Laryngitis**
  (i) *Acute*
  (ii) *Chronic.* That there is a hereditary tendency involved in many cases of laryngitis has long been known, but the precise genetic mechanism is not understood

**Leprosy** (*Hansen's disease*)
Probably irregularly D

**Lymphedema, congenital** (*Milroy's disease*) (Disorder of the lymph vessels, usually confined to lower limbs)
D, with low penetrance
Preponderance of females and female transmitters

**Lymphocytophthisis, essential** (*Hereditary lymphplasmacytic dysgenesia*) (Severe failure of the lymphoreticular system resulting in agammaglobulinemia)
F

**Marcus Gunn phenomenon** (Ptosis, winking when jaw moves)
Irregular D

**Marfan's syndrome** (*Weill-Marchesani syndrome*) (Disorders of connective tissue, short limbs and digits, aortic hypoplasia with necrosis, with small distorted lens of eye)
Incompletely D, possibly R

**Mucoviscidosis** (Abnormal secretion of viscid mucus from exocrine glands, frequently associated with peptic ulcer)
  (i) *Juvenile*
  (ii) *Adult*
D

**Polyserositis, periodic** (Periodic disorder characterized by attacks of fever, abdominal pain, and chest or joint pain. Affects persons of Mediterranean origin)
F

**Renal-hepatic-pancreatic dysplasia** (Abnormal development of kidneys, liver, and pancreas)
F

**Rheumatic fever**
The predisposition may be inherited as an R that will find expression only under the appropriate environmental conditions

**Sexual precocity, familial** (Very early sexual development)
F

**Testicular torsion** (Hypermobility of testes, unilaterally or bilaterally, with subsequent torsion of spermatic cord)
F

# APPENDIX B
# THE CHANCE OF THE OFFSPRING OF MARRIAGE BETWEEN COUSINS EXHIBITING DEFECTIVE CONDITION

It has been estimated by Muller (1950) that the average person carries about eight deleterious mutations in the recessive state. In a random system of mating it is unlikely that husband and wife would also carry any one of these same recessive genes; each would carry different recessive genes. When second cousins marry, there is 1 chance in 16 that both would be carrying one of the sixteen recessives. Second cousins would be the grandchildren of independent sets of grandparents, one member of each grandparental set standing in the relation of brother or sister to each other. The chance that both intermarrying grandchildren would not carry the same recessive gene is $15/16$, and the chance that both would not carry the same one of any of the sixteen recessive genes originally carried by the grandparents between them is $(15/16)^{16}$, or 0.356. The probability, then, that any two of the grandchildren would carry any one of the sixteen recessive genes shared between the grandparents is equal to $1-(15/16)^{16}$, or 0.644. The probability that the first child of the union between such grandchildren would be homozygous for any one of the recessives carried by the parents is one-fourth of 0.644, or 0.161. Should the offspring of such unions amount to four individuals, then 0.644 of these families, on the average, would have at least one individual who would be homozygous for a deleterious recessive. In other words, nearly 65 percent of such families would have at least one member who was in some way unfavorably affected by the defective recessive genes carried by him in the homozygous state.

However, the truth is that not all eight deleterious genes are necessarily related to congenital or childhood abnormalities. The mutant genes involved may exert their effect at any time between conception until terminal age, say up to a hundred years or more. If, then, with Sheldon Reed, who has considered this problem in relation to the practical exigencies of counseling, we arbitrarily assume that one-fifth of the sixteen deleterious genes of the grandparents may produce congenital or childhood abnormalities, one-fifth of 0.161, or 0.032 (1 in 31) of the cousin marriages should result in a congenital or childhood abnormality in the first child, or 1 in 8 of cousin marriages should result in such an anomaly when at least four children are born as the result of such a cousin union.

# APPENDIX C
# GENE AND GENOTYPE EQUILIBRIA, POPULATIONS AND GENES, THE HARDY-WEINBERG LAW*

ONE BRANCH of genetics studies the distribution and movement of genes (gene flow) in populations. The population geneticist is concerned with all those conditions that govern gene distributions and their changes. The "conditions" refer to such factors as size of population, degree of isolation from other populations, forms of mating or marriage-regulations, differential migration (emigration and immigration), mutation, selection, and hybridization.

In the course of the history of virtually every population all these factors tend to be operative, and it is highly probable that every one of these factors, to varying extents, has contributed to the differences in gene frequencies that exist for some traits in different populations.

In the absence of the modifying conditions mentioned above, and frequently in their presence, the genetic structure of a population, that is to say, the frequency distribution of its genes, tends to remain stable. The same proportions of the same genes tend to reappear generation after generation. Such a genetically stable population is said to be in *equilibrium*.

In large populations mating is usually random in respect of any particular gene, and this is what is meant when we speak of random mating—and *not* that individuals choose their mates at random, a condition that applies in no human society. The technical term for random mating is *panmixia*; a random-mating population is said to be *panmictic*.

In a large panmictic population in which the pressures of mutation and selection, or other factors having similar effects upon gene frequencies, are absent or low, genotype frequencies, after the first generation of random mating, will remain indefinitely unchanged. This phenomenon is known as the Hardy-Weinberg Law (or Hardy-Weinberg Equilibrium), a law independently worked out in 1908 by the English mathematician G. H. Hardy and the German physician W. Weinberg.

In spite of three somewhat unreal assumptions, namely, infinite populations, low mutation, and little selection, the Hardy-Weinberg Law is found to work with remarkable precision in real finite populations. Not only this, it is a tool of considerable heuristic value, for when it is found not to work, it may at once be suspected that we are dealing with a population in which special conditions pre-

* From Montagu, Ashley, *An Introduction to Physical Anthropology*, 3rd ed., 1960. Courtesy of Charles C Thomas, Publisher, Springfield, Illinois.

# GENE AND GENOTYPE EQUILIBRIA

vail, such as isolating barriers between segments of the population, that prevent free interbreeding within it. With the removal of such reproductively isolating barriers the Hardy-Weinberg Law would be found to hold true.

We shall take as a simple example of the manner in which the Hardy-Weinberg Law works a population in which there is random mating in respect of two autosomal alleles, $A$ and $a$. The frequencies of these alleles is 50 percent each. Remembering that it is not phenotypes that are so much involved as the gametes, that is, the eggs and sperm, half of the eggs will carry $A$ and half $a$, and so will the sperm. The results of the random combination of such eggs and sperm in the zygotes may be determined from Table XXXIV. From this will be seen that the frequencies of the alleles $A$ and $a$ remain unchanged at 50 percent each, ¼ of the individuals will be $AA$, ½ $Aa$, and another ¼ $aa$.

## TABLE XXXIV

### THE MAINTENANCE OF GENOTYPE EQUILIBRIUM. THE HARDY-WEINBERG LAW

| SPERM / EGGS | $A$ / $p$ | $a$ / $q$ |
|---|---|---|
| $A$ / $p$ | $AA$ / $p^2$ | $Aa$ / $pq$ |
| $a$ / $q$ | $Aa$ / $pq$ | $aa$ / $q^2$ |

$$= p^2 AA + 2pq Aa + q^2 aa,$$

or, in the example given in the text,

$$= ¼\, AA + ½\, Aa + ¼\, aa$$

Matings at random between the members of this population, and the offspring of such marriages, with respect to the genes $A$ and $a$, will yield nine possible matings, and precisely the same proportions of genotypes as existed in the parental population, as shown in Table XXXV.

With special exceptions the Hardy-Weinberg Law applies to all populations, however unequal the frequencies of certain genes may be. It can, therefore, be generalized in the binomial $(p+q)^2$, where p equals the frequency of $A$, and q equals the frequency of its allele $a$, and $p + q = 1$. The genotype frequencies in each generation equal

## TABLE XXXV

### THE PROPORTIONS OF GENOTYPES IN 9 POSSIBLE MATINGS OF INDIVIDUALS WITH GENES $A$ AND $a$.

| 9 Possible Matings | Offspring of 9 Matings | | |
|---|---|---|---|
| 1/16 $AA \times AA$ | 1/16 $AA$ | | |
| 1/8 $AA \times Aa$ | 1/16 $AA$ | 1/16 $Aa$ | |
| 1/8 $Aa \times AA$ | 1/16 $AA$ | 1/16 $Aa$ | |
| 1/16 $AA \times aa$ | | 1/16 $Aa$ | |
| ¼ $Aa \times Aa$ | 1/16 $AA$ | 1/8 $Aa$ | 1/16 $aa$ |
| 1/16 $aa \times AA$ | | 1/16 $Aa$ | |
| 1/8 $Aa \times aa$ | | 1/16 $Aa$ | 1/16 $aa$ |
| 1/8 $aa \times Aa$ | | 1/16 $Aa$ | 1/16 $aa$ |
| 1/16 $aa \times aa$ | | | 1/16 $aa$ |
| Proportion of same genotypes | ¼ $AA$ + | ½ $Aa$ + | ¼ $aa$ |

$(pA + qa)^2 = p^2AA + 2pqAa + q^2aa$, which is precisely what we found in Tables XXXIV and XXXV for the special case of $p = q = \frac{1}{2}$. A similar formula may be derived for loci with more than two alleles.

The Hardy-Weinberg Law explains how it comes about that the genotypic frequencies for such traits as brown eyes and blue eyes, for tasting and nontasting, the blood groups, and all similar traits dependent upon contrasting alleles are likely to be maintained in the same proportions generation after generation.

It should at once be evident, then, why statements to the effect that brown eyes being dominant over blue, the former must eventually swamp the latter, or that the higher frequency of brachycephaly in certain populations is due to the dominance of the alleles for this trait, are erroneous. The error committed is the assumption that the numerical frequency of any trait in a population represents a reflection of the number of dominant alleles conditioning the trait.

The fact, however, is that phenotypic frequency has no necessary relation either to the dominance or to the recessiveness of alleles. Dominance and recessiveness refer only to the expression of the alleles in individuals possessing them. The frequency of any phenotype in a *population* is related to the frequency of the allele controlling it, regardless of that allele's dominance or recessiveness. There are numerous conditions in man that are due to a dominant allele, but those conditions are rare in man simply because the dominant gene is much less frequent than its normal alternative. Examples are partial albinism, the sickle-cell trait, achondroplasia, and parietal foramina of the abnormal type.

In the case of the alleles for tasting and nontasting, tasting is dependent upon a single dominant gene, $T$, and nontasting upon two recessive genes, $tt$. The distribution of these genes in any population will invariably be found to follow the Hardy-Weinberg Law.

Dominant alleles in any population do not tend to replace recessive alleles. The frequency of an allele in a population is not determined by its being either dominant or recessive, but largely by the fact of its being of greater or lesser selective value.

The Hardy-Weinberg Law states that regardless of past history, random mating for one generation yields genotypic frequencies in the proportions $p^2 : 2pq : q^2$.

Gene frequency can be at equilibrium in the sense that p does not change from generation to generation in a number of circumstances, such, for example, as the following:

1. *No selection. Mutation in opposite directions.* In the absence of selection the frequency of a gene will depend upon relative rates of mutation in favorable and unfavorable directions.

2. *Selection in opposite direction from mutation.* A gene may be in process of elimination from a population, but persists within that population because the rate of elimination of the gene by selection is equal to the rate at which the gene is being reintroduced into the population by mutation.

3. *Heterozygote superior to both homozygotes.* This is well illustrated by the case of the sickle-cell trait, in which the heterozygote $AS$ is superior to the normal homozygote $AA$ and the sickle-cell anemia homozygote $SS$.

4. *Diversity of niches or environments.* If the environment is not uniform over the area occupied by a population, different genotypes may have different fitnesses in each phase of the environment. In which case polymorphism (gene equilibrium) may be preserved pro-

vided the heterozygotes are more efficient over a wider range of environments than the homozygotes.

5. *Compensation effect: more children born in families with defectives.* Where the homozygote recessive is defective, there appears to be a tendency in the families involved to have more children than normal, and more of these tend to be heterozygotes than are produced in the general population; a balanced polymorphism can be thus maintained.

6. *Dependence of fitness upon gene frequency.* When the attractiveness of blonds depends upon their rarity in the population, they are at a selective advantage and therefore increase in number. Should blondes increase to be as common as the air, then Phoebe ceases to be esteemed fair and is then at a selective disadvantage and tends to decrease in frequency. At some point in this process genetic equilibrium is reached.

As another illustration of the operation of this factor reference may be made to disease. In a large population of susceptible individuals an epidemic will spread quite rapidly, and the susceptibles will be at a selective disadvantage compared to the immune individuals. However, if most of the population is immune, the disease will not spread readily and the minority (the susceptibles) will be protected by the high frequency of immunes, and equilibrium will thus be maintained.

7. *Sex differences.* If selection is different in the two sexes, gene equilibrium may be achieved. For example, in a population in which fat women are preferred to svelte ones, and lithe men are preferred to somewhat overupholstered ones, a balanced polymorphism for both sexes will be reached.

# APPENDIX D
# LOCATION, NAME OF INSTITUTION, AND PRINCIPAL COUNSELOR OF SOME HEREDITY CLINICS (BY STATE)

Washington, D.C. Genetic Counseling and Research Center, George Washington University Hospital, N. C. Myrianthopoulos.
Tempe, Arizona. Arizona State University, C. M. Woolf.
Berkeley, California. Department of Zoology, University of California, Curt Stern.
Los Angeles, California. Los Angeles Medical Center, University of California, Stanley Wright.
Chicago, Illinois. Department of Zoology, University of Chicago, H. W. Strandskov.
Chicago, Illinois. Genetics Clinic, Children's Memorial Hospital, Northwestern University Medical School, D. Yi-Yung Hsia.
New Orleans, Louisiana. The Medical School, Tulane University, H. W. Kloepfer.
Baltimore, Maryland. Johns Hopkins Hospital, Victor A. McKusick.
Boston, Massachusetts. Department of Immunochemistry, Boston University Medical School, W. C. Boyd.
Ann Arbor, Michigan. Department of Medical Genetics, University of Michigan, J. V. Neel.
East Lansing, Michigan. Department of Zoology, Michigan State University, J. V. Higgins.
Minneapolis, Minnesota. Dight Institute, University of Minnesota, S. C. Reed.
Minneapolis, Minnesota. Human Genetics Unit, State Board of Health, L. E. Schacht.
Rochester, Minnesota. The Mayo Clinic, J. S. Pearson.
New York, New York. Albert Einstein College of Medicine, Yeshiva University, S. Gluecksohn-Waelsch.
New York, New York. New York State Psychiatric Institute, F. J. Kallmann.
New York, New York. Rockefeller Institute, A. G. Bearn.
Winston-Salem, North Carolina. Department of Preventive Medicine, Bowman Gray School of Medicine, C. N. Herndon.
Cleveland, Ohio. Department of Zoology, Western Reserve University, A. G. Steinberg.
Columbus, Ohio. Starling Loving Hall, Ohio State University.

# LOCATION, NAME OF HEREDITY CLINICS 381

Norman, Oklahoma. Department of Zoology, University of Oklahoma, P. R. David.

Providence, Rhode Island. Institute for Research in Health Services, Brown University, G. W. Hagy.

Austin, Texas. The Genetics Foundation, University of Texas, C. P. Oliver.

Logan, Utah. Department of Zoology, Utah State Agricultural College, E. J. Gardner.

Salt Lake City, Utah. Laboratory of Human Genetics, University of Utah, F. E Stephens.

Charlottesville, Virginia. School of Medicine, University of Virginia, R. F. Shaw.

Richmond, Virginia. Department of Biology and Genetics, Medical College of Virginia, B. L. Hanna.

Seattle, Washington. School of Medicine, University of Washington, A. G. Motulsky.

Madison, Wisconsin. Department of Medical Genetics, The Medical School, University of Wisconsin, J. F. Crow.

Edmonton, Alberta. Genetic Counselling Service, University of Alberta, Margaret W. Thompson.

Winnipeg, Manitoba. Hospital for Sick Children, Irene Uchida.

Toronto, Ontario. Hospital for Sick Children, N. F. Walker.

Montreal, Quebec. Department of Genetics, McGill University, F. C. Fraser.

# APPENDIX E

# GLOSSARY

*Acquired character.* A character that is the result of the environment. Acquired characters are *not* inherited.

*Adaptation.* The process, and the result of the process, by which members of a genetic group, either as individuals or in part or as groups, are fitted to past or present changes in their environment. An adaptation is any trait of the organism that contributes to its survival or the survival of the group of which it is a member in the environment it inhabits.

*Allele* or *Allelomorph.* Any of the various forms of a gene. Alleles occupy the same position (*locus*) on a given chromosome pair, influencing in different ways the same developmental process.

*Amino acid.* One of a number of organic acids containing the amino radical ($NH_2$), from which proteins are formed.

*Anaphase.* The stage of mitosis (or meiosis) following metaphase, when the daughter chromosomes are separating toward the poles of the spindle.

*Androgen.* A hormone secreted in both sexes, but more abundantly in males, that influences the development of maleness (in structure, form, and behavior). The principal male sex hormone is testosterone, which is of testicular origin. Androsterone is another male sex hormone.

*Anthropometry.* The measurement of the human body by weight, size, and proportions.

*Antibody.* One of certain chemical substances formed by the body when a foreign material (antigen) is introduced into the bloodstream. The antibody usually combats the ill effects of the foreign material.

*Antigen.* Any foreign material that when introduced into the bloodstream causes the production of antibody.

*Asexual.* Designating any mode of reproduction that does not involve the union of male and female reproductive cells.

*Atavism.* The wholly erroneous notion that long-extinct ancestral traits can reappear in a descendant.

*Autosome.* Any of the twenty-two pairs of chromosomes that are not sex chromosomes. All human beings carry twenty-two pairs of autosomes and one pair of sex chromosomes. The abnormal exceptions may carry more or less.

*Backcross.* The cross of a hybrid with one of its parents.
*Barr body.* See *Chromatin body.*
*Base.* An electropositive element or radical that unites with an acid to form a salt. The four bases in DNA are adenine,

382

thymine, guanine, and cytosine. In RNA thymine is replaced by uracil.

*Bivalent.* A pair of homologous chromosomes united at the first meiotic division, usually by chiasmata.

*Blood.* The red fluid and its constituent parts circulating in the arteries, capillaries, and veins. The fluid portion of the blood, the *plasma* (*serum* when coagulated), carries numerous cells, *erythrocytes* (red cells), *leukocytes* (white cells), and *platelets* (blood plates). Red cells carry oxygen on their hemoglobin surfaces. White cells defend the body against bacteria. When tissues are injured, platelets help make blood clotting possible.

*Catalyst.* A substance that causes or hastens a chemical change without itself being changed. Genes are autocatalytic.

*Caucasoid.* One of the major groups of mankind or a member of that group, characterized most generally by white skin (though many Caucasoids have dark skin), a long nose, and straight or curly hair.

*Cell.* The living active unit of all plants and animals, consisting of many specialized parts. In the nucleus of the cell lie the chromosomes.

*Cell differentiation.* Specialization of the cells for different kinds of function, differing from ordinary growth (increase in size) in that the cells become unlike the parent cells.

*Centriole.* A small body in the centrosome situated just outside the nuclear membrane. The centriole organizes both the spindle and the centromere for movement.

*Centromere.* The small body that holds together at their middle the double strands forming each chromosome: best seen in prophase. In metaphase the centromeres split into two, thus entering upon anaphase, when each of the centromeres with its chromosome becomes attached to a spindle, the centromeres finally moving apart to attach one to each centriole.

*Centrosome.* The self-propagating body that during mitosis divides into two parts, each lying at either pole of the spindle. The centrosome appears to determine the formation of the spindle.

*Character.* A property of an organism in regard to which observations of genetic or other similarities or differences are made. Also called *trait*.

*Chiasma.* A crosswise exchange of partners in a system of paired chromatids observed just before the beginning of anaphase in meiosis. Plural, *chiasmata*.

*Chromatid.* A longitudinal half chromosome between early prophase and metaphase of mitosis, and between diplotene and second metaphase of meiosis.

*Chromatin body.* (*Barr body*). A dark stainable body usually situated against the inner surface of the nuclear membrane of the cell. Present in females and absent in males. Represents the coiled, inactive X chromosome; hence the rule that there is always one less chromatin body than there are X chromosomes.

*Chromosome.* One of a number of thread-shaped bodies situated in the nucleus of animal and plant cells and carrying the genes. A code center. Usually visible only during mitosis or meiosis.

*Codon.* A set of bases that codes one amino acid.

*Conceptus.* The organism from conception to birth.

*Concordant.* Agreeing in traits: used mostly in twin studies.

*Congenital.* Existing at birth as a result either of one's genotype or of one's prenatal environment.

*Corpus luteum.* The yellowish body on the surface of the ovary marking the place through which a ripe ovum has passed. The corpus luteum func-

384 **HUMAN HEREDITY**

tions as a temporary gland secreting progesterone, but only under the influence of the organizing action of the pituitary hormone luteotrophin. See *Progesterone*.

*Cortex.* The outer layers of an organ, in contrast to its inner substance, the *medulla*.

*Cross.* A mating between two individuals.

*Crossing-over.* The exchange of material between homologous chromosomes at meiosis, containing corresponding genes at the same loci as were present on each homologous chromosome. Hence, the shifting of genes from one chromosome to another.

*Culture.* That part of the environment that is learned; the man-made part of the environment. The way of life of a people.

*Cytoplasm.* All the protoplasm of a cell except that of the nucleus, which is called *nucleoplasm*.

*Deoxyribonucleic acid (DNA).* The principal constituent of the gene, believed to be the material of heredity itself. DNA is thought to carry the master plans, or code, containing the information that determines the order in which the amino acids fall into place in the protein molecule for which it is responsible.

*Development.* Increase in complexity.

*Diploid.* Containing the chromosomes in pairs, the members of which are homologous, i.e., containing two sets of chromosomes or twice the haploid number: characteristic of all the cells of the body except the gametes.

*Diplotene.* The stage of meiosis following the division of the chromosomes and chiasma formation at pachytene.

*Discordant.* Having a different trait or traits: used mainly in twin studies.

*Disease.* An acquired morbid change in any tissue or tissues of an organism, or in an organism as a whole, of specific microorganismal causation with characteristic symptoms. See *Disorder*.

*Disorder.* A disturbance of structure or function or both due to a genetic or embryological failure in development or as the result of exogenous factors, such as certain chemical substances, injury, or disease. May be inborn or acquired. See *Disease*.

*Dizygotic.* Derived from two fertilized eggs (zygotes).

*Dominant gene.* A gene that in a heterozygote overrides the effect of its recessive allele. A gene that always expresses itself in the individual who possesses it.

*Down's syndrome (Trisomy 21, Mongolism, Mongolian idiocy).* A developmental disorder, associated with a genetic constitution in which there are forty-seven instead of forty-six chromosomes, there being three instead of two chromosomes No. 21, occurring significantly more frequently in children of mothers who were more than thirty-five years when they conceived. Down's syndrome is characterized by severe mental and physical retardation, the presence of epicanthic folds, a large space between the first and second digits of the hands and feet, stubby hands and fingers, a large tongue, etc.

*Drosophila.* The generic name of the banana fly, vinegar fly, or fruit fly.

*Dysfunction.* Abnormal or incomplete functioning of an organ.

*Dysgenesis.* This term has three distinct but interrelated meanings: (1) maldevelopment, (2) sterility, and (3) detrimental to the heredity of the offspring, the opposite of eugenesis or eugenics.

*Embryo.* An animal in the process of development from the zygote. From the time of conception to the end of the second week of development, the human organism is sometimes said to be in the *ovular* phase. From the end of the second week to the beginning of the eighth week, it is in the *embryonal* phase, and from the beginning of the third month to birth, it is in the *fetal* phase.

**Embryology.** The study of the early phases of development of an organism.

**Endocrine gland.** Any of certain glands of internal secretion that secrete directly into the bloodstream and not through ducts, hence, also called *ductless gland*.

**Environment.** The external conditions that are acting upon or have acted upon an organism. The interaction with the genotype determines the phenotype. See *Internal environment*.

**Enzyme.** Any of various organic compounds secreted by the body cells that act as catalysts causing the chemical processes of the body to be carried on.

**Epicanthic fold.** A fold of skin from the upper eyelid lying over the inner angle of the eye or extending over the whole of the upper eyelid. Sometimes called the "mongoloid fold."

**Epididymis.** The first, convoluted portion of the excretory duct of the testis.

**Epigenesis.** The obsolete theory that the germ cell is structureless and that the embryo develops as a new creation through the action of the environment. Opposed to *preformation*.

**Epistasis.** The suppression of the visible action of one gene by another that is not its allele. Also used to refer to any interaction between nonallelic genes.

**Estrogen.** Any of the female sex hormones, so called because they are capable of inducing estrus in the female, whether mature or immature, spayed or not. The estrogens are *estrone*, *estriol*, and *estradiol*.

**Estrus.** The period of physiological changes in the reproductive organs of the female when in heat.

**Ethnic group.** An arbitrarily recognized population that, having a more or less distinctive assemblage of physical and cultural traits through a common heredity, is distinguishable from other populations within the species.

**Eugenics.** The science of the improvement of human fitness and the reduction of unfitness through the control of heredity.

**Expectation of life.** A measure or statement of the number of years an individual may, on the average, expect to live, from birth or from any age taken as a datum point. It is computed on the basis of the actual ages at death of the population in any given age range at any given age. Often used roughly in the same sense as *longevity*.

**Expressivity.** The manifestation of a trait produced by a gene. When the manifestation differs from individual to individual, the gene is said to have *variable expressivity*, for example, the dominant gene for allergy may take such forms as asthma, eczema, hay fever, angioneurotic edema, or urticarial rash.

**Fertilization.** The union of egg and sperm.

**Fetus.** The intrauterine organism from the beginning of the third month to birth. See *Embryo*.

**Fraternal twins.** Twins developed from separate eggs; two-egg, or dizygotic, twins.

**Gamete.** A mature germ cell; the *ovum* of the female or the *spermatozoon* of the male.

**Gene.** The physical unit of heredity, a small region in a chromosome, consisting of a giant molecule or part of such a molecule, believed to consist mainly of deoxyribonucleic acid (DNA). It is estimated that man has a total of about 30,000 genes on his forty-six chromosomes.

*Gene dosage.* The number of times a given gene is present in the nucleus of a cell or organism.

*Genetic drift* (*Sewall Wright effect*). The nonselective random distribution, extinction, or fixation of genes in a population.

*Genetic equilibrium.* The condition of a population in which successive generations consist of the same genotypes with the same frequencies, with respect to particular genes or arrangements of genes.

*Genetics.* The branch of biology concerned with the manner in which inherited differences and resemblances come into being between similar organisms.

*Genogroup.* A breeding population that differs from others in the frequency of one or more genes.

*Genosex.* The genetic sex of the organism. This may not correspond to the apparent sex, the *phenosex.*

*Genotype.* The genetic constitution, determined by the number, types, and arrangement of genes.

*Germ plasm.* The hereditary material present in germ cells.

*Glutamic acid.* An amino acid occurring as a decomposition product of protein.

*Gonad.* The sex gland, which produces gametes; the *ovary* in the female or the *testis* in the male.

*Gonosomes.* The sex chromosomes.

*Growth.* Increase in size.

*Haploid.* Containing a single set of unpaired chromosomes: characteristic of gametes.

*Heterogametic sex.* The sex that forms two types of gametes in equal proportions; the male, because half the spermatozoa carry an X and half carry a Y chromosome in addition to the autosomes.

*Heterozygous.* Carrying different alleles of a gene on both homologous chromosomes.

*Homogametic sex.* The sex that forms only one type of gamete; the female, because all the ova carry only the X chromosome in addition to the autosomes.

*Homologous chromosomes.* Partner chromosomes, usually identical in form and in the number and types of genes they contain. The exceptions are the sex chromosomes, X and Y, which differ from each other in form and gene content.

*Homozygous.* Carrying the same alleles of a gene on both homologous chromosomes.

*Hormone.* A chemical substance produced in small quantity in an organ and transported to other parts where it exerts specific effects.

*Hybrid.* The offspring of two genetically different parents. Commonly, the offspring of parents of different ethnic or major-group origin.

*Hybrid vigor.* Increased vigor of growth, fertility, or other traits in a hybrid of two different stocks, as compared with either of the parental lines.

*Hydrocephaly.* A condition characterized by great enlargement of the head, generally as a result of increased fluid in the lateral ventricles of the brain. The condition is usually congenital.

*Identical twins.* Twins that develop from the same egg; one-egg, or monozygotic, twins.

*Idiogram.* The diagrammatic representation of a karyotype (*q.v.*), which may be based on measurements of the chromosomes in several or many cells.

*Immunity.* The capacity to resist, or freedom from susceptibility to, disease-producing organisms or poisons.

*Incomplete dominance.* The relationship between two alleles that in the heterozygote produces an intermediate effect. Negroid skin-color genes are incompletely dominant.

*Intelligence.* The individual's total repertory of those problem-

## GLOSSARY 387

solving and thinking discriminatory responses that are usual and expected at a given age level. The "usual and expected" response is one of which 65 to 75 percent of the given population is capable.

**Internal environment.** The conditions within the organism acting upon it.

**Isochromosome.** A chromosome with two identical arms.

**Isolation.** The condition in which potential mating groups are separated by ecological or social barriers and thus prevented from mating.

**Karyotype.** A systematized array of the chromosomes of a single cell prepared by photography or by drawing, with the extension in meaning that the chromosomes of a single cell can typify the chromosomes of an individual or even of a species.

**Lethal gene.** A mutant gene, the effect of which is to kill an organism at any stage from egg to adult.

**Linkage.** Genes situated on the same chromosome are said to be linked. *Close linkage* refers to the association of nonallelic genes that act as if they were inseparable, usually because they are located at or near the same locus on the same chromosome. The linkage is broken by crossing-over.

**Locus.** A particular place on a particular chromosome that always contains one kind of gene or one of a particular set of alleles. Homologous chromosomes usually have identical sets of loci.

**Longevity.** The span of years lived. Often used roughly in the same sense as *expectation of life*.

**Lysosomes.** Cytoplasmic bodies containing digestive enzymes that break down large molecules, such as carbohydrates, fats, proteins, and nucleic acids, into smaller constituents that can be oxidized by the oxidative enzymes of mitochondria.

**Maturation division.** See *Meiosis*.

**Medulla.** The inner substance of an organ in contrast to its outer substance, the *cortex*.

**Meiosis.** The two successive cell divisions, from a diploid mother cell, preceding formation of the gametes. Both divisions resemble mitosis, except that while there are two divisions of the nucleus, in the second meiotic division the chromosomes are not duplicated, hence resulting in the formation of haploid gametes from diploid mother cells. See also *Mitosis*.

**Melanin.** A complex, dark brown pigment, which, in various concentrations, affects the color of hair, skin, and eyes.

**Menarche.** The first menstruation.

**Mendel's laws.** The three principles of chromosome behavior at meiosis: 1. *The Law of Segregation* refers to the behavior of one pair of genes. When the homologous chromosomes separate, the two alleles segregate into different gametes. 2. *The Law of Independent Assortment* or *Free Recombination* refers to the behavior of two or more pairs of alleles carried on different chromosomes. Each pair of chromosomes separates into different gametes independently of other pairs. The result is that the gametes contain all possible combinations of the genes constituting the different pairs. *Linked genes* are the exception to this law; recombination of such genes occurs through *crossing-over*. 3. *The Law of Dominance and Recessiveness* refers to the fact that some genes are capable of suppressing the expression of their alleles. Genes of the first class are called *dominant;* genes whose expression is suppressed are called *recessive*. Recessive genes do not or only minimally express themselves in heterozygous condition, but do, more or less, fully express themselves in homozygous condition.

*Metabolism.* The building up or breaking down of protoplasm within an organism. E. B. Wilson wrote in 1896 that "inheritance is the recurrence, in successive generations, of like forms of metabolism." More generally, the sum of the chemical changes whereby nutrition is effected.

*Metaphase.* The stage following prophase in mitosis (or meiosis) when the chromosomes arrange themselves at the equator of the spindle.

*Mitochondria.* The cytoplasmic particles found in all cells, which constitute the power plants that by oxidation and respiration extract energy from the chemical bonds in the cell nutrients. Each of the power plants makes its energy available to the energy-consuming processes of the cell in the form of adenosine triphosphate (ATP). There are about 2,000 mitochondria in each cell. Diameter: 0.5 micron; length: up to 7 micra.

*Mitosis.* A process of cell division during which the genes and chromosomes reduplicate (*prophase*), migrate to an equatorial plane (*metaphase*), and separate to opposite poles (*anaphase*), and the cell splits and forms two new cells (*telophase*). See also *Meiosis.*

*Modification.* A nonheritable change.

*Modifier.* A gene that modifies the effects of the phenotype controlled by one or more pairs of genes.

*Molecule.* The smallest possible unit of any substance that can exist and in the free state retain the characteristics of that substance. The molecules of elements consist of one atom or two or more similar atoms; those of compounds consist of two or more different atoms.

*Mongolism* or *Mongolian idiocy.* See *Down's syndrome.*

*Mongoloid.* One of the major groups of mankind or a member of that group, generally characterized by a flattened face, high cheekbones, marked overbite of upper teeth, shovel-shaped incisor teeth, epicanthic folds, very slight yellowish tinge to the skin, and lank black hair.

*Mongoloid fold.* See *Epicanthic fold.*

*Monozygotic.* Derived from one fertilized egg (zygote).

*Mosaic.* The constitution of the body by two or more lines of cells, each line being of different chromosomal structure.

*Multiple alleles.* A series of three or more alleles of one gene.

*Multiple factors.* See *Polygenes.*

*Mutation.* A failure of precision in the basic property of self-copying in a gene, resulting in a transmissible hereditary modification in the expression of a trait. The effects of most mutations in any one generation are usually not detectable.

*Mutation pressure.* The measure of the action of mutation in tending to alter the frequency of a gene in a given population.

*Natural selection.* A shorthand phrase for the effects of the differential reproduction of different types. Defined by Darwin: "As many more individuals of each species are born than can possibly survive; and as, consequently, there is a frequently recurring struggle for existence, it follows that any being, if it vary however slightly in any manner profitable to itself, under the complex and sometimes varying conditions of life, will have a better chance of surviving, and thus be *naturally selected.* From the strong principle of inheritance, any selected variety will tend to propagate its new and modified form." *Origin of Species,* 1859, p. 5. See also *Selection pressure.*

*Negroid.* One of the major groups of mankind or a member of that group, characterized by dark-pigmented skin; kinky, fuzzy, or woolly hair; and, generally, a broad nose.

**Nondisjunction.** The failure of separation of paired chromosomes at meiosis or mitosis, and their passage to the same pole. *Primary nondisjunction* is the production of eggs with two X chromosomes or none by an XX individual. *Secondary nondisjunction* is the production of eggs with two X chromosomes or a Y chromosome by an XXY individual.

**Nucleic acid.** Family of substances of large molecular weight found in chromosomes, mitochondria, and viruses.

**Nucleoplasm.** The protoplasm constituting the nucleus of a cell.

**Nucleotide.** Nucleic acid consisting of three components: phosphoric acid, carbohydrates, and purine (adenine, guanine) and/or pyrimidine (cytosine, thymine) bases combined in one unit (see Plate 1). Nucleotides are situated on one side of the helical chain of the DNA molecule; what follows shows only four of the thousands of nucleotides:

phosphoric acid—sugar—base
/
phosphoric acid—sugar—base
/
phosphoric acid—sugar—base
/
phosphoric acid—sugar—base

Observe that the asymmetric structure of the nucleotide conforms well with the spiral structure of the DNA molecule. Deoxyribonucleic acid (DNA) is a nucleotide. The nucleic acid may be built up out of 1,000 such tetranucleotides.

**Ontogeny.** The life development of a single organism.

**Oocyte.** The egg maternal cell with diploid nucleus, which by meiosis results in four haploid nuclei.

**Organ.** Any part of the body exercising a specific function.

**Ovulation.** The discharge of the ripe egg from the ovarian follicle onto the surface of the ovary, whence the egg is then usually taken up by the fimbriated ends of the fallopian tube.

**Pachytene.** Stage in prophase of meiosis in which paired homologous chromosomes are seen to consist of double strands; it is at this stage that crossing-over occurs.

**Paramutation.** A heritable change from one allele to another produced by a particular allele with which it is brought into association. Such paramutations are not entirely stable and under certain conditions will revert partially, but not completely, to the former normal state.

**Parity.** The state of a female with regard to her having borne children. *Nulliparity* is the condition of having borne no children, *primiparity* of having borne one child, *secundiparity* two children, and *multiparity* three or more.

**Parthenogenesis.** Reproduction by unfertilized ova.

**Penetrance.** The regularity with which a gene produces its effect. When a gene regularly produces the same effect, it is said to have *complete penetrance*. When the trait is not manifested in some individuals, the gene is said to have *reduced penetrance*. Dominant genes with reduced penetrance may be mistaken for recessives. When the penetrance of an autosomal gene is completely reduced in one sex, the gene is sex-limited.

**Phenocopy.** A nonhereditary variation or trait phenotypically indistinguishable from the hereditary one.

**Phenosex.** The apparent sex of the organism. May not correspond to genosex.

**Phenotype.** The manifest characteristics of the organism, including anatomical, physiological, and psychological traits. The product of the joint action of environment and genotype.

*Pinocytosis.* The process by which external materials are engulfed by the cell membrane and passed through it into the cytoplasm.

*Pleiotrophy.* The effects of a gene on more than one character.

*Polygenes.* A series of two or more genes or loci affecting the same phenotypic character, usually in a quantitative (additive or multiplicative) way. Also called *multiple factors.*

*Polymorphism.* Having several different forms. Used of traits, such as the blood groups, occurring in fairly constant proportion among the members of a population, genogroups, or species.

*Polyploid.* Having more than two sets (*diploid*) of chromosomes: *triploid* (three), *tetraploid* (four), *pentaploid* (five), *hexaploid* (six), etc.

*Polysome.* An aggregate of ribosomes, probably held together by a single thread of messenger-RNA, in which protein synthesis is carried out.

*Population.* Any contiguously distributed grouping of a single species that is characterized by both genetic and social continuity through one or more generations.

*Position effect.* The difference in effect of two or more genes according to their spatial relations in the chromosomes.

*Preadaptation.* Mutational or already existing genes carried in an individual or a group that enable their bearer or bearers to fill a new niche in the environment when, by chance, it is presented.

*Preformation.* The obsolete theory that the entire diversity of structure is contained in the embryo, and that development consists merely in increase in size (growth). Opposed to *epigenesis.*

*Progesterone.* A hormone secreted by the corpus luteum of the ovary and during pregnancy by the placenta. Prepares the uterus for receiving and developing the fertilized ovum.

*Prophase.* The first stage of mitosis (or meiosis) during which chromosomes appear, and in meiosis undergo pairing.

*Propositus* or *Proband.* An individual whose condition leads to the ascertaining of a pedigree.

*Protein.* Any of certain nitrogenous substances consisting of a complex union of amino acids and containing carbon, hydrogen, nitrogen, oxygen, and frequently sulfur. A chief constituent of chromosomes and of plant and animal bodies.

*Protoplasm.* The essential substance in all plant and animal cells.

*Purine.* The parent substance, $C_5H_4N_4$, of the uric-acid group of compounds. Adenine and guanine are purines, *aminopurines.*

*Pyrimidine.* $C_4H_4N_3$, a crystalline, organic compound providing the fundamental form of a group of bases. Cytosine and thymine are pyrimidine derivatives.

*Race.* A word widely and frequently misused, hence better not used at all. By geneticists, used to mean any population that differs from other populations in the frequency of its genes. By anthropologists, a population whose physical characters, through a common heredity, distinguish it from other populations. The author prefers to use the noncommittal "ethnic group."

*Random mating.* Mating that is largely determined by chance, rather than by careful choice. More especially, chance mating with respect to any particular gene.

*Recessive gene.* A gene that does not show its major effect in the heterozygote.

*Ribonucleic acid (RNA).* Ribonucleic acids of different composition occur everywhere in the cell. Each *transfer-RNA* molecule picks up from the cell fluid an amino acid molecule, which is then carried to a ribosome, where it is transferred to *messenger-RNA* or *intermediary-RNA*, which then instructs *ribosomal-RNA* in the ordering and linking of amino acids to produce finished proteins.

*Ribosome*. A morphologic unit of the cell machinery, which links molecules of amino acids together into proteins.

*Selection pressure*. The measure of the action of selection in tending to alter the frequency of a gene in a given population.

*Selective advantage*. The genotypic condition of an organism or group of organisms that increases its chances, relative to others, of representation in later generations.

*Sex chromatin*. See *Chromatin body*.

*Sex chromosome*. The chromosome that determines the sex of the organism. In mammals and many other organisms the female has two X chromosomes, and the male an X and a Y. In the fruit fly, *Drosophila*, sex is determined by whether or not a single or double complement of certain genes carried in the X chromosomes is balanced against certain genes carried in the autosomes. The double complement produces a female, the single complement a male. But in man and mouse, and probably in all other mammals, it is the Y chromosome that carries the male-determining factors. The X chromosome probably carries some female - determining factors, whereas others are situated on the autosomes.

*Sex-influenced trait*. Any trait conditioned by genes carried in the autosomes, and hence inherited equally by both sexes. Sex, however, controls the dominance. The gene that is dominant in one sex is recessive or intermediate in the other, and vice versa.

*Sex-limited trait*. Any of certain traits expressed in one sex but not in the other, conditioned by genes carried either in autosomes or in sex chromosomes, the expression of sex-limited traits depending largely upon the presence or absence of one or more sex hormones, or, strictly speaking, upon the amount of such hormones circulating within the body of the organism.

*Sex-linked gene*. One of the genes carried on the X chromosome, for which the human Y carries no allele. Sex-linked genes can occur singly in males, on the chromosome derived from the mother, and can occur as a pair in females.

*Somatic cell*. Any of the cells of the body except the germ cells.

*Species*. A group of actually or potentially interbreeding natural populations that is reproductively isolated from other such groups. In man the populations are the ethnic groups, the aggregation of groups is the species.

*Synapsis*. The pairing, point by point, of homologous chromosomes before the maturation division of the gametes.

*Telophase*. The terminal stage of mitosis (or meiosis) during which a nucleus is re-formed around the chromosomes, and the latter uncoil and resume their elongated, threadlike appearance.

*Testosterone*. A male sex hormone produced by the testis and in small amounts by the cortex of the adrenal gland and probably also by the ovary.

*Tetrad*. The paired chromosomes of meiosis after each chromosome has duplicated itself and the pair is visibly four-stranded. Also, a quartet of cells formed by meiosis in a mother cell.

*Throwback*. See *Atavism*.

*Trait*. See *Character*.

*Valine unit.* An amino acid component of protein, as in the hemoglobin molecule.

*Variation.* The occurrence of differences in characters. *Discontinuous variation,* gradations of difference that are perceptible in the phenotype. *Continuous variation,* gradations of difference that are imperceptible in the phenotype.

*Vas deferens.* The upward continuation of the epididymis.

*Viability.* The ability to survive.

*Virus.* A minute, living organism that can be seen only by the electron microscope. Viruses are responsible for such diseases as smallpox, shingles, poliomyelitis, and numerous others.

*Vitamin.* An organic substance present in minute amounts in foodstuffs, necessary to normal metabolism, growth, and development.

*Zygote.* The fertilized ovum, the product of the union of the ovum and a spermatozoon.

# APPENDIX F
# GENETICS IN THE U.S.S.R.:
# A POLITICAL "SCIENCE"

IN THE U.S.S.R. a dogmatic form of "genetics" has been officially decreed as the only politically acceptable version of that subject. The type of genetics described in the present volume is harassed and condemned by the Russians as "reactionary" or "capitalist Mendelian-Morganism." The matter is mentioned here only in order to answer the questions of those readers who have heard something of this development in Russia and who desire to have a plain statement of the facts.

This unfortunate subject can be dealt with only briefly here. There are two excellent books to which the reader may be referred for a more detailed account. These are Professor Conway Zirkle's *Death of a Science in Russia* (University of Pennsylvania Press, Philadelphia, 1949) and Sir Julian Huxley's *Heredity East and West* (Abelard-Schuman, New York, 1949).

The Russian version of genetics is known as "Michurinism" or "Lysenkoism," after the clever Russian gardener and plant breeder Ivan Vladimirovich Michurin (1855-1935) and Trofim Lysenko b. 1898), plant breeder and the principal contemporary proponent of Michurinism. Both men were without scientific training. Lysenko has stated the essence of Michurinism in the following words: "The materialist theory of the evolution of living nature involves recognition of the necessity of hereditary transmission of individual characteristics acquired by the organism under the conditions of its life; it is unthinkable without recognition of the inheritance of acquired characters." As evidence in support of this, Lysenko cites the example of "vernalization" of winter cereals. By treating the seeds with moisture at a low temperature cereals can be made to flower in winter that would otherwise flower in spring. This is not in the least a new discovery and has been known for a long time with respect to the modification of the flowering time of such seeds in a single generation, but Lysenko's claim that winter cereals can have their heredity permanently changed by such treatment so that they will behave like spring cereals in winter has not been substantiated by any scientist outside the U.S.S.R.

Lysenko has also claimed that heredity can be, as he calls it, "shattered." Shock treatment of various kinds, alterations in the environment, changes in metabolism, can, according to Lysenko, permanently change the heredity of the plant or organism. As a result of this "shattering," the heredity is said to become more plastic and malleable.*

Durrant at Aberstwyth has recently shown that special treatment of the soil in which flax is grown with fertilizer combining nitrogen, phosphorus, and potassium may result in the effects of such environmental changes being transmitted to the offspring. The inherited effects, principally expressed in a substantial increase in the weight of the plants, have held good for the four generations thus far investigated. Crossing and grafting experiments discount the effects of nutrition and maternal environment and support the hypothesis that a genetic factor is involved. Whether or not Lysenko may have obtained similar transmissible changes in plants cannot be verified because of the unscientific manner of his pronouncements.

Lysenko claims that evolution occurs as a consequence of environmentally induced changes in organisms and that it is this process of adaptation, and not natural selection, that is the operative factor. Such a belief is entirely contrary to the evidence of the last hundred years.

Is there any scientific evidence for any of Lysenko's claims? The answer is: None whatever. No scientist has been able to corroborate any of the claims made by the Lysenkoists. Indeed, all of these claims had long before the Michurinists, in one form or another, been demonstrated to be without foundation. Why, then, it will be asked, do the Russians insist upon maintaining this dogma of false ideas? Why have they destroyed so many geneticists who adhered to the scientifically demonstrated facts—men like Vavilov, Karpechenko, Levit, Ferry, Agol, Chetverikov, Serebrovsky, Ephroimson, Levitsky, and others?

The answer to this question appears to be that the Russians wanted a theory that maximized the role of environment in heredity and minimized the role of the genes—even to the extent of denying the very existence of genes!

It being a cardinal tenet of the communist creed that the uniqueness of man lies in his social rather than in his biological character, communist evolutionary thinking has tended to emphasize evolution through social rather than through biological changes. In placing the emphasis on social evolution, modification through environmental change becomes a principle applicable to all living things.

Theoretically there is nothing unsound in such a viewpoint. It becomes so only when it develops into an extreme environmentalism. And this is what has happened in the U.S.S.R.

Work of genuine interest, which is best approached with the tools of genetics, is often presented in a philosophical-political manner, together with claims that it is difficult to interpret, still less to substantiate. The customary settling of scientific disputes in an atmosphere of acrimony by political administrators, the dependence upon such administrators by scientists, has resulted in a rash of opportunism, in the lowering of the quality of research, and in a sterilization of brilliant minds in what was once a leading center of genetic activity.

Soviet genetics, for the most part, then, is not so much a scientific discipline as an instrument of political expediency, a party-line ideology. To what purpose?

One of the principal purposes of the Russian government appears to be to lead their subjects to believe that hereditary differences

---

* For a recent illuminating interview with Lysenko, see C. H. Waddington, "Talking to Russian Biologists," *The Listener* (London), January 17, 1963, pp. 119–21.

between ethnic groups and between individuals are insignificant and can be easily and permanently changed through modification of the environment. This idea seems particularly important to the Soviet leaders as a unifying principle in bringing the many different ethnic groups within the Soviet territories into the common fold. Cultural, physical, and mental differences are thus to be turned into cultural, physical, and mental likenesses. Men and women in the U.S.S.R. are to be not only politically equal but also biologically equal—by fiat of the state.

In his presidential address to the Eighth International Congress of Genetics held at Stockholm in July, 1948, the Nobel Prize winner in genetics, Professor H. J. Muller, related how before the Russians called off the Seventh International Congress, which was to be held in the U.S.S.R. in 1937, it had been proposed to proceed with it "provided all sessions and papers dealing with man were eliminated. At about the same time, the plans for a volume which was being prepared by leading geneticists of the world, designed to refute the Nazi racist doctrines, were abandoned. The clue was given when one of the men highest in the administration of applications of biological science, the head of agriculture in the Party, still asserted, even after having presided at most of the 1936 controversy on genetics, that evolution could not have occurred without the inheritance of acquired characteristics. On hearing this remark, the present writer asked this administrator whether this doctrine would not imply that the colonial, minority, and primitive peoples, those who had had less chance for mental and physical development, were not also genetically less advanced than the dominant ones. 'Ah, yes,' he replied in confidential manner, and after some hesitation, 'Yes, we must admit that this is after all true. They are in fact inferior to *us* biologically in every respect, including their heredity. And that,' he added, 'is in fact the *official* doctrine.' The word 'official' here was his, although the italics are ours. And at this point we may interject the question, just who could be more official on the particular subject than this particular individual? The answer to this seems clear. 'But,' continued our authority, 'after two or three generations of living under conditions of socialism, their *genes* would have so improved that *then* we would all be equal.'" \*

Still another reason why the political authorities would like to disseminate the view that acquired characters are inherited is that the genetically uninformed citizen is led to believe that by this means it becomes possible to alter the hereditary constitution of cultivated plants and domestic animals far more rapidly in desired directions than would be the case if it were necessary to select for mutations that have no adaptive relation to the conditions under which they arise. Vavilov was accused of trying to sabotage crop improvement because he said that it would probably take ten years or more to accomplish the results that Lysenko claimed could be achieved in a year or two.

The motivation for the promotion of Michurinism in the U.S.S.R. is clear. Whatever further motives may be involved, it is clear that Michurinism is not science, and even though Lysenko has been appointed by his rulers as its prophet, neither the doctrine of Michurin nor the claims of Lysenko have any scientific merit whatever.

---

\* H. J. Muller, "Genetics in the Scheme of Things," *Hereditas*, Supplementary Volume, 1949, p. 107.

# BOOKS FOR FURTHER READING

ALEXANDER, PETER, *Atomic Radiation and Life*, Penguin Books, Inc., Baltimore, Md., 1957.
A very readable account of the effects of radiation upon living organisms with especial reference to men.

ANFINSEN, CHRISTIAN B., *The Molecular Basis of Evolution*. John Wiley & Sons, Inc., New York, 1959.
An excellent examination of the evolutionary process as the integrated form of genetics and protein chemistry.

ASIMOV, ISAAC. *The Genetic Code*. The Orion Press, Inc., New York, 1962; The New American Library of World Literature, Inc. (Signet Books), New York, 1963.
A brilliantly clear account of the elementary chemistry and mechanics of the genetic code—assuming no knowledge of chemistry.

AUERBACH, CHARLOTTE, *Genetics in the Atomic Age*. Oxford University Press, Inc. (Essential Books), New York, 1956.
A charmingly written little book by an expert lucidly setting out the facts relating to the effects of radiation upon heredity.

BARNETT, ANTHONY, *The Human Species*. Penguin Books, Inc., Baltimore, Md., 1957.
An admirable survey of the human species from the genetic standpoint, accompanied by a thorough discussion of the social implications. Excellently illustrated.

BOYD, WILLIAM C., *Genetics and the Races of Man*. Little, Brown & Company, Boston, Mass., 1950.
A thoroughly sound and stimulating treatment of raciation in man from the standpoint of genetics.

BRINKHOUS, KENNETH M. (editor), *Hemophilia and Hemophilioid Diseases*. University of North Carolina Press, Chapel Hill, N.C., 1957.
An excellent symposium that reports the latest advances in the study of hemophilia and related disorders.

BUTLER, J. A. V., *Inside the Living Cell*. Basic Books, Inc., New York, 1959.
A delightful book by a physical chemist on what goes on in the living cell. Admirable on heredity, aging, cancer, and the brain.

CARTER, C. O., *Human Heredity*. Penguin Books, Inc., Baltimore, Md., 1962.
A thoroughly up-to-date and original book.

CRAGG, J. B., and PIRIE, N. W. (editors), *The Numbers of Man and Animals*. The Macmillan Company, New York, 1957.
A most valuable symposium, participated in by thirteen authorities whose papers on the problems of human population and its control, together with discussion of them by other participants, are printed here.

CROW, JAMES F., *Genetics Notes*. 4th ed. Burgess Publishing Co., Minneapolis, Minn., 1960.
A good working introduction.

DALE, ALAN, *An Introduction to Social Biology*. Rev. 3rd ed. William Heinemann (Medical Books) Ltd., London, 1953; Charles C. Thomas, Publisher, 1959.
Social biology is concerned with the contribution of biological knowledge to the solution of social problems and the promotion of human welfare, and this book is one of the best elementary introductions to the subject in the English language.

DAVIES, S. POWELL, *The Mentally Retarded in Society*. Columbia University Press, New York, 1959.
Most valuable for its solid emphasis on social factors. A very practical book.

DOBZHANSKY, THEODOSIUS, *Genetics and the Origin of Species*. 3rd ed. Columbia University Press, 1951.
A fundamental book written with admirable lucidity on the theories of modern genetics as they apply to the mechanism of organic evolution. The discussion of the problems of human evolution is exemplary.

———. *Evolution, Genetics and Man*. John Wiley & Sons, Inc., New York, 1955.
A delightful and authoritative discussion of the evolution of man in the light of genetic principles.

———. *Mankind Evolving*. Yale University Press, New Haven, Conn., 1962.
Man as a going evolutionary concern. An indispensable book.

———. *The Biological Basis of Human Freedom*. Columbia University Press, New York, 1956.
A consideration of human biological development and ethical striving by a distinguished geneticist.

DUBLIN, LOUIS I., and SPIEGELMAN, MORTIMER, *The Facts of Life from Birth to Death*. The Macmillan Company, New York, 1951.
A virtual short encyclopedia answering everyday questions, from birth to death, about longevity, disease, birth, death, marriage, divorce, and innumerable others.

DUNN, L. C. (editor), *Genetics in the 20th Century*. The Macmillan Company, New York, 1951.
Twenty-six essays by twenty-seven authorities on the progress of genetics during its first fifty years. A fascinating book.

DUNN, L. C., and DOBZHANSKY, TH., *Heredity, Race and Society*. Revised ed. The New American Library of World Literature, Inc. (Mentor Books), New York, 1952.
An excellent paperback.

FRANCIS, ROY G. (editor), *The Population Ahead*. University of Minnesota Press, Minneapolis, Minn., 1958.
A stimulating volume on the population problem.

FRANCOIS, JULES, *Heredity in Ophthalmology*. The C. V. Mosby Co., St. Louis, Mo., 1961.
The best book on the subject.

GATES, REGINALD R., *Human Genetics*. 2 vols. The Macmillan Company, New York, 1946.
A monumental work on human genetics. Illustrated.

GOLDSTEIN, PHILIP, *Genetics Is Easy*. 2nd ed. Lantern Press, Inc., New York, 1955.
A masterly introduction to the subject.

HALDANE, J. B. S., *Heredity and Politics*. W. W. Norton & Company, Inc., New York, 1938.
An extremely readable and interesting book by a great authority.

HAMMONS, H. G. (editor), *Heredity Counseling.* Hoeber-Harper, New York, 1959.
A useful book by a variety of authorities.

HUXLEY, JULIAN, *Heredity East and West.* Abelard-Schuman Limited, New York, 1949.
An admirably clear and judicial examination of the facts of "Lysenkoism" in Russian genetics.

JENNINGS, HERBERT S., *Prometheus; or Biology and the Advancement of Man.* E. P. Dutton and Company, Inc., New York, 1925.
This little book probably represents the best short statement on the meaning of heredity.

KALLMANN, FRANZ J., *Heredity in Health and Mental Disorder.* W. W. Norton & Company, Inc., New York, 1953.
An important and original work, lucidly written and well illustrated.

LAWLER, S. D., and LAWLER, L. J., *Human Blood Groups and Inheritance.* Harvard University Press, Cambridge, Mass., 1957.
A short and very readable account of the history, techniques, blood transfusion, and genetics of the blood groups.

MASLAND, R. L., SARASON, S. B., and GLADWIN, T., *Mental Subnormality.* Basic Books, Inc., New York, 1958.
The best book on the subject, considering the biological, psychological, and cultural factors.

MCELROY, WILLIAM and GLASS, BENTLEY, (editors), *The Chemical Basis of Heredity.* The Johns Hopkins Press, Baltimore, Md., 1957.
A valuable and readable symposium on the chemical basis of heredity. Indispensable for anyone wishing to obtain a firm grasp of the subject.

MCLEISH, JOHN, and SNOAD, BRIAN, *Looking at Chromosomes.* St Martin's Press, Inc., New York, 1958.
Excellent photographs of chromosomes, illustrating the manner in which they behave during the growth of the living organism, with a brief text.

MEIER, RICHARD L., *Modern Science and the Human Fertility Problem.* John Wiley & Sons, Inc., New York, 1959.
An able and timely exploration of the social, economic, and political implications of the physiological control of conception.

MILLER, MARY DURACK, and RUTTER, FLORENCE, *Child Artists of the Australian Bush.* George C. Harrap & Co., Ltd., London, 1952.
An account of the extraordinary artistic abilities of Australian aboriginal children, together with many interesting examples of their work.

MONTAGU, ASHLEY, *An Introduction to Physical Anthropology.* 3rd ed. Charles C Thomas, Publisher, Springfield, Ill., 1960.

———. *Man: His First Million Years.* The World Publishing Company, Cleveland and New York, 1957; The New American Library of World Literature, Inc. (Signet Books), New York, 1958.
The first of these books by the author of the present volume is restricted to an account of the origin and evolution of man. The second of the books is a more popular version of the physical and cultural evolution of man.

———. *Prenatal Influences.* Charles C Thomas, Publisher, Springfield, Ill. 1962.
A discussion of the influences that can affect the human organism in the womb, and the postnatal effect of such influences.

———. (editor), *Genetic Mechanisms in Human Disease.* Charles C Thomas, Publisher, Springfield, Ill., 1961.
The classic papers on the relation between chromosomal aberrations and disease.

MOORE, RUTH, *The Coil of Life*. Alfred A. Knopf, Inc., New York, 1961.
An excellent and highly readable account of great discoveries in the life sciences.

MULLER, H. J., LITTLE, C. C., and SNYDER, L. H., *Genetics, Medicine, and Man*. Cornell University Press, Ithaca, N.Y., 1947.
An introduction to the facts and principles of heredity, and the application of this knowledge to problems of human health and welfare.

NEEL, JAMES V., and SCHULL, WILLIAM J., *Human Heredity*. University of Chicago Press, Chicago, Ill., 1954.
An admirable book, written at university standard, full of the most valuable discussions and material.

NEWMAN, H. H., *Multiple Human Births*. Doubleday, Doran & Company, Inc., New York, 1940.
A popular book on twins and twinning.

NEWMAN, H. H., FREEMAN, F. N., and HOLZINGER, K. J., *Twins: A Study of Heredity and Environment*. University of Chicago Press, Chicago, Ill., 1937.
The authoritative account in the English language of the subject.

OSBORN, FREDERICK, *Preface to Eugenics*. Rev. ed. Harper & Brothers, New York, 1951.
The best and most sensible discussion of eugenics available.

OSBORNE, R. H., and DE GEORGE, V. F., *Genetic Basis of Morphological Variation*. Harvard University Press, Cambridge, Mass., 1960.
An application of the twin methods to the detection of genetic variability, and the analysis of genetic-environmental interaction.

PAULING, LINUS, *No More War!* Dodd, Mead & Co., New York, 1958.
An extremely valuable and important book on the effects of radiation and its effects upon human beings.

PENROSE, LIONEL S., *The Biology of Mental Defect*. Grune & Stratton, Inc., New York, 1949.
The best book on the subject.

———. *Outline of Human Genetics*. William Heinemann, Ltd., London, 1959.
A masterly little book.

REED, SHELDON C., *Counseling in Medical Genetics*. W. B. Saunders Co., Philadelphia, Pa., 1955.
Specially written for physicians and others who may be called upon to counsel in human genetics, this is a most valuable little book.

RIFE, DAVID C., *Heredity and Human Nature*. Vantage Press, Inc., New York, 1960.
A very attractively written little book on human heredity, and is among the most readable.

ROLPH, C. H. (editor), *The Human Sum*. The Macmillan Company, New York, 1957.
On the practical problems of family planning and population by a dozen authorities.

SCHEINFELD, AMRAM, *The New You and Heredity*. J. B. Lippincott Company, Philadelphia, Pa., 1950.
The best popular introduction to human heredity.

SCHUBERT, JACK, and LAPP, RALPH E., *Radiation: What It Is and How It Affects You*. The Viking Press, Inc., New York, 1957.
A most readable and authoritative account of the nature of radiation and its effects upon human beings.

SIMPSON, G. G., *The Meaning of Evolution*. Yale University Press, New Haven, Conn., 1949; The New American Library of World Literature, Inc., New York, 1951.

## 400 HUMAN HEREDITY

The best introduction to the study of evolutionary theory.

SIMPSON, G. G., PITTENDRIGH, C. S., and TIFFANY, L. H., *Life: An Introduction to Biology*. Harcourt, Brace and Company, New York, 1957.
One of the best books on biology ever written, with particularly good chapters on genetics.

SINNOTT, E. W., DUNN, L. C., and DOBZHANSKY, TH., *Principles of Genetics*. 5th ed. McGraw-Hill Book Company, Inc., New York, 1958.
One of the best textbooks on genetics available.

SORSBY, ARNOLD, *Clinical Genetics*. The C. V. Mosby Co., St. Louis, Mo., 1953.
A technical work of great value on the medical aspects of human genetics.

SRB, A. M., and OWEN, R. D., *General Genetics*. W. H. Freeman and Co., San Francisco, Calif., 1953.
One of the best textbooks on general genetics, with many references to human genetics.

STANBURY, J. B., WYNGAARDEN, J. B., and FREDRICKSON, D., *The Metabolic Basis of Inherited Disease*. McGraw-Hill Book Company, Inc., New York, 1960.
An exhaustive and admirable work.

STERN, CURT, *Principles of Human Genetics*. 2nd ed. W. H. Freeman and Co., San Francisco, Calif., 1960.
This is the standard work on human genetics, and is not likely to be soon superseded.

WHITNEY, DAVID D., *Family Treasures*. The Ronald Press Company, New York, 1942.
Subtitled "A Study of the Inheritance of Normal Characteristics in Man," this is a popular guide to human heredity that is unusually clear and well illustrated.

WIENER, A. S., and WEXLER, I. B., *Heredity of the Blood Groups*. Grune & Stratton, Inc., New York, 1958.
An authoritative study.

WINCHESTER, A. M., *Genetics*. 2nd ed. Houghton Mifflin Company, Boston, Mass., 1958.
A very clear, simple, and helpful survey of the principles of genetics with the emphasis on man.

ZAMENHOF, S., *The Chemistry of Heredity*. Charles C Thomas, Publisher, Springfield, Ill., 1959.
An excellent little book presenting the chemistry of heredity for the general reader.

ZIRKLE, C. (editor), *Death of a Science in Russia*. University of Pennsylvania Press, Philadelphia, Pa., 1949.
The fate of genetics in Russia as described in *Pravda* and elsewhere.

# BIBLIOGRAPHY

IN ADDITION to the preceding list of books, the following works are cited in the text either directly or indirectly.

ACOSTA-SISON, H., "Relation Between the State of Nutrition of the Mother and the Birth Weight of the Fetus," *Journal of the Philippine Islands Medical Association*, vol. 9, 1929, pp. 174–176.
ADAMS, S. H., "The Juke Myth," *Saturday Review*, April 2, 1955. (Included in *Saturday Review Treasury*, Simon and Schuster, Inc., New York, 1957, pp. 515–518.)
AIRD, I., BENTALL, H. H., MEHIGAN, J. A., and ROBERTS, J. A. F., "The Blood Groups in Relation to Peptic Ulceration and Carcinoma of Colon, Rectum, Breast, and Bronchus," *British Medical Journal*, vol. 2, 1954, pp. 315–321.
AIRD, I., BENTALL, H. H., and ROBERTS, J. A. F., "Relationship Between Cancer of the Stomach and the ABO Groups," *British Medical Journal*, vol. 1, 1953, pp. 799–801.
BARR, M. L., BERTRAM, L. F., and LINDSAY, A. H., "The Morphology of the Nerve Cell Nucleus According to Sex," *Anatomical Record*, vol. 107, 1950, pp. 283–298.
ARAI, H., "On the Post-Natal Development of the Ovary," *American Journal of Anatomy*, vol. 27, 1920, pp. 405–462.
BARR, M. L., "Sex Chromatin and Phenotype in Man," *Science*, vol. 130, 1959, pp. 679–685.
BATTARINO, P., and CAPODACQUA, A., "Hyperthyroidism and Pregnancy: Clinical and Statistical Observations on 46 Patients" (in Italian), *Minerva Ginecologia*, vol. 9, 1957, pp. 263–269.
BAUER, JULIUS, *Constitution and Disease*, 2nd ed. Grune & Stratton, New York, 1945.
BEADLE, G. W., "Ionizing Radiation and the Citizen," *Scientific American*, vol. 201, 1959, pp. 219–232.
BERGEMANN, E., "Sex Chromatin Determinations in Newborn Infants," *Schweizer Medizinische Wochenschrift*, vol. 91, 1961, pp. 292–293.
BINNING, G., "Peace Be on Thy House," *Health*, March/April, 1948, pp. 6–7, 28, 30.
*The Biological Effects of Atomic Radiation: A Report to the Public.* National Academy of Sciences–National Research Council, Washington, D. C., 1956.
*The Biological Effects of Atomic Radiation: Summary Reports.* National Academy of Sciences, Washington, D.C., 1956.
BLATZ, H., "Luminous Wrist Watches," *Science*, vol. 129, 1959, pp. 1512–1513.
BLOCK, E., "Quantitative Morphological Investigations of Follicular System in Women," *Acta Anatomica*, vol. 14, 1952, pp. 108–123.

BOAS, F., "Changes in Bodily Form of Descendants of Immigrants," *American Anthropologist*, N. S., vol. 14, 1912, pp. 530–562.

BOWLBY, JOHN, *Maternal Care and Mental Health*. World Health Organization, Geneva, 1951, and Columbia University Press, New York, 1951.

BOYD, W. C., and REGUERA, R. M., "Hemagglutinating Substances for Human Cells in Various Plants," *Journal of Immunology*, vol. 62, 1949, pp. 333–339.

BRINK, R. A., "Paramutation at the *R* Locus in Maize," *Cold Spring Harbor Symposia on Quantitative Biology*, vol. 23, 1958, pp. 379–391 (published February, 1959).

BUNGE, R. G., and SHERMAN, J. K., "Fertilizing Capacity of Frozen Human Spermatozoa," *Nature*, vol. 172, 1953, pp. 767–768.

———. "Frozen Human Semen," *Fertility and Sterility*, vol. 5, 1954, pp. 193–194.

———. KEETEL, W. C., and SHERMAN, J. K., "Clinical Use of Frozen Semen," *Fertility and Sterility*, vol. 5, 1954, pp. 520–529.

BURKE, B. S., BEAL, V. A., KIRKWOOD, S. B., and STUART, H. C., "Nutrition Studies During Pregnancy," *American Journal of Obstetrics and Gynecology*, vol. 46, 1943, pp. 38–52.

BURKE, B. S., and STUART, H. C., "Nutritional Requirements During Pregnancy and Lactation," *Journal of the American Medical Association*, vol. 137, 1948, pp. 119–128.

CAIN, A. J., "Possible Significance of Secretor," *Lancet*, vol. 1, 1957, pp. 212–213.

CANTRIL, HADLEY, *The "Why" of Man's Experience*. The Macmillan Co., New York, 1950.

CARRINGTON, E., REARDON, H. S., and SHUMAN, C. R., "Recognition and Management of Problems Associated with Prediabetes During Pregnancy," *Journal of the American Medical Association*, vol. 166, 1958, pp. 245–249.

CHARGAFF, E., "Chemical Specificity of Nucleic Acids and Mechanism of their Enzymatic Degradation," *Experienta*, vol. 6, 1950, pp. 201 ex seq.

CHASE G. D., and OSOL, A., "Radioactive Wrist Watches," *Science*, vol. 128, 1958, p. 788.

CHESLEY, L. C., and ANNTTO, J. E., "Pregnancy in the Patient with Hypertensive Disease," *American Journal of Obstetrics and Gynecology*, vol. 53, 1947, pp. 372–381.

CLARKE, C. A., and others, "ABO Groups and Secretor Character in Duodenal Ulcer. Population and Sibship Studies," *British Medical Journal*, vol. 2, 1956, pp. 725–731.

COOKMAN, H., "A Remarkable Case of Multiple Pregnancy," *Medical Press and Circular*, vol. 75, 1903, pp. 537–538.

CRAVEN, B., and JOKL, E., "A Note on the Effect of Training on the Physique of Adolescent Boys," *Journal of the Cape Town Post-Graduate Medical Association*, vol. 5, 1946, pp. 18–19.

CROW, J. F., "The Nature of Radioactive Fallout and Its Effects on Man." Hearings Before the Special Subcommittee on Radiation, 85th Congress, Government Printing Office, Washington, D.C., 1957, pp. 1009–1028.

———. *Effects of Radiation and Fallout*. Public Affairs Pamphlet No. 26, New York, 1957.

CUDMORE, S. A., and NEAL, N. A., *A Height and Weight Survey of Toronto Elementary School Children, 1939*. Ministry of Trade and Commerce, Ottawa, 1942.

DAVENPORT, C. B., and STEGGERDA, MORRIS, *Race Crossing in Jamaica*. Carnegie Institution of Washington, Washington, D.C., 1929.

DOBZHANSKY, TH., "What Is Heredity?" *Science*, vol. 100, 1944, p. 406.

———. "The Biological Concept of Heredity as Applied to Man," in *The Nature and Transmission of the Genetic and Cultural Characteristics of Human Populations*, Milbank Memorial Fund, New York, 1956, pp. 11–19.

DOTY, P., MARMUR, J., EIGNER, J., and SCHILDKRAUT, C., "Strand Separation and Specific Recombination in Deoxyribonucleic Acids," *Proceedings of the National Academy of Sciences*, vol. 64, 1960, pp. 461–476.

DURRANT, A., "Environmental Conditioning of Flax," *Nature*, vol. 181, 1958, pp. 928–929.

———. "A New Facet of the Chromosome Theory?" *The New Scientist*, vol. 6, 1959, pp. 293–295.

DÜTTEL, P. J., *De Morbis Foetum in Utero Materno*. C. Henckelli, Magdeburg, 1702.

EAST, EDWARD M., and JONES, DONALD F., *Inbreeding and Outbreeding*. J. B. Lippincott Co., Philadelphia, 1919.

EBBS, J. H., BROWN, A., TISDALL, F. F., MOYLE, W. J., and BELL, M., "The Influence of Improved Prenatal Nutrition upon the Infant," *Canadian Medical Association Journal*, vol. 46, 1942, pp. 6–8.

ECK, RICHARD V., "Genetic Code: Emergence of a Symmetrical Pattern," *Science*, vol. 140, 1963, pp. 477–481.

EHRENBERG, L., VON EHRENSTEIN, G., and HEDGRAN, A., "Gonad Temperature and Spontaneous Mutation-rate in Man," *Nature*, vol. 180, 1957, pp. 1433–1434.

EK, J. I., "Thyroid Function in Mothers of Mongoloid Idiots," *Acta Paediatrica*, vol. 48, 1959, pp. 33–42.

ELDERTON, E. M., "Height and Weight of School Children in Glasgow," *Biometrika*, vol. 10, 1914, pp. 288–340.

EVANS, BERGEN, *The Natural History of Nonsense*. Alfred A. Knopf, Inc., New York, 1958.

FARBER, B., *Effects of a Severely Mentally Retarded Child on Family Integration*. Monographs of the Society for Research in Child Development, vol. 24, no. 2, 1959.

FERGUSON-SMITH, M. A., "The Prepubertal Testicular Lesion in Chromatin-Positive Klinefelter's Syndrome (Primary Micro-Orchidism) as Seen in Mentally Handicapped Children," *The Lancet*, vol. 1, January 31, 1959, pp. 219–222.

FLEMING, R. M., "Physical Heredity in Human Hybrids," *Annals of Eugenics*, vol. 9, 1939, pp. 55–81.

FONIO, A., "Behavior of Thrombocytes in Patients with Hemophilia and in Female Conductors" (in German), *Schweitzer-Medinizinische Wochenschrift*, vol. 89, 1959, pp. 1026–1028.

FORD, C. E., JONES, K. W., MILLER, O. J., PENROSE, L. S., MITTWOCH, U., RIDLER, M., and SHAPIRO, A., "The Chromosomes in a Patient Showing Both Mongolism and the Klinefelter Syndrome," *The Lancet*, vol. 1, April 4, 1959, pp. 709–710.

FORD, C. E., JONES, K. W., POLANI, P. E., DE ALMEIDA, J. C., and BRIGGS, J. H., "A Sex-Chromosome Anomaly in a Case of Gonadal Dysgenesis (Turner's Syndrome)" *The Lancet*, vol. 1, April 4, 1959, pp. 711–713.

FORD, C. E., POLANI, P. E., BRIGGS, J. H., and BISHOP, P. M. F., "A Presumptive Human XXY/XX Mosaic," *Nature*, vol. 183, 1959, pp. 1030–1032.

FRACCARO, M., KAYSER, K., and LINDSTEN, J., "Chromosome Complement in Gonadal Dysgenesis (Turner's Syndrome)," *The Lancet*, vol. 1, April 25, 1959, p. 886.

———. "Chromosomal Abnormalities in Father and Mongol Child," *The Lancet*, vol. 1, April 2, 1960, pp. 724–727.

FRIED, R., and MAYER, M. F., "Socio-Emotional Factors Accounting for Growth Failure of Children Living in an Institution," *Journal of Pediatrics*, vol. 33, 1948, pp. 444–456.

GASPAR, J. L., "Diabetes Mellitus and Pregnancy," *Western Journal of Surgery*, vol. 53, 1945, pp. 21–30.

GENTRY, J. T., PARKHURST, E., and BULIN, G. V., JR., "An Epidemiological Study of Congenital Malformations in New York State," *American Journal of Public Health*, vol. 49, 1959, pp. 1–22.

GLUECK, SHELDON and ELEANOR, *Physique and Delinquency*. Harper & Brothers, New York, 1956.

GOLD, A. P., and MICHAEL, A. F., "Testosterone Induced Female Pseudo-hermaphroditism," *Journal of Pediatrics*, vol. 52, 1958, pp. 279–283.

GORDON, M. J., "The Control of Sex," *Scientific American*, vol. 199, 1958, pp. 87–94.

GREENE, A. R., BURRILL, M. W., and IVY, A. C., "Experimental Intersexuality. The Effect of Antenatal Androgens on Sexual Development of Female Rats," *American Journal of Anatomy*, vol. 65, 1939, pp. 415–472.

GREULICH, W. W., "Growth of Children of the Same Race Under Different Environmental Conditions," *Science*, vol. 127, 1958, pp. 515–516.

HADDOCK, D. R., and McCONNELL, R. B., "Carcinoma of the Stomach and ABO Blood Groups," *The Lancet*, vol. 2, 1956, pp. 146–147.

HAMILTON, J. B., "Patterned Loss of Hair in Man: Types and Incidence," *Annals of the New York Academy of Sciences*, vol. 53, 1951, pp. 708–728.

HANDLIN, OSCAR, *Race and Nationality in American Life*. Little, Brown & Co., Boston, 1957, and Doubleday Anchor Books, Garden City, N. Y., 1958.

HARDIN, GARRETT, "Interstellar Migration and the Population Problem," *Journal of Heredity*, vol. 50, 1959, pp. 68–70.

HARNDEN, D. G., and STEWART, J. S. S., "The Chromosomes in a Case of Pure Gonadal Dysgenesis," *British Medical Journal*, December 12, 1959, pp. 1285–1287.

———. and ARMSTRONG, C. N., "The Chromosomes of a True Hermaphrodite," *British Medical Journal*, December 12, 1959, pp. 1287–1288.

HARRELL, RUTH F., WOODYARD, ELLA, and GATES, ARTHUR I., *The Effect of Mothers' Diets on the Intelligence of Offspring*. Bureau of Publications, Teachers College, Columbia University, New York, 1955.

HARRIS, H., *Human Biochemical Genetics*. Cambridge University Press, New York, 1959.

HAYBRITTLE, J. L., "Radiation Hazards from Luminous Watches," *Nature*, vol. 181, 1958, p. 1422.

HENRY, GEORGE W., *All the Sexes*. Rinehart & Co., New York, 1955.

HEPNER, R., "Maternal Nutrition and the Fetus," *Journal of the American Medical Association*, vol. 168, 1958, pp. 1774–1777.

HOLUB, D. A., GRUMBACH, M. M., and JAILER, J., "Seminiferous Tubule Dysgenesis (Klinefelter's Syndrome) in Identical Twins," *Journal of Clinical Endocrinology*, vol. 18, 1958, pp. 1359–1368.

HOOTON, EARNEST A., *Crime and the Man*. Harvard University Press, Cambridge, 1939.

HOROWITZ, S. L., OSBORNE, R. H., and DEGEORGE, F. V., "Caries Experience in Twins," *Science*, vol. 128, 1958, pp. 300–301.

———. "How Common Is Sex Reversal?" *The Lancet*, vol. 1, January 31, 1959, pp. 237–239.

## BIBLIOGRAPHY

HSIA, D. Y., *Inborn Errors of Metabolism*. Year Book Publishers, Inc., Chicago, 1959.

HU, F., FOSNAUGH, R. P., and LESNAY, P. F., "Studies on Albinism," *Archives of Dermatology*, vol. 83, 1961, pp. 723–727.

HUNGERFORD, D. A., DONNELLY, A. J., NOWELL, P. C., and BECK, S., "The Chromosome Constitution of a Human Phenotypic Intersex," *American Journal of Human Genetics*, vol. 11, 1959, pp. 215–236.

HUXLEY, JULIAN, "Eugenics in Evolutionary Perspective," *Eugenics Review*, vol. 54, 1962, pp. 123–141.

IKKALA, E., *Haemophilia*. Thesis, Helsinki, 1960.

JACOBS, P. A., BAIKIE, A. G., COURT-BROWN, W. M., and STRONG, J. A., "The Somatic Chromosomes in Mongolism," *The Lancet*, vol. 1, April 4, 1959, p. 710.

JACOBS, P. A., and KEAY, A. J., "Somatic Chromosomes in a Child with Bonnevie-Ullrich Syndrome," *The Lancet*, vol. 2, October 31, 1959, p. 732.

JACOBS, P. A., and STRONG, J. A., "A Case of Human Intersexuality Having a Possible XXY Sex-Determining Mechanism," *Nature*, vol. 183, 1959, pp. 302–303.

JACOBS, P. A. et al., "Chromosomal Sex in the Syndrome of Testicular Feminization," *The Lancet*, vol. 2, October 17, 1959, pp. 593–594.

JACOBS, P. A., et al., "Evidence for the Existence of the Human 'Superfemale,'" *The Lancet*, vol. 2, September 26, 1959, pp. 423–425.

JALAVISTO, E., "The Influence of Parental Age on the Expectation of Life," *Revue Médicale de Liège*, vol. 5, 1950, pp. 719–722.

JANKE, L. L., and HAVIGHURST, R. J., "Relations Between Ability and Social Status in a Midwestern Community, II. Sixteen-Year-Old Boys and Girls," *Journal of Educational Psychology*, vol. 36, 1945, pp. 499–509.

JENNINGS, HERBERT S., *The Biological Basis of Human Nature*. W. W. Norton & Co., Inc., New York, 1930.

KALLMANN, FRANZ J., *Heredity in Health and Mental Disorder*. W. W. Norton & Co., Inc., New York, 1953.

——— and REISNER, D., "Twin Studies on the Significance of Genetic Factors in Tuberculosis," *American Review of Tuberculosis*, vol. 47, 1943, pp. 549–560.

KIRMAN, B., "Treatment of the Mentally Subnormal," *The Lancet*, vol. 2, 1962, pp. 1265–1268.

KLEBANOW, D., "Die Gefahr der Keimschädigung bei Rückbildungsvorgängen in den weiblichen Gonaden," *Deutsche Medizinsche Wochenschrift*, vol. 74, 1949, pp. 606–610.

KLEIN, H., "Family and Dental Disease: Dental Disease (DMF) Experience in Parents and Offspring," *Journal of the American Dental Association*, vol. 33, 1946, pp. 735–743.

LAUGHLIN, H., "Analysis of America's Melting Pot." Report Presented to the House Immigration Committee, Washington, D.C., 1923.

LEAKE, CHAUNCEY D., and ROMANELL, PATRICK, *Can We Agree? A Scientist and a Philosopher Argue about Ethics*. University of Texas Press, Austin, 1950.

LEDERBERG, J., "A View of Genetics," *Science*, vol. 131, 1960, pp. 269–276.

LEJEUNE, J. B., GAUTIER, M., and TURPIN, R., "Etude des Chromosomes Somatique de neuf enfants mongoliens," *Comptes rendus . . . l'Academie des Sciences*, vol. 248, 1959, pp. 1721–1722.

LENZ, F., "Morbific Hereditary Factors," in BAUR, E., FISCHER, E., and LENZ, F., *Human Heredity*, The Macmillan Co., New York, 1931.

LERNER, A. B., "Hormones and Skin Color," *Scientific American*, vol. 205, 1961, pp. 99–108.

LITCHFIELD, H., *Emma Darwin, A Century of Family Letters, 1792–1896*. Appleton-Century-Crofts, Inc., New York, 1915, vol. 2, p. 162.

MCCONNELL, R. B., PYKE, D. A., and ROBERTS, J. A. F., "Blood Groups and Diabetes Mellitus," *British Medical Journal*, vol. 1, 1956, pp. 772–776.

MCINTYRE, P. A., HAHN, R., CONLEY, C. L., and GLASS, B., "Genetic Factors in Predisposition to Pernicious Anemia," *Bulletin of the Johns Hopkins Hospital*, vol. 67, 1959, pp. 309–342.

MCKUSICK, VICTOR A., "On the X Chromosome of Man," *The Quarterly Review of Biology*, vol. 37, 1962, pp. 69–175.

MADIGAN, F. C., "Are Sex Mortality Differentials Biologically Caused?" *Milbank Memorial Fund Quarterly*, vol. 35, 1957, pp. 1–22.

MANUILA, A., "Blood Groups and Disease—Hard Facts and Delusions," *Journal of the American Medical Association*, vol. 167, 1958, pp. 2047–2053.

MATHEW, JOHN, *Eaglehawk and Crow: A Study of the Australian Aborigines*. David Nutt, London, 1899.

MILKMAN, R., "Potential Genetic Variability of Wild Pairs of *Drosophila melanogaster*," *Science*, vol. 131, 1960, pp. 225–226.

MONTAGU, ASHLEY, *The Direction of Human Development*. Harper & Brothers, New York, 1955.

———. *The Reproductive Development of the Female*. Julian Press, New York, 1957.

———. "The Dentition of Identical Twins with Particular Reference to an Identical Pathological Condition," *Human Biology*, vol. 5, 1933, pp. 629–645.

MOORE, K. L., "Sex Reversal in Newborn Babies," *The Lancet*, vol. 1, January 31, 1959, pp. 217–219.

MOOREHEAD, R. S., MELLMAN, W. J., and WENAR, C., "A Familial Chromosome Translocation Associated with Speech and Mental Retardation," *American Journal of Human Genetics*, vol. 13, 1961, pp. 32–46.

MORGAN, THOMAS HUNT, *Evolution and Genetics*. Princeton University Press, Princeton, 1925.

MOTULSKY, A. G., and GARTLER, S. M., "Consanguinity and Marriage," *The Practitioner*, vol. 183, 1959, pp. 170–177.

MURPHY, D., "Phenylketonuria Treated from Early Infancy," *Irish Journal of Medical Sciences*, vol. 6, 1959, pp. 425–427.

MURPHY, DOUGLAS P., *Congenital Malformations*, 2nd ed., J. B. Lippincott Co., Philadelphia, 1947.

NEHLS, G., "Caries und Paradentose bei Zwillingen," *Zeitschrift für menschliche Vererbungs-und Konstitutionslehre*, vol. 24, 1940, pp. 235–247.

NIERENBERG, MARSHALL W., and MATTHEI, J. HEINRICH., "The Dependence of Cell-Free Protein Synthesis in *E. coli* upon Naturally Occurring or Synthetic Polyribonucleotides," *Proceedings of the National Academy of Sciences*, vol. 47, 1961, pp. 1588–1602.

NILSSON, I. M., BERGMAN, S., REITALU, J., WALDENSTROM, J., "Haemophilia A in a 'Girl' with Male Sex-Chromatin Pattern," *The Lancet*, vol. 2, September 5, 1959, pp. 264–266.

OBERMAYER, M. E., *Psychocutaneous Medicine*. Charles C Thomas, Springfield, Ill., 1955.

OHNO, S., KLINGER, S. P., and ATKIN, N. B., "Human Oögenesis," *Cytogenetics*, vol. 1, 1962, pp. 42–51.

PASTORE, NICHOLAS, *The Nature-Nurture Controversy*. King's Crown Press, Columbia University, New York, 1949.

PEDERSON, J., and BRANDSHUP, E., "Foetal Mortality in Pregnant

Diabetics. Strict Control of Diabetes with Conservative Obstetric Management," *The Lancet*, vol. 1, 1956, pp. 607–608.
PENROSE, LIONEL S., *Mental Defect*. Sidgwick & Jackson, London, 1933.

———, "The Spontaneous Mutation Rate in Man," in *The Hazards to Man of Nuclear and Allied Radiations*, Her Majesty's Stationery Office, London, 1956, pp. 90–92.

———, "Evidence of Heterosis in Man," *Proceedings of the Royal Society*, B, vol. 144, 1956, pp. 90–92.

POLANI, P. E., BRIGGS, J. H., FORD, C. E., CLARKE, C. M., and BERG, J. M., "A Mongol Girl with 46 Chromosomes," *The Lancet*, vol. 1, April 2, 1960.

PRESSEY, S. L., and ROBINSON, F. P., *Psychology and the New Education*. Harper & Brothers, New York, 1944.

PUCK, T. T., "Radiation and the Human Cell," *Scientific American*, vol. 202, 1960, pp. 142–153.

RABOCH, J., and ŠÍPOVÁ, I., "The Mental Level in 74 Cases of True Klinefelter's Syndrome," *Acta Endocrinologica*, vol. 36, 1961, pp. 404–406.

RACE, ROBERT R., and SANGER, RUTH, *Blood Groups in Man*, 3rd ed. Charles C Thomas, Springfield, Ill., 1958.

Radiological Hazards to Patients: Second Report of Adrian Committee. H. M. Stationery Office, London, 1960.

RATNER, B., "A Possible Causal Factor of Food Allergy in Certain Infants," *American Journal of Diseases of Children*, vol. 36, 1928, pp. 277–288.

RECORD, R. C., "Relative Frequencies and Sex Distribution of Human Multiple Births," *British Journal of Social Medicine*, vol. 6, 1952, pp. 192–196.

RENKONEN, K. O., "Studies on Hemagglutinins Present in Seeds of Some Representatives of the Family Leguminoseae," *Annals of Medicine and Experimental Biology*, vol. 26, 1948, pp. 66–72.

RENWICK, J. H., and LAWLER, S. D., "Linkage Between the ABO and Nail-Patella Loci," *Annals of Human Genetics*, vol. 19, 1955, pp. 312–331.

*Report of the United Nations Scientific Committee on the Effects of Atomic Radiation*. United Nations, New York, 1958.

ROHRER, J. H., "The Test Intelligence of Osage Indians," *Journal of Social Psychology*, vol. 16, 1942, pp. 99–105.

RUNDLE, A., COPPEN, A., and COWIE, V., "Steroid Excretion in Mothers of Mongols." *The Lancet*, vol. 2, 1961.

RUSSELL, W. L., "Genetic Effects of Radiation in Mice and Their Bearing upon the Estimation of Human Hazards," *United Nations Reports*, vol. 11, 1956, p. 382.

——— and RUSSELL, L. B., "Radiation Hazards to the Embryo and Fetus," *Radiology*, vol. 58, 1952, pp. 369–376.

———, RUSSELL, L. B., and GOWER, J. S., "Exceptional Inheritance of Sex-Linked Gene in the Mouse Explained on the Basis That the X/O Sex-Chromosome Constitution Is Female," *Proceedings of the National Academy of Sciences*, vol. 45, 1959, pp. 554–560.

SALDANHA, P. H., "The Genetic Effects of Immigration in a Rural Community of Sao Paulo, Brazil," *Acta Geneticae Medicae et Gemellologiae*, vol. 11, 1962, pp. 158–224.

SANDERS, BARKEV S., *Environment and Growth*. Warwick & York, Inc., Baltimore, 1934.

SCHEINFELD, A., "The Kallikaks After Thirty Years," *Journal of Heredity*, vol. 35, 1944, pp. 259–264.

———, "The Mortality of Men and Women," *Scientific American*, vol. 198, 1958, pp. 22–27.

SHELDON, WILLIAM H., *The Varieties of Human Physique.* Harper & Brothers, New York, 1940.

———, *The Varieties of Temperament.* Harper & Brothers, New York, 1942.

———, *Varieties of Delinquent Youth.* Harper & Brothers, New York, 1949.

SHIMBERG, M. E., "An Investigation into the Validity of Norms with Special Reference to Urban and Rural Groups," *Archives of Psychology,* No. 104, 1929.

SIMPSON, W. J., "A Preliminary Report on Cigarette Smoking and the Incidence of Prematurity," *American Journal of Obstetrics and Gynecology,* vol. 73, 1957, pp. 805–815.

SIMPSON, W. T., "Sextuplets," *British Medical Journal,* vol. 2, 1959, p. 638.

SNYDER, L. H. "The Mutant Gene in Man," in *Genetics, Medicine, and man.* Cornell University Press, Ithaca, N.Y., 1947, pp. 133–134.

———, "Fifty Years of Medical Genetics," *Science,* vol. 129, 1959, pp. 7–13.

SONTAG, L. W., and WALLACE, R. F., "The Effect of Cigarette Smoking During Pregnancy upon the Fetal Heart Rate," *American Journal of Obstetrics and Gynecology,* vol. 29, 1935, pp. 3–8.

SPEISER, P., "Bestehen mathematisch gesicherte Beziehungen der A-B-O Gruppen," *Der Krebsarzt,* vol. 11, 1956, pp. 344–348.

SPITZ, R. A., "Hospitalism," in *The Psychoanalytic Study of the Child,* vol. 1. International Universities Press, New York, 1945, pp. 53–74.

———, "Hospitalism: A Follow-up Report," in *The Psychoanalytic Study of the Child,* vol. 2. International Universities Press, New York, 1947, pp. 113–117.

STEARNS, G., "Nutritional State of the Mother Prior to Conception," *Journal of the American Medical Association,* vol. 168, 1958, pp. 1655–1659.

STEPHENS, J. W., HOLCOMB, B., and PAGE, O. C., "Pregnancy and Diabetes," *Journal of the American Medical Association,* vol. 161, 1956, pp. 224–226.

STERN, C., "The Problem of Complete Y-Linkage," *American Journal of Human Genetics,* vol. 9, 1957, pp. 147–166.

STEWART, J. S. S., "The Chromosomes in Man," *The Lancet,* vol. 1, April 18, 1959, p. 833.

STRUTHERS, D., "ABO Groups of Infants and Children Dying in the West of Scotland (1949–1951)" *British Journal of Social Medicine,* Vol. 5, 1951, pp. 223–228.

SZONTÁGH, F. E., JAKOBOVITS, A., and MÉHES, C., "Primary Embryonal Sex Ratio in Normal Pregnancies Determined by the Nuclear Chromatin," *Nature,* vol. 192, 1961, p. 476.

TANNER, J. M., PRADER, A., HABICH, H., and FERGUSON-SMITH, M. A., "Genes on the Y Chromosome Influencing the Rate of Maturation in Man," *The Lancet,* vol. 2, August 22, 1959, pp. 141–144.

TELLER, W. M., ROSEVEAR, J. W., and BURKE, E. C., "Identification of Heterozygous Carriers of Gargoylism," *Proceedings of the Society of Experimental Biology,* vol. 108, 1961, pp. 267–268.

TJIO, J. H. and LEVAN, A., "The Chromosome Number of Man," *Hereditas,* vol. 42, 1956, p. 16.

——— and PUCK, T. T., "The Somatic Chromosomes in Man," *Proceedings of the National Academy of Sciences.*

———, PUCK, T. T., and ROBINSON, A., "The Somatic Chromosomal Constitution of Some Human Subjects with Genetic Defects," *Proceedings of the National Academy of Sciences*, vol. 45, 1959, pp. 1008–1016.

TOBIAS, P. V., "On a Bushman-European Hybrid Family," *Man*, vol. 54, 1955, pp. 179–182.

TOUGH, I. M., BUCKTON, K. E., BAIKE, A. G., and COURT-BROWN, W. M., "X-Ray-Induced Chromosome Damage in Man," *The Lancet*, vol. 2, 1960, pp. 849–851.

WALLACE, B., and DOBZHANSKY, TH., *Atoms, Genes, and Man*. Henry Holt, New York, 1959.

WALLACE, J., "Blood Groups and Disease," *British Medical Journal*, vol. 2, 1954, p. 534.

WATSON, B. B. (editor), *The Delayed Effects of Whole-Body Radiation: A Symposium*. The Johns Hopkins Press, Baltimore, Md., 1960.

WATSON, J. D. and CRICK, F. H. C., "Molecular Structure of Nucleic Acids. A Structure for Deoxyribose Nucleic Acid," *Nature*, vol. 171, 1953, p. 737.

WEBSTER, E. W., "Hazards of Diagnostic Radiology: A Physicist's Point of View," *Radiology*, vol. 72, 1959, pp. 493–507.

WELSHSON, W. J., and RUSSELL, L. B., "The Y-Chromosome as the Bearer of Male Determining Factors in the Mouse," *Proceedings of the National Academy of Sciences*, vol. 45, 1959.

WESTWOOD, GORDON, *Society and the Homosexual*. E. P. Dutton & Co., New York, 1952.

WIENER, A. S., and WEXLER, I. B., "Blood Group Paradoxes," *Journal of the American Medical Association*, vol. 161, 1956, p. 1474.

———, *Heredity of the Blood Groups*. Grune & Stratton, New York, 1958.

ZABRISKIE, J. B., "Effect of Cigaret Smoking During Pregnancy: Study of 2000 Cases," *Obstetrics and Gynecology*, vol. 21, 1963, pp. 405–411.

ZIMÁNYI, S., MÁTTYUS, A., KOZÁK, E., NAGY, T., and SZEGEDI M., "Clinical and Laboratory Observations on the Effects of Ribonucleic Acid," *The Lancet*, vol. 2, 1962, p. 1283.

# INDEX

A-B-O hemolytic disease, 98, 347
Aborigines, 259
  Australian, 321–22
Acanthrocytosis, 347
Acanthosis nigricans, 370
Acatalasemia, 362
Acatalasia, 347
Achondroplastic dwarfism, 269–70
Acosta-sison, H., 104
Acoustic neuroma, bilateral, 366
Acrobrachycephalia, 361
Acrocephalosyndactylia, 361
Acrodermititis onteropathica, 371
Acroosteolysis, 351
Addison's disease, 359
Adenine, 31, 33, 39
Aebi et al., 308
Afibrogenemia, 347
Agammaglobulinemia, 347
Age, maternal, 88–94
Agglutinins, 276–78
  Anti-A, 277
  Anti-B, 277
Agglutinogens, 276–77, 280, 285–86
Aird, I., 293–94
Albers-Schönberg disease, 352
Albinism, 197, 235–39, 246, 248, 317, 341, 371
  deaf-mutism with, 354
  ocular, 239, 354

Alder's anomaly, 348
Aldrich's syndrome, 351
Alkaptonuria, 197, 362
Allele(s), 59, 61, 63, 65
Allergies, 81, 104, 347
Alopecia, familial congenital, 360
Alpha particles, 327
Alport's syndrome, 358
Altzheimer's syndrome, 366
Amaurosis, congenital, 354
Amaurotic idiocy, 197, 369
  infantile, 321, 369
Amino acids, 39
  determination of, 34
  free, 37
  protein molecules and, 30
Amputations, congenital, 351
Amyloidosis, familial primary, 362
Amyotrophic, lateral sclerosis, 366
Analgesia, congenital generalized, 366
*Analysis of America's Melting Pot* (Davenport), 300
Andalusian fowl, cross breeding of, 60–61
Androgens, 194
Anemia, 347–48
Anencephaly, 103, 366
Angiokeratoma corporis diffusum, 371
Angioneurotic edema, 347
Aniridia, 355
Ankyglossia, 372

INDEX 411

Anophthalmos, 354
Anosmia, 370
Antibody, 98, 277, 284
  definition of, 98*n*.
Antigen, 98, 277
  definition of, 98*n*.
  P, 292
  Rh, 284
Apert's syndrome, 361
Apical dystrophy, 357
Arachnodactyly, 352, 357
Aristotle, 20, 71
Arrhinencephalia unilateral, 361
Arteriosclerosis, 348
Arthralgia, periodic, 367
Arthritis, chronic rheumatoid, 351
Arthronychodysplasia, 351
*Ascaris megalocephala*, 27
Asthma, 347
Atavisms, 160
Ataxia-telangieclasia, 366
Ateliotic dwarfism, 270
Atom(s), 327–28, 338–39
Atom bomb, heredity and, 325
ATP (adenosine triphosphate), 31, 37
Atresia of auditory meatus, 353
Atrial septal defect, 348
Auditory imperception, congenital, 366
Auricular appendages, 354
Auricular fistula, 354
Auricular septal defect, 349
Australoids, 221, 289
Autosome(s), 55, 198, 202, 205–6

Bacteria, 105–6
"Bad blood," 70
*Bad Seed, The* (March), 19

Baldness, 191–96
Banting, Frdereick G., 81
Barbiturates, 107
Barr, Murray L., 200
Batterino, P., 96
"Bayonet" hand, 352
BCG (bacillus Calmette-Guérin), 168
Behavioral potentialities, 118
Beiser, S. M., 44
Bell's palsy, 367
Bendich, A., 44
Bentall, H. H., 293
Beregmann, E., 210
Best, C. H., 81
Beta-liproprotein deficiency, 362
Beta rays, 327
Beutler, Ernest, 202–3
Bilirubin, 99
*Bill of Divorcement, A* (Dane), 19
Binning, Griffith, 116
*Biological Effects of Atomic Radiation, The*, 343
"Biologistic fallacy," 80
Birthmarks, 103
Biryukov, D. A., 341
Bivalent (dyad), 50, 53
Blood, chemical composition of, 72
  heredity and, 67–75, 276–96
  myths about, 67, 71–73
  secreting factor of, 281–82
Blood group(s), A, 98
  AB, 276–77
  A-B-O, 276–81, 294, 296
  agglutinogens and agglutinins of, 276–78
  B, 98
  determination of, tables, 277–78
  disease and, 293–96
  disorders of, 347–51

## 412 INDEX

Duffy type, 392
in Europe, 289
   illustrated, 279
genetic constitution and, table, 282
Henshaw type, 292
Hunter type, 292
Kell type, 292
Kidd type, 292
Lewis type, 292
Lutheran type, 292
M-N-S series, 282–84, 292
   tables of, 284–85
O, 98, 280, 282, 295
Rh. *See* Rh factor.
Sutter type, 292
*Blood and Honor* (Rosenberg), 72
"Blood relationship," 68, 70–72
"Blood royal," 68
Blood serum, 277–78
Blood transfusion, 73, 277–78, 284
"Blue blood," 68
Boas, Franz, 122–24
Body cells, division of, 25–26
Body form, 266–75
   environment and, 122–24, 266
   variety in, 267–68
Body type, 267–69
   temperament and, 268–69
Bone growth, 116
Bones, diseases of, 351–53
Bonner, James, 47
Bonnevie-Ullrich syndrome, 205–6, 368
Bourneville's disease, 369
Bowlby, John, 114
Boyd, W. C., 295
Brachycardia, 348
Brachydactylia, 357
Brachymesophalangy, 357

Brachomorphy with spherophacia, 357
Brachytelephalangy, 357
Brain degeneration, 366
Branchial defects, 361
Braun-Bayer syndrome, 359
Brink, Alexander, 65–66
Bronchial pneumonia, 293
Bronchiectasis, 362
Bronchitis, 361
Bronze diabetes, 363
Buerger's disease, 351
Bulbar paralysis, progressive, 368
Bunions, 352
Buphthalmia, 355
Burke, B. S., 104
Bushmen, 221, 252
   interbreeding with, 223–24
Butler, Samuel, quoted, 54


Caesarian section, 96
Cain, A. J., 295–96
Calmette-Guérin bacillus, 168
Camptodactyly, 357
Cancer, 106, 176, 293–94
   of breast, 373
Capodacqua, A., 96
Carbon 14, 344
Carboxylase deficiency disease, 362
Cardiomegaly, familial, 348
Cardiovascular, dysplasia, hereditary, 348
Caries, 259–60, 372
Carotid-body tumors, 373
Carrington, E., 96
*Castinidium variable*, 27
Cataract, 354, 369
Cat's ear, 354
Caucasoids, 218, 221
Cell(s), body, 25–26
   chromosome, 32

INDEX 413

mother, 53
typical, illustrated, 36
Cell division, 48–50
Cell functions, 47–48
Cerebral diplegia, congenital, 368
Cerebral sclerosis, 191
diffuse, 366
Cerebromacular degeneration, 366
Characteristics, "inherited," 83
Charcot-Marie-Tooth disease, 308, 368
Charcot's disease, 366
Chargaff, E., 31
Chase, G. D., 329
Cherubism, 351
Chest, disorders of, 362
Chin, 258
receding, 258, 372
Choanal atresia, 370
Cholesterol ester storage disease, familial, 363
Chondrodystrophia calcificans congenita, 351
Chondrodystrophy, 351
Chorioid defects, 354
Christmas disease, 348
Chromatids, 50
Chromatin, 35–36, 47
Chromatin body, 200–3, 206–7
Chromosomal complexes, abnormal, 203–9
illustrated, 208
Chromosome(s), 25–36
Denver classification of, 93
extra, 91, 93
function of, 33–34
of human female, illustrated, 92
of human male, illustrated, 92

nature of, 31
number of, in body cell, 26–27
table of, 29
radiation damage to, 339–40
Rh factor and, 286–89
sex, 33, 203, 205–6, 210
size of, 27
X. See X chromosomes.
Y. See Y chromosomes.
Chromosome rearrangement, 340
Chronic lymphatic leukemia, 350
Chronic myeloid leukemia, 350
Cigarette smoking, 107–9
Clarke, C. A., 296
Class differences, intelligence and, 138–39
Cleft palate, 110, 260–61, 372
Cleidocranial dysostosis, 361
Clermont, 53
Clothing, male, mutations and, 334–35
Clubfoot, 103
congenital, 351
Codon, 39, 41
Color-blindness, inheritance of, 187, 189–91, 354–55
illustrated, 190
Congenital arrhythmia, 349
Congenital hemolytic jaundice, 350
Congenital virilizing adrenal hyperplasia, 362
Congenital word blindness, 367
Conjugation, 50
Conklin, Edwin Grant, 298
Consanguinity, 315
Constitution, definition of, 164
genes and, 164–167

"Constitutional idiopathy," 80
Convex vagus foot, 353
Coppen, A., 90
Cornea defects, 355
Coronary xanthomatosis, 348
Correns, Karl, 61
Cosmic rays, 328, 333
*Counseling in Medical Genetics* (Reed), 313
Cousins, marriage between, 315, 317–19, 375
Cowie, V., 90
Cowlicks, 250–51
Craniofacial dysostosis, 361
Craven, B., 121
Cretinism, 111–12, 271, 369
Crick, F. H. C., 32
Crigler-Najjar syndrome, 364
Crime, cause of, 158–59
  definition of, 156
  environment and, 131–33, 158–63
  heredity and, 156–63
  in twins, 156–58
*Crime and the Man* (Hooton), 161
Crossbreeding, 65–66
Cross-eyes, 356
Crossing-over, 52–53
  illustrated, 52
Crouzon's disease, 361
Crow, James F., 336
Cryptorchism, 358
Cudmore, S. A., 122
Cultural environment, 134–36
Cystic kidney, 358
Cystinuria, 362
Cytoplasm, 25
  RNA in, 35
Cytosine, 31, 33, 39

Dacryocystitis, 361
Dacryostenosis, congenital, 361

Darwin, Charles, genetic endowment of, 126
  marriage of, 319
  pedigree of, 320
Davenport, C. B., 225–26, 300
Day blindness, 355
Deaf-mutism, 354
*Death of a Science in Russia* (Zirkle), 393
Deep acetabulum and intrapelvic protrusion, 351
Deformity, physical agents and, 109–10
DeGeorge, F. V., 260
Dehn, Paul, quoted, 332
Dehydroepiandrosterone (DHA), 90
Delinquency, 159
Dentin dysplasia, 372
Dentinogenesis imperfecta, 372
Dentition, 372–73
  of identical twins, 150–53
Depression, 369
Developmental process, genetic, 23–56
  *See also* Embryo; Fetus.
DeVries, Hugo, 61
Dextrocardia, 348
DHA (dehydroepiandrosterone), 90
Diabetes, 81, 95–96, 197
  bronze, 363
  constitution and, 166
  false, 359
Diabetes insipidus, 359, 363
Diabetes mellitus, 295, 359
Diaphyseal aclasis, 351
Diet, maternal, 101–4
Differences, 18
Dinter, Artur, 75
Disease, inherited, 346–374
  susceptibility, to, sex and, table, 186

Disorders, human, list of inherited, 346–74
Disseminated sclerosis, 366
DNA (deoxyribonucleic acid), 30–38, 42–47, 180, 309
  helical structure of, 32–33, 340
  nucleic acids of, 31
  synthetic, 44
Dominance and recessiveness, law of, 63–66
Dominants, 63
Doty, Paul M., 45
Down, Langdon, 89
Down's syndrome, 89–91, 93, 103, 204, 314, 369
Drosophilia melanogaster, 22, 28–29, 56
  sex determination in, 181
Drugs, effect of, on embryo and fetus, 107–9
Duchenne-Greisinger disease, 308
Dugdale, Richard L., 131–32
Durrant, A., 394
Dwarf virus, 47n.
Dwarfism, 269–70
Dyad, 50, 53
Dyck et al., 308
Dysautonomia, familial, 367
Dysfunction, maternal, 95–96
Dyskeratosis, 355
Dysplasia epiphysialis punctata, 351
Dystonia musculorum deformans, 368


Ear(s), 265
  disorders of, 353–54
  hypertrichosis of, 250
  malformed, 354
Earlobes, 265
  attached, inheritance of, illustrated, 264

East, Edward M., 319
Ebbs, J. H., 101–2
Ectodactylia, 357
Ectoderm, 254n.
Ectodermal dysplasia, 373
Eczema, 371
Egg. See Ovum.
Egg cells, primordial, 23
Ehlers-Danlos syndrome, 371
Ehrenberg, L., 334
Einstein, Albert, genetic endowment of, 126
Ek, J. I., 90
Elderton, E. M., 120–21
Electrons, 338
Elliptocytosis, 350
Embryo, drugs and, 107
  emotional disturbances and, 110
  infections and, 105–6
  organ-forming period of, 105
  physical agents and, 110
Emotional disturbances, effect of, on embryo and fetus, 110–11
Emphysema of lungs, 362
Enamel, defective development of, 373
Enchondromatosis, 351
Endocardial fibrosis, 349
Endoderm, 254
Endometriosis, 358
Energy, nucleotides and, 31
Environment(s), 20–21
  adjustment to, 298
  after birth, 113–40
  body form and, 119–26, 266
  of cells, 26
  control of, 84
  crime and, 158–63
  cultural, 134–36
  definition of, 26

## 416 INDEX

effects of, on embryo and fetus, 86–112
extrauterine, 94
geographic, changes in body form and, 119–26, 266
heredity and, 76–85, 117
intercellular, 84
longevity and, 124–26
mental abilities and, 126–40
socioeconomic, growth and, 119–22
uterine, 79, 84
Enzyme(s), 37, 174–175, 202
Eosinophilia, familial, 349
Epicanthus, 355
Epidermodysplasia verruciformis, 371
Epidermolysis bullosa, 191, 371
Epigenesis, 78
Epilepsy, 154–55, 367
Epiloia, 369
Epistasis, 199
Epitaxis, 370
Epithelial adenoides cysticum, 373
Erythemato-keratotic-phacomatosis, 371
Erythemia nodosum, familial, 371
Erythroblastosis, 97–99
treatment of, 99
Erythroblastosis fetalis, 349
Erythroglucose deficiency disease, 365
Estabrook, Arthur H., 132
Ethics, science and, 297–99
Ethnic group(s), 218–24
definition of, 220
illustration of, 222
Rh genes in, 289
table of, 290–91

*Eugenical Sterilization in the United States* (Laughlin), 301
Eugenics, negative, 321–22
positive, 322–23
Eunuchoidism, 358
Evolution of man, 218–21, 223–24, 299
Exophthalmic goiter, 359
Exotosis, multiple hereditary, 351
Expressivity, 199
Eye(s), disorders of, 354–56
Eye color, 244–46
Eyebrows, 249, 255
Eyelashes, 249–50, 255

Fabry's disease, 371
Facial hair, 250, 255
Facial hemiatrophy, 361
Fallopian tubes, 24
Fallout, 333–35
Familial hemolytic anemia, 347
Familial hypophosphatemia, 353
Familial intestinal polypsis, 357
Familial nonhemolytic jaundice, 350
Fanconi's syndrome, 347
Farber, Bernard, 314
Farsightedness, 356
Favism, 363
Features, the, 256–65
Feeblemindedness, 369
Female(s), baldness in, 193–95
birth rate of, 184
mortality rates of, 184
tables of, 182–83
sex determination of, 180–81
sex role of, 178–80

survival rate of, 182–87
　*See also* entries under Maternal; Mother(s); Mothering.
Ferguson-Smith, M. A., 211
Ferri, Enrico, 160
Fertility, hybrid vigor and, 227–28
Fertilization, 24–26
Fetal deaths, 184
Fetal hemoglobin, persistent, 349
Fetal malformations, 94, 106
Fetus, drugs, and 107–9
　emotional factors and, 110
　environment and, 86–112
　infections and, 105–6
　nutrition and, 100–4
　physical agents and, 109–10
Fibrocystic pulmonary dysplasia, familial, 362
Fibrous dysplasia of the jaws, 351
Filum terminale brevis, 367
Finger(s), defects of, 357
　hair on, 251–252
　supernumary, 103
Finger smudges, 243
First-born children, 97–98
Fisher, R. A., 286–88
Flatfoot, 353
Fleming, Miss R. M., 226–27
Flocculi iridis, 355
Fonio, A., 173–74
Ford, C. E., 204
Forehead, 273–74
Fraccaro, M., 91, 93
Fraenkel-Conrat, H., 44
Fragilitas ossium, 352
Fraternal twins, 90, 154, 167
　criminal behavior of, 156–58
　frequency of, 141
Freckles, 243

*Free martin(s)*, 197
Freeman, F. N., 142
Frenulum, hypertrophied, 373
Fried, Ralph, 116
Fructose intolerance, 363
Fructosuria, essential, 363
Fruit fly, 22, 28–29, 56, 334
"Full blood," 69
Fundulus heteroclitus, 86
Fundus dystrophy, 355
Funnel chest, 362

G-6-PD, 202–3
Galactosemia, 363
Galactosuria, 363
Gamete(s), 54
Gamma rays, 327, 329
Gargoylism, 308, 373
Gaspar, J. L., 96
Gastrointestinal bleeding, familial, massive, 357
Gastrointestinal tract, disorders of, 357–58
Gates, A. I., 104
Gaucher's disease, 373
Gauguin, Paul, 127–28
Gauthier, M., 91
Gene(s), 25, 27–28
　action of, 82–85
　allelic pairs of, 59
　allelic Rh series of, table, 289
　baldness and, 192–96
　blood groups and, 279–89, 293
　chemical defects in, 341
　constitution and, 164–77
　criminal tendencies and, 158, 160–63
　D, C, and E, 286–88
　defective, 90
　definition of, 33
　dominant, 199, 239, 264–65

environment and, interaction of, 130–31
function of, 33–35
heterozygous. *See* Heterozygous genes.
homozygous. *See* Homozygous genes.
inalterability of, 65
lethal. *See* Lethal genes.
linkage of, 63
and mental defectives, 310–11
multiple, 240
mutant, 35, 198
mutation rates of, table, 337
number of, in chromosome, 28
overvaluation of, 78–82
penetrance, expressivity, and viability, 198–212
pleiotropic, 198
potentialities and, 127
populations and, 376–79
radiation damage to, 338–41
recessive, 199, 236–39, 264–65, 317
size of, 28
of twins, environment and, 141–55
in X chromosome, 187
Gene deficiencies, 153, 155
Gene frequency, 378–79
*Generation of Animals, The* (Aristotle), 71
Genetic code, 31, 34, 39
  elements of, illustrated, 43
Genetic counseling, 311, 380
  list of institutions giving, 380–81
Genetic damage, 333
  mechanism of, 338–341
Genetic drift, 220

Genetics, 62
  Russian, 66, 393–95
Genitourinary system, disorders of, 358–59
Genius, 126–27, 129
Genotype (s) 153, 165, 195, 210, 239, 242, 245
  blood groups and, 280–82, 286–89
Genotype equilibrium, 376–79
German measles, 105–6
Gigantism, 271–72
Gilbert's disease, 350
Gingival hyperplasia, 373
Glandular disorders, 359–60
Glanzmann's thromboasthenia, 351
Glaucoma, 355
Glomerubonephritis, familial, 358
Glucoglycinuria, 363
Glucose-6-phosphate dehydrogenase deficiency, 202–03, 363
Glueck, Eleanor and Sheldon, 162
Glycosuria, renal, 363
Goddard, Henry H., 132–33
Goiter, 350–60
Goiter-deafness syndrome, 374
Gold, A. P., 211
Goldacre, J. R., 32
Goldscheider's disease, 371
Gonosomes, 55
Goodenough "Draw-a-Man" test, 136
Gordon, M. J., 211
Gout, 196–97, 363
Gower, Josephine S., 181
Graafian follicles, 24
Grant, Madison, 300
Granular dystrophy, 355

INDEX 419

Grave's disease, 359
Greene, A. R., et al., 211
Gröndblad-Strandberg syndrome, 372
Growth, bone, 116
See also Development.
Growth rates, environment and, 119–22
Grumbach, M. M., 211
Gruneberg, 274
Guanine, 31, 33, 39
Guthrie, Robert, 175
Gynecomastia, 360

Haddock, D. R., 294
Hageman trait, 349
Hair, 246–52
  adaptive value of, 254–55
  body, 251–52
  disorders of, 360
  facial, 250, 255
  graying of, 248–49
Hair color, 246–49
Hair form, 252–53
  inheritance of, illustrated, 252
Hairlessness, 249
Haldane, J. B. S., quoted, 306
"Half blood," 69
"Half-caste," 69
Hallermann-Streiff syndrome, 361
Hallervorden and Spatz syndrome, 368
Hallux rigidus, 352
Hallux valgus, 352
Hammertoe, 352
Hand(s), hair on, 251–52
Handlin, Oscar, 300n.
Hansen's disease, 373
Hardin, Garrett, 212
Hardy, G. H., 376
Hardy-Weinberg Law, 376–78

Harelip, 260–61, 373
Harrell, Ruth F., 104
Hartnup disease, 363
Hashimoto's disease, 360
Havighurst, R. J., 138
Haybrittle, J. L., 329
Head, the, 272–75
  disorders of, 361
Hearing, defective, 81
  disorders of, 353–54
Heart, disorders of, 348–49
  enlargement of, 348
Heart disease, congenital, 349
  rheumatic, 350
Heberden's nodes, 191
Hedgran, A., 333
Heller, 53
Hellin's rule, 141
Hemangioma, of cerebellum and retina, 367
  of small intestine, 349
Hemeralopia, 355
Hemochromatosis, 363
Hemoglobin, 30
Hemoglobin M disease, 365
Hemolysis, 277
Hemolytic anemia, familial, 349
Hemolytic disease of the newborn, 97, 349
Hemophilia, 81, 180, 185–87, 304–6, 308, 349
  cause of, 168–69
  and Christmas disease, 348
  constitution and, 168–74
  family lines affected by, 171–73
  and parahemophilia, combined, 348
  transmission of genes of, illustrated, 170
Hemorrhagic disease, autosomal, 349
Henry, George W., 152
Heparinemia, 349

# 420 INDEX

Hepatic fibrosis, congenital, 374
Hepatolenticular degeneration, 364
Hereditary capillary purpura, 349
"Hereditary condition," 80
Hereditary hemolytic icturus, 176
Hereditary hemorrhagic diathesis, 191
Hereditary iron-loading anemia, 347
Hereditary labile factor V deficiency, 349
Hereditary nephritis, 348
Hereditary, 116
 crime and, 156-63
 definition of, 78-79, 84
 and environment, 76-85, 117
 fallacies about, 80
 questions about, 297
*Heredity East and West* (Huxley), 393
Heredopathia atactica polyneuritiformis, 368
Hermaphroditism, 210, 358
Herrick, C. Judson, 298
Heterozygous genes, 60, 236-39, 249, 280-81, 286, 332
Hill, J. B. and H. D., 51
Hip, congenital dislocation of, 352
Hippocrates, 75n.
Hirschhorn, Kurt, 92
Hirshsprung's disease, 357
Histidinosis, familial, 364
Histone, 47
Hitler, 159
 racist theories of, 20, 72, 75
Hives, 347
Holcomb, B., 96

Holmes, 192
Holub, D. A., 209, 211
Holzinger, K. J., 142
*Homo sapiens*, 45, 220
 See also Man.
Homosexuality, environment and, 153
Homozygous genes, 60, 236-39, 249, 280, 286-87, 332
Homozygous hemoglobin C disease, 349
Hooton, Ernest, 156, 161
Hormonal imbalance, 153
Hormones, 83
 female sex, 195
 male sex, 194
Horowitz, S. L., 260
Huang, Ru-chih H., 47
Huntington's chorea, 367
Hurler-Pfaundler syndrome, 373
Huxley, Aldous, quoted, 23
Huxley, Julian, 383
Hybrid vigor (heterosis), 227-28
Hybridization, 220
Hydrocephalus, 93-94, 103, 314, 361
Hydrogen bomb, heredity and, 325
Hypercalcemia, 364
Hyperchronic anemia, 347
Hyperelastosis cutis, 371
Hyperostosis, infantile cortical, 352
Hyperoxaluria, 364
Hyperprolinemia, familial, 364
Hypertelorism, 361
Hypertension, 95, 176
 essential, 349
 and nephrosclerosis, 359
Hyperthyroidism, 96
 familial, 360

## INDEX 421

Hypertipemia, familial, 364
Hypertrichosis of the ears, 250, 354
Hypertrophic polyneuritis, progressive, 368
Hyperuricemic nephropathy, familial, 364
Hypochromic, microcytic anemia, 348
Hypogammaglobulinemia, 350
Hypoglycemia, spontaneous, 364
Hypogonadism and ichthyosis, 360
Hypokalemia, congenital, 364
Hypolipemia, 364
Hypophosphatasia, 363, 364
Hypoplastic anemia of childhood, 348
Hypoprothrombinemia, 350
Hypospadias, 358
Hypostasis, 199
Hypotension, 350

Ichthyosis, 191, 371
Identical twins, 154, 165, 167, 265
  criminal behavior of, 156–158
  dentition of, 150–152
  Downs, syndrome, 90
  environment and, 142–51
  frequency of, 141
  reared apart, 140, 142–148
Idiocy, 369
Idiopathic megacolon, 357
Imbecility, 369
Immigrants to United States, changes in body form of descendants of, 122–24
  tables of, 123–24
  crime among, 159

Immigration laws, 21, 300
Inbreeding, 315, 318–19
*Inbreeding and Outbreeding* (East and Jones), 319
Independent assortment, law of, 62
  illustrated, 64
Indian corn, hybrid, 65–66
Indians, American, 236, 239
  intelligence of, 135–36
  intermarrying with, 57
  Asiatic, 239
  South American, blood groups of, 280
Infant mortality, 96, 106
Infections, effect of, on uterus, 105–6
Inferiority, biological, 161
Inheritance, laws of, 57–66
  modes of, 198
Insulin, 81, 95
Intelligence, of American Indians, 135–36
  "race" and, 230–32
Intelligence tests, 129–30, 135–39
Interaction, 76–77, 82–83
  of genes and environment, 164
International Commission on Radiation Protection, 330
Intersexuality, 211
Involutional psychosis, 154, 155
Iodine deficiency, 111–12
Ion(s), 328
Ionization, 328–29
IQ (intelligence quotient), diet and, 104
  environment and, 135–38, 148
Iris, 244–46
  defects of, 355
Isolation, 219–20
Isotope(s), 327

## 422  INDEX

Jacobs, Patricia, 205–6, 209
Jailer, J., 211
Jalavisto, E., 94
Janke, L. L., 138
Jaundice, 350, 364
Jaws, the, 257–60, 373
Jenner, Edward, 105
Jennings, Herbert S., 226
Joints, disorders of, 351–53
Jokl, E., 121
Jones, Donald F., 319
"Jukes, the," 126, 131–34
*Jukes, The: A study in Crime . . .* (Dogdale), 131
Juvenile Tay-Sachs disease, 366

"Kallikaks, the," 126, 131–34
Kallmann, Franz J., 147, 151–53, 167, 304n.
Kartagener's syndrome, 362
Keay, A. J., 205
Keloids, multiple, 371
Keratosis, 371
Kirman, B., et al., 307
Klebanow, D., 103
Klein, H., 259
Klinefelter's syndrome (testicular dysgenesis), 203–4, 209, 211, 358
Klippel-Feil syndrome, 361
Knowledge, of facts of heredity, 20

Labyrinthine deafness, 354
Landsteiner, Karl, 99, 276, 282
Landsteiner Rule, 277
Larsell, Olaf, 298
Laryngitis, 374
Latticelike dystrophy, 355
Laughlin, Harry, 300–2

Laurence-Moon-Bardet-Biedl syndrome, 369
Lawler, S. D., 293
Laws, of dominance and recessiveness, 63–66
  of independent assortment, 62–63
  of inheritance, 57–66
  Mendel's. *See* Mendel.
  of segregation, 59–62
Leake, Chauncey, 298
Legg-Perthes disease, 352
Lejeune, J. B., 91
Lens defects, 355
Lenz, Fritz, 223
Leonardo, genetic endowment of, 126
Leprosy, 374
Lethal genes, 199, 317
  conditions due to, table of, 201
Leukemia, 350
Levine, Phillip, 282–84
Levulosuria, 363
Lewandowsky-Lutz disease, 371
Lewin, Sherry, 211
Life expectancy, 124–26
  of females, 185
  of males, 185
  table of, 127–28
Likenesses, 18
Lindau's disease, 367
Lipoidosis, 364
Lipomatosis, multiple, 364
Livanoy, M. N., 341
Lobstein's disease, 352
Lobster hand, 357
"Lofenalac," 175
Lombroso, Cesare, 156, 160
Longevity, environment and, 124–26
  marriage and, 185
Lowe, Terrey and MacLachlan syndrome, 365

Lower-urinary-tract obstruction, 358
Lowe's syndrome, 356
Lubarsch-Pick syndrome, 361–362
Lumostabil, 330–31
Lungs, disorders of, 362
Lupus vulgaris, 371
Lutenbacher's syndrome, 348
Lymphedema, congenital, 374
Lymphocytophthisis, essential, 374
Lysenko, T. D., 66, 393–95
Lysosomes, 36


McArdle's syndrome, 364
McCarran-Walter Act of 1952, 300
McConnell, R. B., 294–95
McKusick, V. A., 188
Macular dystrophy, 355
Madelung's deformity, 352
Major ethnic groups, 221
Malaria, 106
Males, mortality rate of, 184
  tables of, 182–83
  sex determination of, 180
  survival rate of, 182–87
Malformations, congenital, 106
Malnutrition, effect of, on embryo and fetus, 102–3
Man, evolution of, 218–21, 223–24
  *See also* Females; *Homo sapiens;* Males.
Mandibulofacial dysostosis, 361
Manic-depressive psychoses, 154, 155, 197
Manuila, A., 295
Maple sugar urine disease, 362
Marble bones, 352

Marcus Gunn phenomenon, 374
Marfan's syndrome, 352, 357, 374
Marmur, Julius, 45
Marriage, of close relatives, 315
  Church bann on, 316
  in Egyptian dynasties, 315–16
  between cousins, 315, 317–19, 375
  table of, 317
  longevity and, 185
  between whites and Negroes, 57, 59, 225–26, 241
Marshall's syndrome, 361
Maternal age, 88–94
*Maternal Care and Mental Health* (Bowlby), 114
Maternal dysfunction, 95–96
Maternal (uterine) environment, 79, 84
Maternal parity, 94
Maternal sensitization, 97–100
Mathew, the Rev. John, 232
Matthaei, J. Heinrich, 39
Mayer, M. F., 116
MBS (mercaptobenzoselenazole), 198
Meckel's diverticulum, 357
Mediterranean fever, familial, 363
Megalocornea, 355
Meiosis, 50–54
  illustrated, 51, 55
Melanasians, 289
Melanin, 235, 240
Melanocytes, 235
Mendel, Gregor, 61–63, 65
  first law of, 59
  second law of, 63

## 424 INDEX

study of peas by, 62–63
third law of, 63
Mental abilities, environment and, 126–40
race and, 229–30
*Mental Defect* (Penrose), 306
Mental defectives, 306–10
problem of, 310–11, 313–24
Mental deficiencies, list of, 369–70
Mental disorders, list of, 370
in twins, 150–55
Mental health, 114–15
Mesoderm, 254n.
Messenger-(template)-RNA, 37–38, 41, 47–48
Metabolic disorders, 362–66
Metabolism, 341
errors in, 202
Metaphyseal dysostosis, familial, 352
Methemoglobinemis, 364
Michael, A. F., 211
Michurin, Ivan V., 393
Michurinism, 393–95
Microcephaly, 361, 370
Microcornea, 355
Micrognathia glossoptosis, 361
Micron, definition of, 24n.
Microphthalmos, 355
and mental deficiency, 370
Microtia, 354
Migraine, 367
Migration, 219–20
Milroy's disease, 374
Minnows, effect of environment on, 86
Miosis, congenital, 355
Mirror reading, 355
Miscarriage, 111
Mitosis, 49–50
Mohr, O. L., 253
Molars, absence of, 373

Molecular genetics, summary of, 52
Molecular weight, of DNA molecule, 32
Molecule(s), DNA, 31–34, 45–47
fragmented, 338
hybrid, 45
in nucleotides, 31
protein, 30
Mongolism, 369
Mongoloids, 89, 218, 221, 252, 255, 289
"Monsters," 82–83
Montague, Ashley, 89n., 160n., 186, 287
Moore, K. L., 210
Morgan, Thomas Hunt, quoted, 134
Morphinism, in mothers, 107
Morquio-Brailsford disease, 352
Mortality, infant, 96, 106
Mortality rates, sex and, tables of, 182–83
socioeconomic status and, 120
Mosaics and mosaicism, 205
Moses, Grandma, 128
Mother(s), pregnant, 71
*See also* entries under Maternal.
Mothering, importance of, 113–19
Mouth, disorders of, 372–73
Mozart, genetic endowment of, 126, 129
Mucovisidosis, 374
Mulattoes, 57–58, 69, 241, 243
Muller, H. J., 375, 395
Multiple congenital contractures, 352
Muscular dystrophy, 308, 368

INDEX 425

Muscular weakness, 368
Mutation(s), 35, 44, 219–20
   radiation and, 326–28
   rates of, of human genes, table of, 337
   spontaneous, 332
   male dress and, 334–35
Mutation pressure, 81
Myopia, 355
Myotonia congenita, 368
Myxedemia, 360


Nail-deafness syndrome, 360
Nail nevi, 360
Nail-patella syndrome, 293, 360
   blood groups and, illustrated, 294
Nails, disorders of, 360
Natural selection, 220, 331
*Nature-Nurture Controversy, The* (Pastore), 21
Nazis, blood myths and, 72–73
   false theories of, 20
Neal, N. A., 122
Nearsightedness, 355
Neck, disorders of, 361
Negroes, 69, 236, 239, 254, 259
   blood of, 73
   IQ scores of, 136–37
   mating of, with other groups, 223–26
   with whites, 57, 59, 69, 225–27, 241
   multiple births among, 141–42
   Nilotic, 272
   pigmentation of, 240–43
Negroids, 218, 221, 253, 255
Nehls, G., 260
Nephritis, familial, 358
Nephropathy, 358

Nephrophthisis, 358
Nephrosclerosis, 359
Nephrosis, 359
Nervous system, disorders of, 366–70
Neurofibromatosis, 367
Neurological disorders, 366–70
Neutropenia, 350
Nevi, 372
   Nail, 360
Newman, H. H., 142–44
Newton, genetic endowment of, 126
Niemann-Pick disease, 197
Night blindness, 355
Nillson, Inga M., 169
Nirenberg, Marshall W., 39
Nitrogenous bases of nucleotides, 31
Noise, effect of, on embryo and fetus, 110
Nose, 256–57
   disorders of, 370
"Nothing-but fallacy," 80
Nuclear war, 325
Nucleic acids, 30–31
   nucleotides in, 31
Nucleolus (nucleoli), 35
Nucleotide(s), 31
   and genetic instructions, 33
   linkage of, 31
   number of, in fertilized egg, 33
Nucleus (nuclei), of chromosome cell, 32, 35–36
   of egg, 25
   of sperm, 25
Nutrition, effect of, on embryo and fetus, 100–4
Nyctalopia, 355
Nystagmus, 355


Obermayer, M. E., 193

## 426 INDEX

Occipital protuberance, 274
Ochronosis, hereditary, 356
Ocular paralysis, 356
Oculo-cerebro-renal syndrome, 356, 365
Oguchi's disease, 191
Ollier's disease, 351
Oogonia, 23
Optic nerve atrophy, 356
Organism, definition of, 34
Osborne, R. H., 260
Osol, A., 329
Osteitis deformans, 352
Osteoarthropathy of fingers, familial, 352
Osteochondritis deformans juvenilis, 352
Osteochondritis dissicans, 352
Osteochondrodystrophia deformans, 352
Osteogenesis imperfecta, 352
Osteopetrosis, 352
Osteopoikilosis, 353
Otitis media, 354
Otomandibular dysostosis, 361
Otosclerosis, 354
Ovalocytosis, 350
Ovary (ovaries), 23
  number of eggs in, at birth, 23
Ovum (ova), 178–80
  fertilized, information in, 32
  number of nucleotides in, 33
  number of, in ovary at birth, 23
  size of, 24–25
Oxalosis, familiar, 365
Oxaluria, 365
Oxycephaly, 361
Oxygen reduction, brain damage and, 99

Pachyonychia congenita, 372
Page, O. C., 96
Paget's disease, 352
Pain insensitivity, 367
Pancreatic cyctic fibrosis, 360
Parahemophilia, 348, 350
Paralysis agitans, 368
Paramytonia, 368
Parathyroid adenomas, familial multiple, 360
Parietal foramina, enlarged, 353
Parity, maternal, 94
Parkinsonism, familial, 368
*Passing of the Great Race, The* (Grant), 300
Pastore, Nicholas, 21
Patent ductus arteriosis, 349
"Pathetic fallacy," 80
Pauling, Linus, qouted, 344
Pavlovsky, A., 306*n*.
  quoted, 305–6
Peas, study of, by Mendel, 62–53
Pedigree charts, 311
  illustrated, 312
Pelger's nuclear anomaly, 350
Penetrance, 199
Penrose, L. S., 154, 306
  quoted, 228, 322
Pentosuria, 365
Peptic ulcer, 357
Performance hypothesis, 77
Periodic paralysis, familial, 365
Pernicious anemia, 176, 295, 348
Peroneal muscular atrophy, 368
Personalization, 114
Pes planus, 353
Peutz-Jegher's syndrome, 357
"Phenistix," 175
Phenotype(s), 57, 198, 210, 239, 280, 282, 286–87

# INDEX 427

Phenylalanine 40–41, 174
Phenylketonuria, 308, 365
  constitution and, 174–76
Phenylpyruvic oligophrenia, 365
Pheochronocytoma, familial, 360
Phi X virus, 47n.
Phlebectasia, 350
Phlebitis, 351
Phosphoric acid, 31
Photosensitivity, familial, 368
Physical development, environment and, 119–22
Pick's disease, 367
Piebaldness, 372
Pigmentary cirrhosis, 363
Pigmies, 221, 270–71
  interbreeding with, 223
Pili torti, 360
Pityriasis rubia pilaris, 372
Planned breeding, 323–24
Plasma thromboplastin antecedent deficiency, 350
Pleiotropy, 63
Pneumonia, 293
Pneumothorax, 362
Polani, P. E., 91
Polar bodies, 54
Poliomyelitis, 106
Politics, heredity and, 20–21
Polydactylism, 357
Polyendocrine adenomatosis, 360
Polygenes, 240
Polynesians, 289
Polyposis of large intestine, 357
Polyseroseritis, 365
Polyserositis, periodic, 374
Polyspondily, 353
Population(s), control of, 212
  genes and, 376–79
Porokeratosis, 372
Porphyria, 365

"Potato nose," 370
Potter, R. V., 47n.
Prediabetes, 95–96
Pregnancy. *See* developmental process, genetic; Embryo; Fetus; Uterus.
Presbyopia, 356
Progressive ophthalmoplegia, 356
Protein, 30
  defective, 30
  production of, 37
Protein molecules, 30
Protein synthesis, 35, 39
  illustrated, 38, 40
Pseudo-cholinerterase deficiency, 365
Pseudoglaucoma, 356
Pseudoglioma, 356
Pseudohermaphrodite, 210, 358
Pseudoxanthoma elasticum, 372
Psoriasis, 372
PTC (phenylthiocarbamide),
  ability to taste, 261–63, 265
  table of, by population, 262–63
Pterygium, 356
Ptosis, 356, 374
Puck, T. T., 180, 339
Pupillary membrane, congenital, 356
  persistent, 355
"Pureblood," 69
Purines, 31–32
Pyke, D. A., 295
Pyloric stenosis, congenital, 358
Pyrimidines, 31, 33

Raboch, J., 204
Race, 18–19

## 428  INDEX

false ideas of, 217–18
heredity and, 217–32
intelligence and, 230–32
mental capacities and, 229–230
"Race"-crossing, 224–28
*Race Crossing in Jamaica* (Davenport), 225, 300
Racism, 301
Radial reduction, 353
Radiation, 110–11
  background, 333
  effects of, 341–45
  high-energy, 339
  ionizing, 327
  mutation and, 326–38
Radioactivity, 328–37, 342
  of radium watch dials, 329–31, 342
Radiohumeral, ulnahumeral synostosis, familial, 353
Radiology, diagnostic, 330
Radioulnar defect, 353
Radium, 329–31, 342
Ratner, B., 100
Reardon, H. S., 96
Recessiveness, dominance and, law of, 63–66
Recessives, 63, 65
Rectal stenosis, 358
Recurrent bulbous eruption of feet, 372
Redheads, 236, 244, 247–49
Reduction division, 52
Reduplication, 45
Reed, Sheldon, 313, 375
Refsum's disease, 368
Reguera, R. M., 295
Reisner, D., 167
Renal dysplasia, 359
Renal glycosuria, 359
Renal-hepatic-pancreatic dysplasia, 374
Renal sclerosis, familial, 359

Renal tubular acidosis, familial, 359
Rendu-Osler-Weber disease, 351
Renkonen, K. O., 295
Renwick, J. H., 293
Reproduction, 33, 45, 53
  optimum age for, in females, 88–94
  *See also* Cell division.
Retinal glioma, 356
Retinitis pigmentosa, 191, 197, 356
Retinoblastoma, 356
Retinoschisis, hereditary, 356
Rh antibodies, 97–99
Rh factor, 72, 97, 283, 311
  negative, 97–99
  positive, 97–99
Rheumatic fever, 374
Rheumatic heart disease, 350
Rhinitis, 370
Rhinophyma, 370
Ribosomal-RNA, 41
Ribosome(s), 36–37
Rickets, 364
  due to maternal defect, 353
  vitamin-D resistant, 353
RNA (ribonucleic acid), 35–44, 47, 307
  function of, 35
  synthetic, 39–40
Roberts, Fraser, 274
Roberts, J. A. F., 293, 295
Robin's syndrome, 361
Roentgen(s), 328–30, 344
Roentgen, Wilhelm Konrad 328n.
Rohrer, J. H., 136
Rosenberg, Alfred, quoted, 72–73
Rosenkranz, H., 44
"Royal blood," 68–69
Rundle, A., 90
Russell, Liane Brauch, 181

Russell, W. L., 181, 336

Sacral spot, 243
St. Vitus dance, 367
Saldanha, P. H., 228
Scapulae, elevated, 353
Scheinfeld, A., 133, 271
Schizophrenia, 154–55, 303–4, 370
Schofield, 192
Schröder, V. N., 211
Science, definition of, 19
   ethics and, 297–99
   national policy and, 299–306
Secondary sex characteristics, 197
Segregation, law of, 59
Senses, learning use of, 118
Sensitization, maternal, 97–100
Sex, 178–216
   determination of, 55–56, 180–81, 199–201, 211–12
   female role in, 178–80
   mortality rates and, table of, 182–83
   survival rates and, 182–87
Sex-influenced traits, 191–92
Sex-limited traits, 192–98
Sex-linkage, 180
   traits and, 187–91
   table of, 188
Sex reversal, 206, 211
Sexual precocity, familial, 374
Shaking palsy, 368
Sheldon, W. H., 162
Shettles, L. B., 211
Shilder's disease, 366
Shimberg, Myra, 137
Shuman, C. R., 96
Sickle-cell anemia, 30–31, 33, 348

Sickness rates, socioeconomic status and, 120
Simpson, W. J., 108–9
Sinsheimer, R. L., 47n.
Sinuses, 274–75
Šíprová, I., 204
Sjögren's disease, 366
Skeletal muscle phosphorylase deficiency, 366
Skin, 235
   disorders of, 370–72
   pigmentation of, 220, 232–39
   general, 239–43
   of Negroes, 240–43
Smallpox, 105
Social changes, 21
Social inadequates, 301–2
Socioeconomic status, growth and, 119–20
   maternal environment and, 101–4
   sickness and mortality rates and, 120
Solnitzky, Othmas, 208
Somotype(s), 267–69
   temperament and, 268–69
Sontag, L. W., 108
Spastic diplegia, 369
Spastic paraplegia, 191, 369
Specific dyslexia, 367
Speech, potentialities for, 117
Speiser, P., 294
Sperm, 55–56, 211
   formation of, 178
   illustrated, 54
Spermatids, 53
Spermatogenesis, 53
Spermatozoa, 24, 178
   Hartsoeker's drawing of, 79
   as homunculus, 79
   nucleus of, 53
   number of, at ejaculation, 183

## INDEX

preservation of, 215–16
production of, past sixty, 179
separation of X-bearing from Y-bearing, 211
size of, 24–25
*See also* Sperm.
Spherocytosis, 350
Spielmeyer-Vogt disease, 366
Spina bifida, 353
Spinocerebellar ataxia, 369
Spitz, René, 114
Split foot, 353
Split hand, 357
Spodylitis, ankylosing, 353
Spotted bones, 353
Sprengel's deformity, 353
Steggerda, Morris, 225–26
Stephens, J. W., 96
Sterility, 111
Sterilization, 301–6, 322
Sterilization laws, 21, 301–2
states having, table of, 303
Stigmata of crime, 156, 161
Stillbirth, 97
Strabismus, 356
Strontium, 334
Struthers, D., 293
Stuart-Prower clotting defect, 350
Sulzberger, 196
Survival rates, differential, 182–87
Sydenham's chorea, 197, 367
Symphalangism, 357
Synapsis, 50
Syndactyly, 357
Syphilis, 106, 197
Systemic lupus erythematosus, 372

T-substance anomaly, 365
Tangier disease, 366
Taste, ability to, 261–63, 265
Teeth, 258–60
decay of, 259–60
disorders of, 372–73
malocclusion of, 258
Telangiectasis, hereditary hemorrhagic, 351
Telegony, 73–75
Teller, W. M., et al., 308
Testicular feminization, 359
Testicular torsion, 374
Tetrad, 50
Tetralogy of Fallot, 349
Thalassemia, 348
Thalidomide, 107
*Theory of Generation, A,* (Wolff), 78
Thomsen's disease, 368
Thromboangiitis obliterans, 351
Thrombocytes, 174
Thrombocytopathic purpura, 351
Thrombocytopenia, familial, 351
Thrombocytopenic purpura, idiopathic, 351
Thromboplastin, 174
Thymine, 31, 33
Thyroxine, lack of, 111
Tibial defect, 353
Tjio, J. H., 180
Tobias, Phillip, 223
Toe(s), defects of, 357
supernumerary, 103
Tongue, the, 261
Tongue-tie, 372
Tooth decay, 372
Torsion dystonia, 369
Torticollis (wryneck), 109
Torus palatinus, 361
Toxemias, material, 96
Traits, 219
of ethnic groups, illustrated, 222
inheritance of, 273, 323

## INDEX

sex-influenced, 191–92
sex-limited, 191–98
sex-linked, 187–91
   table of, 188
Transfer-RNA, 37–38, 40–41
Translocation, 340
Treacher-Collins syndrome, 361
Trinuclides (triplets), 39, 41
   illustrated, 39
Trisomy, 21, 369
Tschermak-Seysenegg, Erich von, 61
Tsugita, A., 44
Tuberculosis, 81, 106
   constitution and, 167–68
   death rate and, 167
   hereditary disposition to, 362
   resistance to, 167
   of the skin, 371–72
Tuberous sclerosis, 369
Turner's syndrome, 203–5, 207, 209, 359
Turpin, R., 91
*Twins* (Newman et al.), 142
Twins, criminal behavior of, 156–58
   table of, 157
   fraternal. (*See* Fraternal) twins.
   genes, environment and, 141–55
   identical. *See* Identical twins
   mental disorder in, 152–55
   reared apart, environment and, 140, 142–48
Tylosis, 372
Tyrosinase, 235
Tyrosine, 174, 235

Ulcers, stomach, 295
Ulnar defects, 353

United Nations Scientific Commission on Effects of Atomic Radiation, 341
United States Immigration Act of 1924, 300
Unpigmented eyes, 356
Uracil, 37, 39
Uricaria, 347
Uterus (womb), 24
   absence or malformation of, 359
   effects of environment in, 71, 79, 86–112
Uticaria pigmentosa, 372

Vaccination, 105, 168
Valence, 338
Van den Bosch syndrome, 356
Van der Hoeve's syndrome, 352
Variation, 17
   crossing-over and, 52
   in molecules, 31–32
Varicose veins, 350
Vascular system, diseases of, 347–51
Viability, of genes, 199
Viruses, 105
Vision, defective, 81
   *see also* Eyes,
Vitamin B, 101
Vitamin C, 101
Vitamin D, 101, 240
   rickets resistant to, 353
Vitamins, 101, 104
Von Ehrenstein, G., 333
Von Recklinghausen's disease, 367
Von Willebrand's disease, 349

Waardenburg-Klein syndrome, 356, 360

## 432 INDEX

Waddington, C. H., 394n.
Wallace, J., 294
Watson, J. D., 32
Watson-Crick model of DNA molecule, 32
Webster, Edward W., 330, 335
Weight, at birth, 26
  of DNA molecule, 32
  at fertilization, 26
Weill-Marchesoni syndrome, 374
Weinberg, W., 376
Weisberger, Austin S., 44
Welshons, W. J., 181
Werlhof's disease, 351
Westwood, Gordon, 152
Wexler, I. B., 285, 295
White forelock, 191, 248, 360
White-mouth syndrome, hereditary, 361
Whites, 221, 225–26
  light-pigmented genes of, 240–41
  mating of, with Negroes, 52, 59, 69, 225–27, 241
  *See also* Caucasoids.
Whorls, 250–51
Wiener, A. S., 97, 285–86, 289, 295
Wilkins, M. F. H., 32
Wilson's disease, 364
Wolff, Kaspar, 78
Womb. *See* Uterus.
Woodworth-Mathews test, 143
Woodyard, Ella, 104

X chromosome(s), 48, 55–56, 180–83, 186–87, 189–91, 193, 202–7, 209–10
  chromatin body and, 202–3
  hemophilia and, 169–71, 180
X rays, 110, 327, 329
  radiation danger from, 333–35, 342–43
Xanthelasma, 372
Xanthomatosis, 372
Xeroderma pigmentosum, 191, 372
XO chrosome combination, 181, 202–5, 207
XX chromosome combination, 190, 202–3, 205–7

XXX chromosome combination, 203, 205–7
XXY chromosome combination, 181, 203, 207
XY chromosome combination, 202, 205–6
XYY chromosome combination, 205

Y chromosome(s), 28, 48, 55–56, 169–70, 180–185, 187, 189–91, 205, 207, 209–10

Zimányi, Stephen, 307
Zinsser-Cole-Engman syndrome, 355
Zirkle, Conway, 393
Zollinger-Ellison syndrome, 360
Zygote(s), 54, 205